The Behaviour and Design of Steel Structures to EC3

Also available from Taylor & Francis

Steel Structures: Practical Design Studies
3rd Edition
H. Al Nagiem and T. MacGinley   Hb: ISBN 978–0–415–30156–5
                                          Pb: ISBN 978–0–415–30157–2

Limit States Design of Structural Steelwork
D. Nethercot   Hb: ISBN 978–0–419–26080–6
                                          Pb: ISBN 978–0–419–26090–5

Fracture Mechanics
M. Janssen et *al.*   Pb: ISBN 978–0–415–34622–1

Assessment and Refurbishment of Steel
Structures
Z. Agocs et *al.*   Hb: ISBN 978–0–415–23598–3

Information and ordering details

For price, availability and ordering visit our website **www.tandf.co.uk/builtenvironment**

Alternatively our books are available from all good bookshops.

The Behaviour and Design of Steel Structures to EC3

Fourth edition

N.S. Trahair, M.A. Bradford,
D.A. Nethercot, and L. Gardner

Taylor & Francis
Taylor & Francis Group

LONDON AND NEW YORK

First edition published 1977
Second edition 1988, revised second edition published 1988
Third edition – Australian (AS4100) published 1998
Third edition – British (BS5950) published 2001

Fourth edition published 2008
by Taylor & Francis
2 Park Square, Milton Park, Abingdon, Oxon OX14 4RN

Simultaneously published in the USA and Canada
by Taylor & Francis
711 Third Avenue, New York, NY 10017

*Taylor & Francis is an imprint of the Taylor & Francis Group,
an informa business*

© 1977, 1988, 2001, 2008 N.S. Trahair, M.A. Bradford,
D.A. Nethercot, and L. Gardner

Typeset in Times New Roman by
Newgen Imaging Systems (P) Ltd, Chennai, India
Printed and bound in Great Britain by
TJ International Ltd, Padstow, Cornwall

British Library Cataloguing in Publication Data
A catalogue record for this book is available from the British Library

Library of Congress Cataloging in Publication Data
     The behaviour and design of steel structures to EC3 / N.S. Trahair
... [et al.]. – 4th ed.
       p. cm.
     Includes bibliographical references and index.
     1. Building, Iron and steel. I. Trahair, N.S.

   TA684.T7 2008
   624.1′821–dc22   2007016421

ISBN10: 0–415–41865–8 (hbk)
ISBN10: 0–415–41866–6 (pbk)
ISBN10: 0–203–93593–4 (ebk)

ISBN13: 978–0–415–41865–2 (hbk)
ISBN13: 978–0–415–41866–9 (pbk)
ISBN13: 978–0–203–93593–4 (ebk)

Contents

Preface

This fourth British edition has been directed specifically to the design of steel structures in accordance with *Eurocode 3 Design of Steel Structures*. The principal part of this is *Part 1-1:General Rules and Rules for Buildings* and this is referred to generally in the text as EC3. Also referred to in the text are *Part 1-5: Plated Structural Elements*, and *Part 1-8: Design Of Joints*, which are referred to as EC3-1-5 and EC3-1-8. EC3 will be accompanied by National Annexes which will contain any National Determined Parameters for the United Kingdom which differ from the recommendations given in EC3.

Designers who have previously used BS5950 (which is discussed in the third British edition of this book) will see a number of significant differences in EC3. One of the more obvious is the notation. The notation in this book has been changed generally so that it is consistent with EC3.

Another significant difference is the general absence of tables of values computed from the basic design equations which might be used to facilitate manual design. Some designers will want to prepare their own tables, but in some cases, the complexities of the basic equations are such that computer programs are required for efficient design. This is especially the case for members under combined compression and bending, which are discussed in Chapter 7. However, the examples in this book are worked in full and do not rely on such design aids.

EC3 does not provide approximations for calculating the lateral buckling resistances of beams, but instead expects the designer to be able to determine the elastic buckling moment to be used in the design equations. Additional information to assist designers in this determination has been given in Chapter 6 of this book. EC3 also expects the designer to be able to determine the elastic buckling loads of compression members. The additional information given in Chapter 3 has been retained to assist designers in the calculation of the elastic buckling loads.

EC3 provides elementary rules for the design of members in torsion. These are generalised and extended in Chapter 10, which contains a general treatment of torsion together with a number of design aids.

The preparation of this fourth British edition has provided an opportunity to revise the text generally to incorporate the results of recent findings and research. This is in accordance with the principal objective of the book, to provide students and practising engineers with an understanding of the relationships between structural behaviour and the design criteria implied by the rules of design codes such as EC3.

N.S. Trahair, M.A. Bradford, D.A. Nethercot, and L. Gardner
April 2007

Units and conversion factors

Units

While most expressions and equations used in this book are arranged so that they are non-dimensional, there are a number of exceptions. In all of these, SI units are used which are derived from the basic units of kilogram (kg) for mass, metre (m) for length, and second (s) for time.

The SI unit of force is the newton (N), which is the force which causes a mass of 1 kg to have an acceleration of 1 m/s^2. The acceleration due to gravity is 9.807 m/s^2 approximately, and so the weight of a mass of 1 kg is 9.807 N.

The SI unit of stress is the pascal (Pa), which is the average stress exerted by a force of 1 N on an area of 1 m^2. The pascal is too small to be convenient in structural engineering, and it is common practice to use either the megapascal (1 MPa = 10^6 Pa) or the identical newton per square millimetre (1 N/mm^2 = 10^6 Pa). The newton per square millimetre (N/mm^2) is used generally in this book.

Table of conversion factors

To Imperial (British) units			To SI units		
1 kg	=	0.068 53 slug	1 slug	=	14.59 kg
1 m	=	3.281 ft	1 ft	=	0.304 8 m
	=	39.37 in.	1 in.	=	0.025 4 m
1 mm	=	0.003 281 ft	1 ft	=	304.8 mm
	=	0.039 37 in.	1 in.	=	25.4 mm
1 N	=	0.224 8 lb	1 lb	=	4.448 N
1 kN	=	0.224 8 kip	1 kip	=	4.448 kN
	=	0.100 36 ton	1 ton	=	9.964 kN
1 N/mm^{2*†}	=	0.145 0 kip/in.^{2} (ksi)	1 kip/in.^{2}	=	6.895 N/mm^2
	=	0.064 75 ton/in.^{2}	1 ton/in.^{2}	=	15.44 N/mm^2
1 kNm	=	0.737 6 kip ft	1 kip ft	=	1.356 kNm
	=	0.329 3 ton ft	1 ton ft	=	3.037 kNm

Notes

* 1 N/mm^2 = 1 MPa.

$\dagger$ There are some dimensionally inconsistent equations used in this book which arise because a numerical value (in N/mm^2) is substituted for the Young's modulus of elasticity E while the yield stress f_y remains algebraic. The value of the yield stress f_y used in these equations should therefore be expressed in N/mm^2. Care should be used in converting these equations from SI to Imperial units.

Glossary of terms

Actions   The loads to which a structure is subjected.

Advanced analysis   An analysis which takes account of second-order effects, inelastic behaviour, residual stresses, and geometrical imperfections.

Beam   A member which supports transverse loads or moments only.

Beam-column   A member which supports transverse loads or moments which cause bending and axial loads which cause compression.

Biaxial bending   The general state of a member which is subjected to bending actions in both principal planes together with axial compression and torsion actions.

Brittle fracture   A mode of failure under a tension action in which fracture occurs without yielding.

Buckling   A mode of failure in which there is a sudden deformation in a direction or plane normal to that of the loads or moments acting.

Buckling length   The length of an equivalent simply supported member which has the same elastic buckling load as the actual member.

Cleat   A short-length component (often of angle cross-section) used in a connection.

Column   A member which supports axial compression loads.

Compact section   A section capable of reaching the full plastic moment. Referred to in EC3 as a Class 2 section.

Component method of design   A method of joint design in which the behaviour of the joint is synthesised from the characteristics of its components.

Connection   A joint.

Dead load   The weight of all permanent construction. Referred to in EC3 as permanent load.

Deformation capacity   A measure of the ability of a structure to deform as a plastic collapse mechanism develops without otherwise failing.

Design load   A combination of factored nominal loads which the structure is required to resist.

Design resistance   The capacity of the structure or element to resist the design load.

Distortion   A mode of deformation in which the cross-section of a member changes shape.

Effective length   The length of an equivalent simply supported member which has the same elastic buckling load as the actual member. Referred to in EC3 as the buckling length.

Effective width   That portion of the width of a flat plate which has a non-uniform stress distribution (caused by local buckling or shear lag) which may be considered as fully effective when the non-uniformity of the stress distribution is ignored.

Elastic buckling analysis   An analysis of the elastic buckling of the member or frame out of the plane of loading.

Elastic buckling load   The load at elastic buckling. Referred to in EC3 as the elastic critical buckling load.

Elastic buckling stress   The maximum stress at elastic buckling. Referred to in EC3 as the elastic critical buckling stress.

Factor of safety   The factor by which the strength is divided to obtain the working load capacity and the maximum permissible stress.

Fastener   A bolt, pin, rivet, or weld used in a connection.

Fatigue   A mode of failure in which a member fractures after many applications of load.

First-order analysis   An analysis in which equilibrium is formulated for the undeformed position of the structure, so that the moments caused by products of the loads and deflections are ignored.

Flexural buckling   A mode of buckling in which a member deflects.

Flexural–torsional buckling   A mode of buckling in which a member deflects and twists. Referred to in EC3 as torsional–flexural buckling or lateral–torsional buckling.

Friction-grip joint   A joint in which forces are transferred by friction forces generated between plates by clamping them together with preloaded high-strength bolts.

Geometrical imperfection   Initial crookedness or twist of a member.

Girt   A horizontal member between columns which supports wall sheeting.

Gusset   A short-plate element used in a connection.

Imposed load   The load assumed to act as a result of the use of the structure, but excluding wind load.

Inelastic behaviour   Deformations accompanied by yielding.

In-plane behaviour   The behaviour of a member which deforms only in the plane of the applied loads.

Joint   The means by which members are connected together and through which forces and moments are transmitted.

Lateral buckling   Flexural–torsional buckling of a beam. Referred to in EC3 as lateral–torsional buckling.

Limit states design   A method of design in which the performance of the structure is assessed by comparison with a number of limiting conditions of usefulness.

The most common conditions are the strength limit state and the serviceability limit state.

Load effects   Internal forces and moments induced by the loads.

Load factor   A factor used to multiply a nominal load to obtain part of the design load.

Loads   Forces acting on a structure.

Local buckling   A mode of buckling which occurs locally (rather than generally) in a thin-plate element of a member.

Mechanism   A structural system with a sufficient number of frictionless and plastic hinges to allow it to deform indefinitely under constant load.

Member   A one-dimensional structural element which supports transverse or longitudinal loads or moments.

Nominal load   The load magnitude determined from a loading code or specification.

Non-uniform torsion   The general state of torsion in which the twist of the member varies non-uniformly.

Plastic analysis   A method of analysis in which the ultimate strength of a structure is computed by considering the conditions for which there are sufficient plastic hinges to transform the structure into a mechanism.

Plastic hinge   A fully yielded cross-section of a member which allows the member portions on either side to rotate under constant moment (the plastic moment).

Plastic section   A section capable of reaching and maintaining the full plastic moment until a plastic collapse mechanism is formed. Referred to in EC3 as a Class 1 section.

Post-buckling strength   A reserve of strength after buckling which is possessed by some thin-plate elements.

Preloaded bolts   High-strength bolts used in friction-grip joints.

Purlin   A horizontal member between main beams which supports roof sheeting.

Reduced modulus   The modulus of elasticity used to predict the buckling of inelastic members under the so called constant applied load, because it is reduced below the elastic modulus.

Residual stresses   The stresses in an unloaded member caused by non-uniform plastic deformation or by uneven cooling after rolling, flame cutting, or welding.

Rigid frame   A frame with rigid connections between members. Referred to in EC3 as a continuous frame.

Second-order analysis   An analysis in which equilibrium is formulated for the deformed position of the structure, so that the moments caused by the products of the loads and deflections are included.

Semi-compact section   A section which can reach the yield stress, but which does not have sufficient resistance to inelastic local buckling to allow it to reach or to maintain the full plastic moment while a plastic mechanism is forming. Referred to in EC3 as a Class 3 section.

Semi-rigid frame   A frame with semi-rigid connections between members. Referred to in EC3 as a semi-continuous frame.

Service loads   The design loads appropriate for the serviceability limit state.

Shear centre   The point in the cross-section of a beam through which the resultant transverse force must act if the beam is not to twist.

Shear lag   A phenomenon which occurs in thin wide flanges of beams in which shear straining causes the distribution of bending normal stresses to become sensibly non-uniform.

Simple frame   A frame for which the joints may be assumed not to transmit moments.

Slender section   A section which does not have sufficient resistance to local buckling to allow it to reach the yield stress. Referred to in EC3 as a Class 4 section.

Splice   A connection between two similar collinear members.

Squash load   The value of the compressive axial load which will cause yielding throughout a short member.

Stiffener   A plate or section attached to a web to strengthen a member.

Strain-hardening   A stress–strain state which occurs at stresses which are greater than the yield stress.

Strength limit state   The state of collapse or loss of structural integrity.

System length   Length between adjacent lateral brace points, or between brace point and an adjacent end of the member.

Tangent modulus   The slope of the inelastic stress–strain curve which is used to predict buckling of inelastic members under increasing load.

Tensile strength   The maximum nominal stress which can be reached in tension.

Tension field   A mode of shear transfer in the thin web of a stiffened plate girder which occurs after elastic local buckling takes place. In this mode, the tension diagonal of each stiffened panel behaves in the same way as does the diagonal tension member of a parallel chord truss.

Tension member   A member which supports axial tension loads.

Torsional buckling   A mode of buckling in which a member twists.

Ultimate load design   A method of design in which the ultimate load capacity of the structure is compared with factored loads.

Uniform torque   That part of the total torque which is associated with the rate of change of the angle of twist of the member. Referred to in EC3 as St Venant torque.

Uniform torsion   The special state of torsion in which the angle of twist of the member varies linearly. Referred to in EC3 as St Venant torsion.

Warping   A mode of deformation in which plane cross-sections do not remain in plane.

Warping torque   The other part of the total torque (than the uniform torque). This only occurs during non-uniform torsion, and is associated with changes in the warping of the cross-sections.

Working load design   A method of design in which the stresses caused by the service loads are compared with maximum permissible stresses.

Yield strength   The average stress during yielding when significant straining takes place. Usually, the minimum yield strength in tension specified for the particular steel.

Notations

The following notation is used in this book. Usually, only one meaning is assigned to each symbol, but in those cases where more meanings than one are possible, then the correct one will be evident from the context in which it is used.

Main symbols

A	Area
B	Bimoment
b	Width
C	Coefficient
c	Width of part of section
d	Depth, or Diameter
E	Young's modulus of elasticity
e	Eccentricity, or Extension
F	Force, or Force per unit length
f	Stress property of steel
G	Dead load, or Shear modulus of elasticity
H	Horizontal force
h	Height, or Overall depth of section
I	Second moment of area
i	Integer, or Radius of gyration
k	Buckling coefficient, or Factor, or Relative stiffness ratio
L	Length
M	Moment
m	Integer
N	Axial force, or Number of load cycles
n	Integer
p	Distance between holes or rows of holes
Q	Load
q	Intensity of distributed load
R	Radius, or Reaction, or Resistance
r	Radius

s	Spacing
T	Torque
t	Thickness
U	Strain energy
u	Deflection in x direction
V	Shear, or Vertical load
v	Deflection in y direction
W	Section modulus, or Work done
w	Deflection in z direction
x	Longitudinal axis
y	Principal axis of cross-section
z	Principal axis of cross-section
α	Angle, or Factor, or Load factor at failure, or Stiffness
χ	Reduction factor
Δ	Deflection
$\Delta\sigma$	Stress range
δ	Amplification factor, or Deflection
ε	Normal strain, or Yield stress coefficient $= \sqrt{(235/f_y)}$
ϕ	Angle of twist rotation, or Global sway imperfection
γ	Partial factor, or Shear strain
κ	Curvature
λ	Plate slenderness $= (c/t)/\varepsilon$
$\overline{\lambda}$	Generalised slenderness
μ	Slip factor
ν	Poisson's ratio
θ	Angle
σ	Normal stress
τ	Shear stress

Subscripts

as	Antisymmetric
B	Bottom
b	Beam, or Bearing, or Bending, or Bolt, or Braced
c	Centroid, or Column, or Compression
cr	Elastic (critical) buckling
d	Design
Ed	Design load effect
eff	Effective
el	Elastic
F	Force
f	Flange
G	Dead load
I	Imposed load

i	Initial, or Integer
j	Joint
k	Characteristic value
L	Left
LT	Lateral (or lateral–torsional) buckling
M	Material
m	Moment
max	Maximum
min	Minimum
N	Axial force
n	Integer, or Nominal value
net	Net
op	Out-of-plane
p	Bearing, or Plate
p, pl	Plastic
Q	Variable load
R	Resistance, or Right
r	Rafter, or Reduced
Rd	Design resistance
Rk	Characteristic resistance
s	Slip, or Storey, or Sway, or Symmetric
ser	Service
st	Stiffener, or Strain hardening
T	Top, or Torsional buckling
t	St Venant or uniform torsion, or Tension
TF	Flexural–torsional (or torsional–flexural) buckling
tf	Tension field
ult	Ultimate
V, v	Shear
W	Wind load
w	Warping, or Web, or Weld
x	x axis
y	y axis, or Yield
z	z axis
σ	Normal stress
τ	Shear stress
0	Initial value
$1\text{–}4$	Cross-section class

Additional notations

A_e	Area enclosed by hollow section
$A_{f,max}$	Flange area at maximum section
$A_{f,min}$	Flange area at minimum section

A_h	Area of hole reduced for stagger
A_{nt}, A_{nv}	Net areas subjected to tension or shear
A_s	Tensile stress area of a bolt
A_v	Shear area of section
C	Index for portal frame buckling
C_m	Equivalent uniform moment factor
D	Plate rigidity $Et^3/12(1 - v^2)$
$\{D\}$	Vector of nodal deformations
E_r	Reduced modulus
E_t	Tangent modulus
F	Buckling factor for beam-columns with unequal end moments
$F_{p,C}$	Bolt preload
F_L, F_T	Weld longitudinal and transverse forces per unit length
$F_{T,Rd}$	Design resistance of a T-stub flange
$[G]$	Stability matrix
I_{cz}	Second moment of area of compression flange
I_m	Second moment of area of member
I_n	$= b_n^3 t_n/12$
I_r	Second moment of area of restraining member or rafter
I_t	Uniform torsion section constant
I_{yz}	Product second moment of area
I_w	Warping torsion section constant
I_{zm}	Value of I_z for critical segment
I_{zr}	Value of I_z for restraining segment
K	Beam or torsion constant $= \sqrt{(\pi^2 EI_w/GJL^2)}$, or Fatigue life constant
$[K]$	Elastic stiffness matrix
K_m	$= \sqrt{(\pi^2 EI_y d_f^2/4GI_t L^2)}$
L_c	Distance between restraints, or Length of column which fails under N alone
L_j	Length between end bolts in a long joint
L_m	Length of critical segment, or Member length
L_r	Length of restraining segment or rafter
L_{stable}	Stable length for member with plastic hinges
LF	Load factor
M_A, M_B	End moments
$M_{b0,y,Rd}$	Design member moment resistance when $N = 0$ and $M_z = 0$
$M_{c0,z,Rd}$	Design member moment resistance when $N = 0$ and $M_y = 0$
$M_{cr,MN}$	Elastic buckling moment reduced for axial compression
$M_{N,y,Rd}, M_{N,z,Rd}$	Major and minor axis beam section moment resistances
M_E	$= (\pi/L)\sqrt{(EI_y GI_t)}$
M_f	First-order end moment of frame member
M_{fb}	Braced component of M_f

M_{fp}	Major axis moment resisted by plastic flanges
M_{fs}	Sway component of M_f
M_I	Inelastic beam buckling moment
M_{Iu}	Value of M_I for uniform bending
M_L	Limiting end moment on a crooked and twisted beam at first yield
$M_{max,0}$	Value of M_{max} when $N = 0$
M_N	Plastic moment reduced for axial force
M_b	Out-of-plane member moment resistance for bending alone
M_{bt}	Out-of-plane member moment resistance for bending and tension
M_{ry}, M_{rz}	Section moment capacities reduced for axial load
M_S	Simple beam moment
M_{ty}	Lesser of M_{ry} and M_{bt}
M_{zx}	Value of M_{cr} for simply supported beam in uniform bending
M_{zxr}	Value of M_{zx} reduced for incomplete torsional end restraint
$\{N_i\}$	Vector of initial axial forces
$N_{b,Rd}$	Design member axial force resistance when $M_y = 0$ and $M_z = 0$
$N_{cr,MN}$	Elastic buckling load reduced for bending moment
$N_{cr,L}$	$= \pi^2 EI / L^2$
$N_{cr,r}$	Reduced modulus buckling load
N_{im}	Constant amplitude fatigue life for ith stress range
$N_{cr,t}$	Tangent modulus buckling load
Q_D	Concentrated dead load
Q_I	Concentrated imposed load
Q_m	Upper-bound mechanism estimate of Q_{ult}
Q_{ms}	Value of Q_s for the critical segment
Q_{rs}	Value of Q_s for an adjacent restraining segment
Q_s	Buckling load for an unrestrained segment, or Lower bound static estimate of Q_{ult}
R	Radius of circular cross-section, or Ratio of column and rafter stiffnesses, or Ratio of minimum to maximum stress
R_H	Ratio of rafter rise to column height
R, R_{1-4}	Restraint parameters
SF	Factor of safety
S_j	Joint stiffness
T_M	Torque exerted by bending moment
T_P	Torque exerted by axial load
V_R	Resultant shear force
V_{Ty}, V_{Tz}	Transverse shear forces in a fillet weld
V_{vi}	Shear force in ith fastener
a	$= \sqrt{(EI_w/GJ)}$, or Distance along member, or Distance from web to shear centre, or Effective throat size of a weld, or Ratio of web to total section area, or Spacing of transverse stiffeners

a_0	Distance from shear centre
$\bar{b}$	c_f or c_w
c	Factor for flange contribution to shear resistance
c_m	Bending coefficient for beam-columns with unequal end moments
d_e	Depth of elastic core
d_f	Distance between flange centroids
d_0	Hole diameter
e_1	End distance in a plate
e_2	Edge distance in a plate
e_{Ny}	Shift of effective compression force from centroid
f	Factor used to modify χ_{LT}
h_w	Clear distance between flanges
$i_{f,z}$	Radius of gyration of equivalent compression flange
i_p	Polar radius of gyration
i_0	$= \sqrt{(i_p^2 + y_0^2 + z_0^2)}$
k	Deflection coefficient, or Modulus of foundation reaction
k_c	Slenderness correction factor, or Correction factor for moment distribution
k_{ij}	Interaction factors for bending and compression
k_s	Factor for hole shape and size
k_t	Axial stiffness of connector
k_v	Shear stiffness of connector
k_σ	Plate buckling coefficient
k_1	Factor for plate tension fracture
k_1, k_2	Stiffness factors
ℓ_{eff}	Effective length of a fillet weld, or Effective length of an unstiffened column flange
ℓ_y	Effective loaded length
m	Fatigue life index, or Torque per unit length
n	Axial compression ratio, or Number of shear planes
p_F	Probability of failure
$p(x)$	Particular integral
p_1	Pitch of bolt holes
p_2	Spacing of bolt hole lines
s	Distance around thin-walled section
s_s	Stiff bearing length
s	Staggered pitch of holes
s_m	Minimum staggered pitch for no reduction in effective area
s_1, s_2	Side widths of a fillet weld
w	$= W_{pl}/W_{el}$
w_{AB}	Settlement of B relative to A
w_c	Mid-span deflection
$\bar{z}$	Distance to centroid

y_p, z_p	Distances to plastic neutral axes
y_r, z_r	Coordinates of centre of rotation
y_0, z_0	Coordinates of shear centre
z_c	Distance to buckling centre of rotation, or Distance to centroid
z_n	Distance below centroid to neutral axis
z_Q	Distance below centroid to load
z_t	Distance below centroid to translational restraint
α	Coefficient used to determine effective width, or Unit warping (see equation 10.35)
α_{bc}	Buckling coefficient for beam columns with unequal end moments
α_{bcl}	Inelastic moment modification factor for bending and compression
α_{bcu}	Value of α_{bc} for ultimate strength
α_d	Factor for plate tear out
α_i	In-plane load factor
α_L	Limiting value of α for second mode buckling
α_{LT}	Imperfection factor for lateral buckling
α_L, α_0	Indices in interaction equations for biaxial bending
α_m	Moment modification factor for beam lateral buckling
α_n	$= (1/A) \int_0^E \alpha t \, \mathrm{d}s$
α_r, α_t	Rotational and translational stiffnesses
α_x, α_y	Stiffnesses of rotational restraints acting about the x, y axes
α_{st}	Buckling moment factor for stepped and tapered beams
α, β	Indices in section interaction equations for biaxial bending
β	Correction factor for the lateral buckling of rolled sections, or Safety index
β_e	Stiffness factor for far end restraint conditions
β_{Lf}	Reduction factor for bolts in long joints
β_m	End moment factor, or Ratio of end moments
β_w	Fillet weld correlation factor
β_y	Monosymmetry section constant for I-beam
$\beta_{2,3}$	Effective net area factors for eccentrically connected tension members
$\Delta\sigma_C$	Reference value for fatigue site
$\Delta\sigma_L$	Fatigue endurance limit
ε	Load height parameter $= (K/\pi)2z_Q/d_f$
Φ	Cumulative frequency distribution of a standard normal variate, or Value used to determine χ
ϕ_{Cd}	Design rotation capacity of a joint
ϕ_j	Joint rotation
$\gamma_F, \gamma_G, \gamma_Q$	Load partial factors
$\gamma_m, \gamma_n, \gamma_s$	Factors used in moment amplification

$\gamma_{1,2}$	Relative stiffnesses
η	Crookedness or imperfection parameter, or Web shear resistance factor ($= 1.2$ for steels up to S460)
$\overline{\lambda}_{c,0}$	Slenderness limit of equivalent compression flange
$\overline{\lambda}_p$	Generalised plate slenderness $= \sqrt{(f_y/\sigma_{cr})}$
λ_1	$= \pi\sqrt{(E/f_y)}$
μ	$= \sqrt{(N/EI)}$
θ	Central twist, or Slope change at plastic hinge, or Torsion stress function
ρ	Perpendicular distance from centroid, or Reduction factor
ρ_m	Monosymmetric section parameter $= I_{zc}/I_z$
ρ_c, ρ_r	Column and rafter factors for portal frame buckling
ρ_0	Perpendicular distance from shear centre
σ_{ac}, σ_{at}	Stresses due to axial compression and tension
σ_{bcy}	Compression stress due to bending about y axis
$\sigma_{bty}, \sigma_{btz}$	Tension stresses due to bending about y, z axes
$\sigma_{cr,l}$	Bending stress at local buckling
$\sigma_{cr,p}$	Bearing stress at local buckling
σ_L	Limiting major axis stress in a crooked and twisted beam at first yield
τ_h, τ_v	Shear stresses due to V_y, V_z
τ_{hc}, τ_{vc}	Shear stresses due to a circulating shear flow
τ_{ho}, τ_{vo}	Shear stresses in an open section
ψ	End moment ratio, or Stress ratio
ψ_0	Load combination factor

Chapter 1

Introduction

1.1 Steel structures

Engineering structures are required to support loads and resist forces, and to transfer these loads and forces to the foundations of the structures. The loads and forces may arise from the masses of the structure, or from man's use of the structures, or from the forces of nature. The uses of structures include the enclosure of space (buildings), the provision of access (bridges), the storage of materials (tanks and silos), transportation (vehicles), or the processing of materials (machines). Structures may be made from a number of different materials, including steel, concrete, wood, aluminium, stone, plastic, etc., or from combinations of these.

Structures are usually three-dimensional in their extent, but sometimes they are essentially two-dimensional (plates and shells), or even one-dimensional (lines and cables). Solid steel structures invariably include comparatively high volumes of high-cost structural steel which are understressed and uneconomic, except in very small-scale components. Because of this, steel structures are usually formed from one-dimensional members (as in rectangular and triangulated frames), or from two-dimensional members (as in box girders), or from both (as in stressed skin industrial buildings). Three-dimensional steel structures are often arranged so that they act as if composed of a number of independent two-dimensional frames or one-dimensional members (Figure 1.1).

Structural steel members may be one-dimensional as for beams and columns (whose lengths are much greater than their transverse dimensions), or two-dimensional as for plates (whose lengths and widths are much greater than their thicknesses), as shown in Figure 1.2c. While one-dimensional steel members may be solid, they are usually thin-walled, in that their thicknesses are much less than their other transverse dimensions. Thin-walled steel members are rolled in a number of cross-sectional shapes [1] or are built up by connecting together a number of rolled sections or plates, as shown in Figure 1.2b. Structural members can be classified as tension or compression members, beams, beam-columns, torsion members, or plates (Figure 1.3), according to the method by which they transmit the forces in the structure. The behaviour and design of these structural members are discussed in this book.

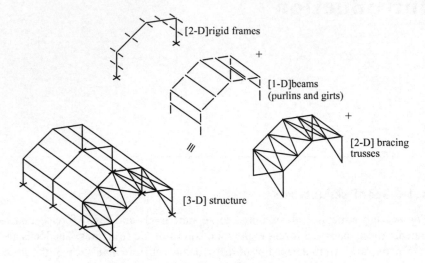

Figure 1.1 Reduction of a [3-D] structure to simpler forms.

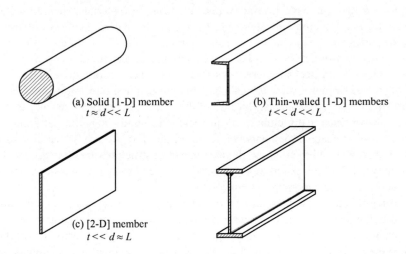

Figure 1.2 Types of structural steel members.

Structural steel members may be connected together at joints in a number of ways, and by using a variety of connectors. These include pins, rivets, bolts, and welds of various types. Steel plate gussets, or angle cleats, or other elements may also be used in the connections. The behaviour and design of these connectors and joints are also discussed in this book.

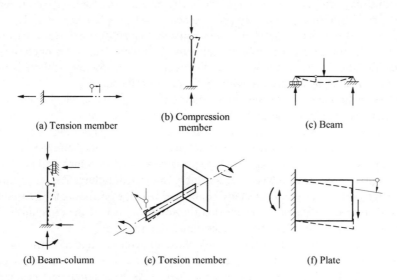

(a) Tension member

(b) Compression
member

(c) Beam

(d) Beam-column

(e) Torsion member

(f) Plate

Figure 1.3 Load transmission by structural members.

This book deals chiefly with steel frame structures composed of one-dimensional members, but much of the information given is also relevant to plate structures. The members are generally assumed to be hot-rolled or fabricated from hot-rolled elements, while the frames considered are those used in buildings. However, much of the material presented is also relevant to bridge structures [2, 3], and to structural members cold-formed from light-gauge steel plates [4–7].

The purposes of this chapter are first, to consider the complete design process and the relationships between the behaviour and analysis of steel structures and their structural design, and second, to present information of a general nature (including information on material properties and structural loads) which is required for use in the later chapters. The nature of the design process is discussed first, and then brief summaries are made of the relevant material properties of structural steel and of the structural behaviour of members and frames. The loads acting on the structures are considered, and the choice of appropriate methods of analysing the steel structures is discussed. Finally, the considerations governing the synthesis of an understanding of the structural behaviour with the results of analysis to form the design processes of EC3 [8] are treated.

1.2   Design

1.2.1   Design requirements

The principal design requirement of a structure is that it should be effective; that is, it should fulfil the objectives and satisfy the needs for which it was created. The structure may provide shelter and protection against the environment by enclosing space, as in buildings; or it may provide access for people and materials, as in bridges; or it may store materials, as in tanks and silos; or it may form part of a machine for transporting people or materials, as in vehicles, or for operating on materials. The design requirement of effectiveness is paramount, as there is little point in considering a structure which will not fulfil its purpose.

The satisfaction of the effectiveness requirement depends on whether the structure satisfies the structural and other requirements. The structural requirements relate to the way in which the structure resists and transfers the forces and loads acting on it. The primary structural requirement is that of safety, and the first consideration of the structural engineer is to produce a structure which will not fail in its design lifetime, or which has an acceptably low risk of failure. The other important structural requirement is usually concerned with the stiffness of the structure, which must be sufficient to ensure that the serviceability of the structure is not impaired by excessive deflections, vibrations, and the like.

The other design requirements include those of economy and of harmony. The cost of the structure, which includes both the initial cost and the cost of maintenance, is usually of great importance to the owner, and the requirement of economy usually has a significant influence on the design of the structure. The cost of the structure is affected not only by the type and quantity of the materials used, but also by the methods of fabricating and erecting it. The designer must therefore give careful consideration to the methods of construction as well as to the sizes of the members of the structure.

The requirements of harmony within the structure are affected by the relationships between the different systems of the structure, including the load resistance and transfer system (the structural system), the architectural system, the mechanical and electrical systems, and the functional systems required by the use of the structure. The serviceability of the structure is usually directly affected by the harmony, or lack of it, between the systems. The structure should also be in harmony with its environment, and should not react unfavourably with either the community or its physical surroundings.

1.2.2   The design process

The overall purpose of design is to invent a structure which will satisfy the design requirements outlined in Section 1.2.1. Thus the structural engineer seeks to invent a structural system which will resist and transfer the forces and loads acting on it with adequate safety, while making due allowance for the requirements of serviceability, economy, and harmony. The process by which this may be achieved is summarised in Figure 1.4.

The first step is to define the overall problem by determining the effectiveness requirements and the constraints imposed by the social and physical environments and by the owner's time and money. The structural engineer will need to consult the owner; the architect, the site, construction, mechanical, and electrical engineers;

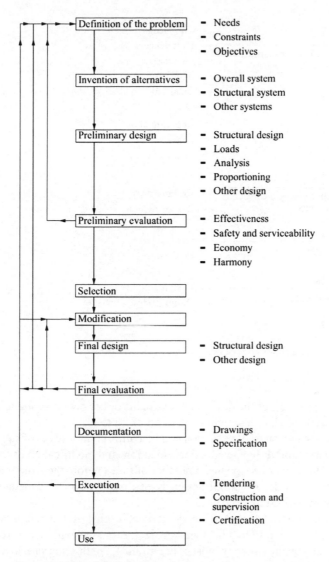

Figure 1.4 The overall design process.

and any authorities from whom permissions and approvals must be obtained. A set of objectives can then be specified, which if met, will ensure the successful solution of the overall design problem.

The second step is to invent a number of alternative overall systems and their associated structural systems which appear to meet the objectives. In doing so, the designer may use personal knowledge and experience or that which can be

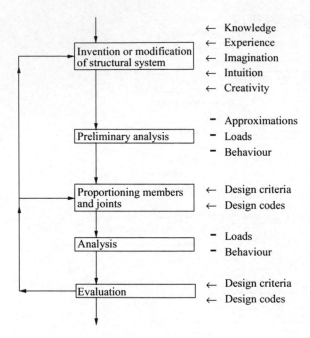

Figure 1.5 The structural design process.

gathered from others [9–12]; or the designer may use his or her own imagination, intuition, and creativity [13], or a combination of all of these.

Following these first two steps of definition and invention come a series of steps which include the structural design, evaluation, selection, and modification of the structural system. These may be repeated a number of times before the structural requirements are met and the structural design is finalised. A typical structural design process is summarised in Figure 1.5.

After the structural system has been invented, it must be analysed to obtain the information required for determining the member sizes. First, the loads supported by and the forces acting on the structure must be determined. For this purpose, loading codes [14, 15] are usually consulted, but sometimes the designer determines the loading conditions or commissions experts to do this. A number of approximate assumptions are made about the behaviour of the structure, which is then analysed and the forces and moments acting on the members and joints of the structure are determined. These are used to proportion the structure so that it satisfies the structural requirements, usually by referring to a design code, such as EC3 [8].

At this stage a preliminary design of the structure will have been completed, however, because of the approximate assumptions made about the structural behaviour, it is necessary to check the design. The first steps are to recalculate the loads and to reanalyse the structure designed, and these are carried out with more precision than was either possible or appropriate for the preliminary analysis. The performance of the structure is then evaluated in relation to the structural requirements, and any changes in the member and joint sizes are decided on. These changes may require a further reanalysis and re-proportioning of the structure, and this cycle may be repeated until no further change is required. Alternatively, it may be necessary to modify the original structural system and repeat the structural design process until a satisfactory structure is achieved.

The alternative overall systems are then evaluated in terms of their service-ability, economy, and harmony, and a final system is selected, as indicated in Figure 1.4. This final overall system may be modified before the design is finalised. The detailed drawings and specifications can then be prepared, and tenders for the construction can be called for and let, and the structure can be constructed. Further modifications may have to be made as a consequence of the tenders submitted or due to unforeseen circumstances discovered during construction.

This book is concerned with the structural behaviour of steel structures, and the relationships between their behaviour and the methods of proportioning them, particularly in relation to the structural requirements of the European steel struc-tures code EC3 and the modifications of these are given in the National Annexes. This code consists of six parts, with basic design using the conventional members being treated in Part 1. This part is divided into 12 sub-parts, with those likely to be required most frequently being:

Part 1.1 General Rules and Rules for Buildings [8],
Part 1.5 Plated Structural Elements [16],
Part 1.8 Design of Joints [17], and
Part 1.10 Selection of Steel for Fracture Toughness and Through-Thickness
    Properties [18].

Other parts that may be required from time to time include: Part 1.2 that covers resistance to fire, Part 1.3 dealing with cold-formed steel, Part 1.9 dealing with fatigue and Part 2 [19] that covers bridges. Composite construction is covered by EC4 [20]. Since Part 1.1 of EC3 is the document most relevant to much of the content of this text (with the exception of Chapter 9 on joints), all references to EC3 made herein should be taken to mean Part 1.1 [8], including any modifications given in the National Annex, unless otherwise indicated.

Detailed discussions of the overall design process are beyond the scope of this book, but further information is given in [13] on the definition of the design problem, the invention of solutions and their evaluation, and in [21–24] on the execution of design. Further, the conventional methods of structural analysis are adequately treated in many textbooks [25–27] and are discussed in only a few isolated cases in this book.

1.3   Material behaviour

1.3.1   Mechanical properties under static load

The important mechanical properties of most structural steels under static load are indicated in the idealised tensile stress–strain diagram shown in Figure 1.6. Initially the steel has a linear stress–strain curve whose slope is the Young's modulus of elasticity E. The values of E vary in the range 200 000–210 000 N/mm^2, and the approximate value of 205 000 N/mm^2 is often assumed (EC3 uses 210 000 N/mm^2). The steel remains elastic while in this linear range, and recovers perfectly on unloading. The limit of the linear elastic behaviour is often closely approximated by the yield stress f_y and the corresponding yield strain $\varepsilon_y = f_y/E$. Beyond this limit the steel flows plastically without any increase in stress until the strain-hardening strain ε_{st} is reached. This plastic range is usually considerable, and accounts for the ductility of the steel. The stress increases above the yield stress f_y when the strain-hardening strain ε_{st} is exceeded, and this continues until the ultimate tensile stress f_u is reached. After this, large local reductions in the cross-section occur, and the load capacity decreases until tensile fracture takes place.

The yield stress f_y is perhaps the most important strength characteristic of a structural steel. This varies significantly with the chemical constituents of the steel, the most important of which are carbon and manganese, both of which increase the yield stress. The yield stress also varies with the heat treatment used and with the amount of working which occurs during the rolling process. Thus thinner plates which are more worked have higher yield stresses than thicker plates of the same constituency. The yield stress is also increased by cold working. The rate of straining affects the yield stress, and high rates of strain increase the upper or first yield stress (see the broken line in Figure 1.6), as well as the lower yield stress f_y. The strain rates used in tests to determine the yield stress of a particular

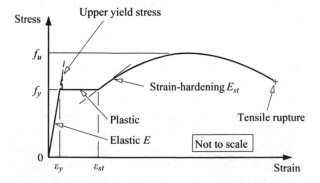

Figure 1.6 Idealised stress–strain relationship for structural steel.

steel type are significantly higher than the nearly static rates often encountered in actual structures.

For design purposes, a 'minimum' yield stress is identified for each different steel classification. For EC3, these classifications are made on the basis of the chemical composition and the heat treatment, and so the yield stresses in each classification decrease as the greatest thickness of the rolled section or plate increases. The minimum yield stress of a particular steel is determined from the results of a number of standard tension tests. There is a significant scatter in these results because of small variations in the local composition, heat treatment, amount of working, thickness, and the rate of testing, and this scatter closely follows a normal distribution curve. Because of this, the minimum yield stress f_y quoted for a particular steel and used in design is usually a characteristic value which has a particular chance (often 95%) of being exceeded in any standard tension test. Consequently, it is likely that an isolated test result will be significantly higher than the quoted yield stress. This difference will, of course, be accentuated if the test is made for any but the thickest portion of the cross-section. In EC3 [8], the yield stress to be used in design is listed in Table 3.1 for hot-rolled structural steel and for structural hollow sections for each of the structural grades.

The yield stress f_y determined for uniaxial tension is usually accepted as being valid for uniaxial compression. However, the general state of stress at a point in a thin-walled member is one of biaxial tension and/or compression, and yielding under these conditions is not so simply determined. Perhaps the most generally accepted theory of two-dimensional yielding under biaxial stresses acting in the $1'2'$ plane is the maximum distortion-energy theory (often associated with names of Huber, von Mises, or Hencky), and the stresses at yield according to this theory satisfy the condition

$$\sigma_{1'}^2 - \sigma_{1'}\sigma_{2'} + \sigma_{2'}^2 + 3\sigma_{1'2'}^2 = f_y^2, \tag{1.1}$$

in which $\sigma_{1'}$, $\sigma_{2'}$ are the normal stresses and $\sigma_{1'2'}$ is the shear stress at the point. For the case where $1'$ and $2'$ are the principal stress directions 1 and 2, equation 1.1 takes the form of the ellipse shown in Figure 1.7, while for the case of pure shear ($\sigma_{1'} = \sigma_{2'} = 0$, so that $\sigma_1 = -\sigma_2 = \sigma_{1'2'}$), equation 1.1 reduces to

$$\sigma_{1'2'} = f_y/\sqrt{3} = \tau_y, \tag{1.2}$$

which defines the shear yield stress τ_y.

1.3.2 Fatigue failure under repeated loads

Structural steel may fracture at low average tensile stresses after a large number of cycles of fluctuating load. This high-cycle fatigue failure is initiated by local damage caused by the repeated loads, which leads to the formation of a small local crack. The extent of the fatigue crack is gradually increased by the subsequent load repetitions, until finally the effective cross-section is so reduced that catastrophic

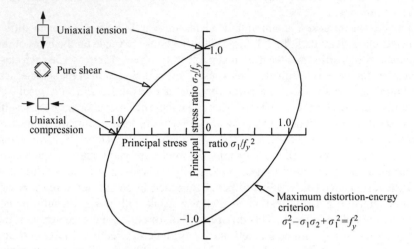

Figure 1.7 Yielding under biaxial stresses.

failure may occur. High-cycle fatigue is only a design consideration when a large number of loading cycles involving tensile stresses is likely to occur during the design life of the structure (compressive stresses do not cause fatigue). This is often the case for bridges, cranes, and structures which support machinery; wind and wave loading may also lead to fatigue problems.

Factors which significantly influence the resistance to fatigue failure include the number of load cycles N, the range of stress

$$\Delta\sigma = \sigma_{max} - \sigma_{min} \tag{1.3}$$

during a load cycle, and the magnitudes of local stress concentrations. An indication of the effect of the number of load cycles is given in Figure 1.8, which shows that the maximum tensile stress decreases from its ultimate static value f_u in an approximately linear fashion as the logarithm of the number of cycles, N, increases. As the number of cycles increases further the curve may flatten out and the maximum tensile stress may approach the endurance limit $\Delta\sigma_L$.

The effects of the stress magnitude and stress ratio on the fatigue life are demonstrated in Figure 1.9. It can be seen that the fatigue life N decreases with increasing stress magnitude σ_{max} and with decreasing stress ratio $R = \sigma_{min}/\sigma_{max}$.

The effect of stress concentration is to increase the stress locally, leading to local damage and crack initiation. Stress concentrations arise from sudden changes in the general geometry and loading of a member, and from local changes due to bolt and rivet holes and welds. Stress concentrations also occur at defects in the member, or its connectors and welds. These may be due to the original rolling of the steel, or due to subsequent fabrication processes, including punching, shearing,

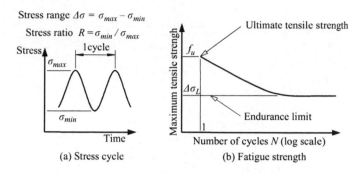

Figure 1.8 Variation of fatigue strength with number of load cycles.

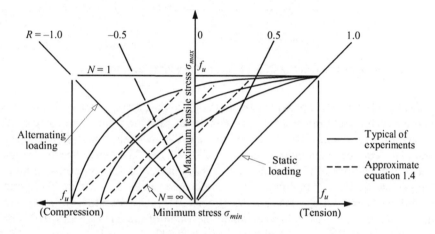

Figure 1.9 Variation of fatigue life with stress magnitudes.

and welding, or due to damage such as that caused by stray arc fusions during welding.

It is generally accepted for design purposes that the fatigue life N varies with the stress range $\Delta\sigma$ according to equations of the type

$$\left(\frac{\Delta\sigma}{\Delta\sigma_C}\right)^m \left(\frac{N}{2 \times 10^6}\right) = K \tag{1.4}$$

in which the reference value $\Delta\sigma_C$ depends on the details of the fatigue site, and the constants m and K may change with the number of cycles N. This assumed dependence of the fatigue life on the stress range produces the approximating straight lines shown in Figure 1.9.

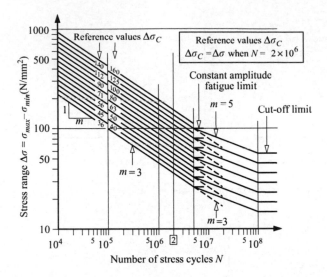

Figure 1.10 Variation of the EC3-1-9 fatigue life with stress range.

EC3-1-1 [8] does not provide a treatment of fatigue, since it is usually the case that either the stress range $\Delta\sigma$ or the number of high amplitude stress cycles N is comparatively small. However, for structures supporting vibrating machinery and plant, reference should be made to EC3-1-9 [28]. The general relationships between the fatigue life N and the service stress range $\Delta\sigma$ for constant amplitude stress cycles are shown in Figure 1.10 for reference values $\Delta\sigma_C$ which correspond to different detail categories. For $N \leq 5 \times 10^6$, $m = 3$ and $K = 1$, so that the reference value $\Delta\sigma_C$ corresponds to the value of $\Delta\sigma$ at $N = 2 \times 10^6$. For $5 \times 10^6 \leq N \leq 10^8$, $m = 5$ and $K = 0.4^{2/3} \approx 0.543$.

Fatigue failure under variable amplitude stress cycles is normally assessed using Miner's rule [29]

$$\sum N_i / N_{im} \leq 1 \tag{1.5}$$

in which N_i is the number of cycles of a particular stress range $\Delta\sigma_i$ and N_{im} the constant amplitude fatigue life for that stress range. If any of the stress ranges exceeds the constant amplitude fatigue limit (at $N = 5 \times 10^6$), then the effects of stress ranges below this limit are included in equation 1.5.

Designing against fatigue involves a consideration of joint arrangement as well as of permissible stress. Joints should generally be so arranged as to minimise stress concentrations and produce as smooth a 'stress flow' through the joint as is practicable. This may be done by giving proper consideration to the layout of a joint, by making gradual changes in section, and by increasing the amount of material used at points of concentrated load. Weld details should also be determined with this in mind, and unnecessary 'stress-raisers' should be avoided. It will also be advantageous to restrict, where practicable, the locations of joints to low stress regions such as at points of contraflexure or near the neutral axis. Further information and guidance on fatigue design are given in [30–33].

1.3.3 Brittle fracture under impact load

Structural steel does not always exhibit a ductile behaviour, and under some circumstances a sudden and catastrophic fracture may occur, even though the nominal tensile stresses are low. Brittle fracture is initiated by the existence or formation of a small crack in a region of high local stress. Once initiated, the crack may propagate in a ductile (or stable) fashion for which the external forces must supply the energy required to tear the steel. More serious are cracks which propagate at high speed in a brittle (or unstable) fashion, for which some of the internal elastic strain energy stored in steel is released and used to fracture the steel. Such a crack is self-propagating while there is sufficient internal strain energy, and will continue until arrested by ductile elements in its path which have sufficient deformation capacity to absorb the internal energy released.

The resistance of a structure to brittle fracture depends on the magnitude of local stress concentrations, on the ductility of the steel, and on the three-dimensional geometrical constraints. High local stresses facilitate crack initiation, and so stress concentrations due to poor geometry and loading arrangements (including impact loading) are dangerous. Also of great importance are flaws and defects in the material, which not only increase the local stresses, but also provide potential sites for crack initiation.

The ductility of a structural steel depends on its composition, heat treatment, and thickness, and varies with temperature and strain rate. Figure 1.11 shows the increase with temperature of the capacity of the steel to absorb energy during

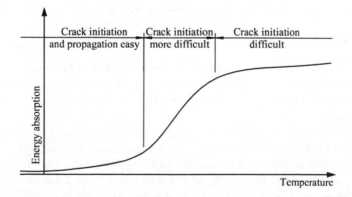

Figure 1.11 Effect of temperature on resistance to brittle fracture.

impact. At low temperatures the energy absorption is low and initiation and propagation of brittle fractures are comparatively easy, while at high temperatures the energy absorption is high because of ductile yielding, and the propagation of cracks can be arrested. Between these two extremes is a transitional range in which crack initiation becomes increasingly difficult. The likelihood of brittle fracture is also increased by high strain rates due to dynamic loading, since the consequent increase in the yield stress reduces the possibility of energy absorption by ductile yielding. The chemical composition of steel has a marked influence on its ductility: brittleness is increased by the presence of excessive amounts of most non-metallic elements, while ductility is increased by the presence of some metallic elements. Steel with large grain size tends to be more brittle, and this is significantly influenced by heat treatment of the steel, and by its thickness (the grain size tends to be larger in thicker sections). EC3-1-10 [18] provides values of the maximum thickness t_1 for different steel grades and minimum service temperatures, as well as advice on using a more advanced fracture mechanics [34] based approach and guidance on safeguarding against lamellar tearing.

Three-dimensional geometrical constraints, such as those occurring in thicker or more massive elements, also encourage brittleness, because of the higher local stresses, and because of the greater release of energy during cracking and the consequent increase in the ease of propagation of the crack.

The risk of brittle fracture can be reduced by selecting steel types which have ductilities appropriate to the service temperatures, and by designing joints with a view to minimising stress concentrations and geometrical constraints. Fabrication techniques should be such that they will avoid introducing potentially dangerous flaws or defects. Critical details in important structures may be subjected to inspection procedures aimed at detecting significant flaws. Of course the designer must give proper consideration to the extra cost of special steels, fabrication techniques, and inspection and correction procedures. Further information on brittle fracture is given in [31, 32, 34].

1.4   Member and structure behaviour

1.4.1   Member behaviour

Structural steel members are required to transmit axial and transverse forces and moments and torques as shown in Figure 1.3. The response of a member to these actions can be described by the load-deformation characteristics shown in Figure 1.12.

A member may have the linear response shown by curve 1 in Figure 1.12, at least until the material reaches the yield stress. The magnitudes of the deformations depend on the elastic moduli E and G. Theoretically, a member can only behave linearly while the maximum stress does not exceed the yield stress f_y, and so the presence of residual stresses or stress concentrations will cause early non-linearity. However, the high ductility of steel causes a local redistribution after this premature yielding, and it can often be assumed without serious error

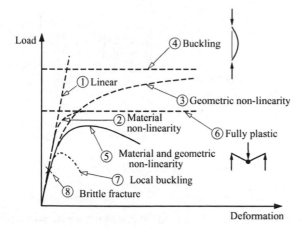

Figure 1.12 Member behaviour.

that the member response remains linear until more general yielding occurs. The member behaviour then becomes non-linear (curve 2) and approaches the condition associated with full plasticity (curve 6). This condition depends on the yield stress f_y.

The member may also exhibit geometric non-linearity, in that the bending moments and torques acting at any section may be influenced by the deformations as well as by the applied forces. This non-linearity, which depends on the elastic moduli E and G, may cause the deformations to become very large (curve 3) as the condition of elastic buckling is approached (curve 4). This behaviour is modified when the material becomes non-linear after first yield, and the load may approach a maximum value and then decrease (curve 5).

The member may also behave in a brittle fashion because of local buckling in a thin plate element of the member (curve 7), or because of material fracture (curve 8).

The actual behaviour of an individual member will depend on the forces acting on it. Thus tension members, laterally supported beams, and torsion members remain linear until their material non-linearity becomes important, and then they approach the fully plastic condition. However, compression members and laterally unsupported beams show geometric non-linearity as they approach their buckling loads. Beam-columns are members which transmit both transverse and axial loads, and so they display both material and geometric non-linearities.

1.4.2 Structure behaviour

The behaviour of a structure depends on the load-transferring action of its members and joints. This may be almost entirely by axial tension or compression, as in the triangulated structures with joint loading as shown in Figure 1.13a.

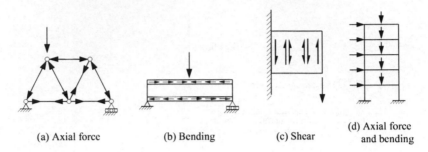

(a) Axial force   (b) Bending   (c) Shear   (d) Axial force and bending

Figure 1.13 Structural load-transfer actions.

Alternatively, the members may support transverse loads which are transferred by bending and shear actions. Usually the bending action dominates in structures composed of one-dimensional members, such as beams and many single-storey rigid frames (Figure 1.13b), while shear becomes more important in two-dimensional plate structures (Figure 1.13c). The members of many structures are subjected to both axial forces and transverse loads, such as those in multistorey buildings (Figure 1.13d). The load-transferring action of the members of a structure depends on the arrangement of the structure, including the geometrical layout and the joint details, and on the loading arrangement.

In some structures, the loading and joints are such that the members are effectively independent. For example, in triangulated structures with joint loads, any flexural effects are secondary, and the members can be assumed to act as if pin-jointed, while in rectangular frames with simple flexible joints the moment transfers between beams and columns may be ignored. In such cases, the response of the structure is obtained directly from the individual member responses.

More generally, however, there will be interactions between the members, and the structure behaviour is not unlike the general behaviour of a member, as can be seen by comparing Figures 1.14 and 1.12. Thus, it has been traditional to assume that a steel structure behaves elastically under the service loads. This assumption ignores local premature yielding due to residual stresses and stress concentrations, but these are not usually serious. Purely flexural structures, and purely axial structures with lightly loaded compression members, behave as if linear (curve 1 in Figure 1.14). However, structures with both flexural and axial actions behave non-linearly, even near the service loads (curve 3 in Figure 1.14). This is a result of the geometrically non-linear behaviour of its members (see Figure 1.12).

Most steel structures behave non-linearly near their ultimate loads, unless they fail prematurely due to brittle fracture, fatigue, or local buckling. This non-linear behaviour is due either to material yielding (curve 2 in Figure 1.14), or member or frame buckling (curve 4), or both (curve 5). In axial structures, failure may

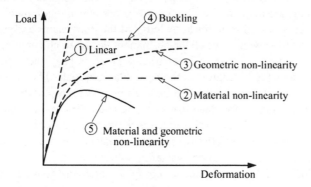

Figure 1.14 Structure behaviour.

involve yielding of some tension members, or buckling either of some compression
members or of the frame, or both. In flexural structures, failure is associated with
full plasticity occurring at a sufficient number of locations that the structure can
form a collapse mechanism. In structures with both axial and flexural actions, there
is an interaction between yielding and buckling (curve 5 in Figure 1.14), and the
failure load is often difficult to determine. The transitions shown in Figure 1.14
between the elastic and ultimate behaviour often take place in a series of non-linear
steps as individual elements become fully plastic or buckle.

1.5   Loads

1.5.1   General

The loads acting on steel structures may be classified as dead loads, as imposed
loads, including both gradually applied and dynamic loads, as wind loads, as earth
or ground-water loads, or as indirect forces, including those due to temperature
changes, foundation settlement, and the like. The more general collective term
Actions is used throughout the Eurocodes. The structural engineer must evaluate
the magnitudes of any of these loads which will act, and must determine those
which are the most severe combinations of loads for which the structure must be
designed. These loads are discussed in the following subsections, both individually
and in combinations.

1.5.2   Dead loads

The dead loads acting on a structure arise from the weight of the structure including
the finishes, and from any other permanent construction or equipment. The dead

loads will vary during construction, but thereafter will remain constant, unless significant modifications are made to the structure or its permanent equipment.

The dead load may be assessed from the knowledge of the dimensions and specific weights or from the total weights of all the permanent items which contribute to the total dead load. Guidance on specific weights is given in [14], the values in which are average values representative of the particular materials. The dimensions used to estimate dead loads should also be average and representative, in order that consistent estimates of the dead loads can be made. By making these assumptions, the statistical distribution of dead loads is often taken as being of a Weibull type [35]. The practice sometimes used of consistently overestimating dimensions and specific weights is often wasteful, and may also be dangerous in cases where the dead load component acts in the opposite sense to the resultant load.

1.5.3 Imposed loads

The imposed loads acting on a structure are gravity loads other than the dead loads, and arise from the weights of materials added to the structure as a result of its use, such as materials stored, people, and snow. Imposed loads usually vary both in space and time. Imposed loads may be sub-divided into two groups, depending on whether they are gradually applied, in which case static load equivalents can be used, or whether they are dynamic, including repeated loads and impact or impulsive loads.

Gradually applied imposed loads may be sustained over long periods of time, or may vary slowly with time [36]. The past practice, however, was to consider only the total imposed load, and so only extreme values (which occur rarely and may be regarded as lifetime maximum loads) were specified. The present imposed loads specified in loading codes [14] often represent peak loads which have 95% probability of not being exceeded over a 50-year period based on a Weibull type distribution [35].

It is usual to consider the most severe spatial distribution of the imposed loads, and this can only be determined by using both the maximum and minimum values of the imposed loads. In the absence of definite knowledge, it is often assumed that the minimum values are zero. When the distribution of imposed load over large areas is being considered, the maximum imposed loads specified, which represent rare events, are often reduced in order to make some allowance for the decreased probability that the maximum imposed loads will act on all areas at the same time.

Dynamic loads which act on structures include both repeated loads and impact and blast loads. Repeated loads are of significance in fatigue problems (see Section 1.3.2), in which case the designer is concerned with both the magnitudes, ranges, and the number of repetitions of loads which are very frequently applied. At the other extreme, impact loads (which are particularly important in the brittle fracture problems discussed in Section 1.3.3) are usually specified by values of extreme magnitude which represent rare events. In structures for which the static loads dominate, it is common to replace the dynamic loads by static force equivalents [14]. However, such a procedure is likely to be inappropriate when the dynamic loads form a significant proportion of the total load, in which case a proper dynamic analysis [37, 38] of the structure and its response should be made.

1.5.4 Wind loads

The wind loads which act on structures have traditionally been allowed for by using static force equivalents. The first step is usually to determine a basic wind speed for the general region in which the structure is to be built by using information derived from meteorological studies. This basic wind speed may represent an extreme velocity measured at a height of 10 m and averaged over a period of 3 seconds which has a return period of 50 years (i.e. a velocity which will, on average, be reached or exceeded once in 50 years, or have a probability of being exceeded of 1/50). The basic wind speed may be adjusted to account for the topography of the site, for the ground roughness, structure size, and height above ground, and for the degree of safety required and the period of exposure. The resulting design wind speed may then be converted into the static pressure which will be exerted by the wind on a plane surface area (this is often referred to as the dynamic wind pressure because it is produced by decelerating the approaching wind velocity to zero at the surface area). The wind force acting on the structure may then be calculated by using pressure coefficients appropriate to each individual surface of the structure, or by using force coefficients appropriate to the total structure. Many values of these coefficients are tabulated in [15], but in special cases where these are inappropriate, the results of wind tunnel tests on model structures may be used.

In some cases it is not sufficient to treat wind loads as static forces. For example, when fatigue is a problem, both the magnitudes and the number of wind fluctuations must be estimated. In other cases, the dynamic response of a structure to wind loads may have to be evaluated (this is often the case with very flexible structures whose long natural periods of vibration are close to those of some of the wind gusts), and this may be done analytically [37, 38], or by specialists using wind tunnel tests. In these cases, special care must be taken to model correctly those properties of the structure which affect its response, including its mass, stiffness, and damping, as well as the wind characteristics and any interactions between wind and structure.

1.5.5 Earth or ground-water loads

Earth or ground-water loads act as pressure loads normal to the contact surface of the structure. Such loads are usually considered to be essentially static.

However, earthquake loads are dynamic in nature, and their effects on the structure must be allowed for. Very flexible structures with long natural periods of vibration respond in an equivalent static manner to the high frequencies of earthquake movements, and so can be designed as if loaded by static force equivalents.

On the other hand, stiff structures with short natural periods of vibration respond significantly, and so in such a case a proper dynamic analysis [37, 38] should be made. The intensities of earthquake loads vary with the region in which the structure is to be built, but they are not considered to be significant in the UK.

1.5.6 Indirect forces

Indirect forces may be described as those forces which result from the straining of a structure or its components, and may be distinguished from the direct forces caused by the dead and applied loads and pressures. The straining may arise from temperature changes, from foundation settlement, from shrinkage, creep, or cracking of structural or other materials, and from the manufacturing process as in the case of residual stresses. The values of indirect forces are not usually specified, and so it is common for the designer to determine which of these forces should be allowed for, and what force magnitudes should be adopted.

1.5.7 Combinations of loads

The different loads discussed in the preceding subsections do not occur alone, but in combinations, and so the designer must determine which combination is the most critical for the structure. However, if the individual loads, which have different probabilities of occurrence and degrees of variability, were combined directly, the resulting load combination would have a greatly reduced probability. Thus, it is logical to reduce the magnitudes of the various components of a combination according to their probabilities of occurrence. This is similar to the procedure used in reducing the imposed load intensities used over large areas.

The past design practice was to use the worst combination of dead load with imposed load and/or wind load, and to allow increased stresses whenever the wind load was included (which is equivalent to reducing the load magnitudes). These increases seem to be logical when imposed, and wind loads act together because the probability that both of these loads will attain their maximum values simultaneously is greatly reduced. However, they are unjustified when applied in the case of dead and wind load, for which the probability of occurrence is virtually unchanged from that of the wind load.

A different and more logical method of combining loads is used in the EC3 limit states design method [8], which is based on statistical analyses of the loads and the structure capacities (see Section 1.7.3.4). Strength design is usually carried out for the most severe combination of actions for normal (termed persistent) or temporary (termed transient) conditions using

$$\sum_{j \geq 1} \gamma_{G,j} G_{k,j} + \gamma_{Q,1} Q_{k,1} + \sum_{i>1} \gamma_{Q,i} \psi_{0,i} Q_{k,i} \qquad (1.6)$$

where Σ implies 'the combined effect of', γ_G and γ_Q are partial factors for the persistent G and variable Q actions, and ψ_0 is a combination factor. The concept is

Table 1.1 Partial load factors for common situations

Ultimate limit state	Permanent actions γ_G		Variable actions γ_Q	
	Unfavourable	Favourable	Unfavourable	Favourable
EQU	1.1	0.9	1.5	0
STR/GEO	1.35	1.0	1.5	0

thus to use all the persistent actions $G_{k,j}$ such as self-weight and fixed equipment with a leading variable action $Q_{k,1}$ such as imposed, snow, or wind load, and reduced values of the other variable actions $Q_{k,i}$. More information on the Eurocode approach to loading for steel structures is given in [39–41].

This approach is applied to the following forms of ultimate limit state:

EQU = loss of static equilibrium of the structure on any part of it

STR = failure by excessive deformation, transformation of the structure or any part of it into a mechanism, rupture or loss of stability of the structure or of any part of it

GEO = failure or excessive deformation of the ground

FAT = fatigue failure

For the most common set of design situations, use of the appropriate values from [41] gives the load factors of Table 1.1.

Using the combination factors of $\psi_0 = 0.7$ and 0.5 of [41] for most variable actions and wind actions, respectively leads to the following common STR load combinations for buildings:

$$1.35\, G + 1.5\, Q_I + 0.75\, Q_W$$

$$1.35\, G + 1.05\, Q_I + 1.5\, Q_W$$

$$-1.0\, G + 1.5\, Q_I, \text{ and}$$

$$-1.0\, G + 1.5\, Q_W.$$

in which the minus signs indicate that the permanent action is favourable.

1.6 Analysis of steel structures

1.6.1 General

In the design process, the assessment of whether the structural design requirements will be met or not requires the knowledge of the stiffness and strength of the structure under load, and of its local stresses and deformations. The term structural analysis is used to denote the analytical process by which this knowledge of the response of the structure can be obtained. The basis for this process is the knowledge of the material behaviour, and this is used first to analyse the behaviour of the individual members and joints of the structure. The behaviour of the complete structure is then synthesised from these individual behaviours.

The methods of structural analysis are fully treated in many textbooks [e.g. 25–27], and so details of these are not within the scope of this book. However, some discussion of the concepts and assumptions of structural analysis is necessary so that the designer can make appropriate assumptions about the structure and make a suitable choice of the method of analysis.

In most methods of structural analysis, the distribution of forces and moments throughout the structure is determined by using the conditions of static equilibrium and of geometric compatibility between the members at the joints. The way in which this is done depends on whether a structure is statically determinate (in which case the complete distribution of forces and moments can be determined by statics alone), or is statically indeterminate (in which case the compatibility conditions for the deformed structure must also be used before the analysis can be completed).

An important feature of the methods of structural analysis is the constitutive relationships between the forces and moments acting on a member or connection and its deformations. These play the same role for the structural element as do the stress–strain relationships for an infinitesimal element of a structural material. The constitutive relationship may be linear (force proportional to deflection) and elastic (perfect recovery on unloading), or they may be non-linear because of material non-linearities such as yielding (inelastic), or because of geometrical non-linearities (elastic) such as when the deformations themselves induce additional moments, as in stability problems.

It is common for the designer to idealise the structure and its behaviour so as to simplify the analysis. A three-dimensional frame structure may be analysed as the group of a number of independent two-dimensional frames, while individual members are usually considered as one-dimensional and the joints as points. The joints may be assumed to be frictionless hinges, or to be semi-rigid or rigid. In some cases, the analysis may be replaced or supplemented by tests made on an idealised model which approximates part or all of the structure.

1.6.2 Analysis of statically determinate members and structures

For an isolated statically determinate member, the forces and moments acting on the member are already known, and the structural analysis is only used to determine the stiffness and strength of the member. A linear elastic (or first-order elastic) analysis is usually made of the stiffness of the member when the material non-linearities are generally unimportant and the geometrical non-linearities are often small. The strength of the member, however, is not so easily determined, as one or both of the material and geometric non-linearities are most important. Instead, the designer usually relies on a design code or specification for this information. The strength of the isolated statically determinate members is fully discussed in Chapters 2–7 and 10.

For a statically determinate structure, the principles of static equilibrium are used in the structural analysis to determine the member forces and moments, and the stiffness and strength of each member are then determined in the same way as for statically determinate members.

1.6.3 Analysis of statically indeterminate structures

A statically indeterminate structure can be approximately analysed if a sufficient number of assumptions are made about its behaviour to allow it to be treated as if determinate. One method of doing this is to guess the locations of points of zero bending moment and to assume there are frictionless hinges at a sufficient number of these locations that the member forces and moments can be determined by statics alone. Such a procedure is commonly used in the *preliminary analysis* of a structure, and followed at a later stage by a more precise analysis. However, a structure designed only on the basis of an approximate analysis can still be safe, provided the structure has sufficient ductility to redistribute any excess forces and moments. Indeed, the method is often conservative, and its economy increases with the accuracy of the estimated locations of the points of zero bending moment. More commonly, a preliminary analysis is made of the structure based on the linear elastic computer methods of analysis [42, 43], using approximate member stiffnesses.

The accurate analysis of statically indeterminate structures is complicated by the interaction between members: the equilibrium and compatibility conditions and the constitutive relationships must all be used in determining the member forces and moments. There are a number of different types of analysis which might be made, and some indication of the relevance of these is given in Figure 1.15 and in the following discussion. Many of these can only be used for two-dimensional frames.

For many structures, it is common to use a *first-order elastic analysis* which is based on linear elastic constitutive relationships and which ignores any geometrical non-linearities and associated instability problems. The deformations determined by such an analysis are proportional to the applied loads, and so the principle of superposition can be used to simplify the analysis. It is often assumed that axial and shear deformations can be ignored in structures whose action is predominantly flexural, and that flexural and shear deformations can be ignored in structures whose member forces are predominantly axial. These assumptions further simplify the analysis, which can then be carried out by any of the well-known methods [25–27], for which many computer programs are available [44, 45]. Some of these programs can be used for three-dimensional frames.

However, a first-order elastic analysis will underestimate the forces and moments in and the deformations of a structure when instability effects are present.

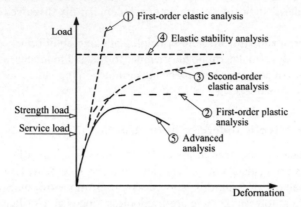

Figure 1.15 Predictions of structural analyses.

Some estimate of the importance of these in the absence of flexural effects can be obtained by making an *elastic stability analysis*. A *second-order elastic analysis* accounts for both flexure and instability, but this is difficult to carry out, although computer programs are now generally available [44, 45]. EC3 permits the use of the results of an elastic stability analysis in the amplification of the first-order moments as an alternative to second-order analysis.

The analysis of statically indeterminate structures near the ultimate load is further complicated by the decisive influence of the material and geometrical non-linearities. In structures without material non-linearities, an *elastic stability analysis* is appropriate when there are no flexural effects, but this is a rare occurrence. On the other hand, many flexural structures have very small axial forces and instability effects, in which case it is comparatively easy to use a *first-order plastic analysis*, according to which a sufficient number of plastic hinges must form to transform the structure into a collapse mechanism.

More generally, the effects of instability must be allowed for, and as a first approximation the nominal first yield load determined from a second-order elastic analysis may be used as a conservative estimate of the ultimate load. A much more accurate estimate may be obtained for structures where local and lateral buckling is prevented by using an *advanced analysis* [46] in which the actual behaviour is closely analysed by allowing for instability, yielding, residual stresses, and initial crookedness. However, this method is not yet in general use.

1.7   Design of steel structures

1.7.1   Structural requirements and design criteria

The designer's task of assessing whether or not a structure will satisfy the structural requirements of serviceability and strength is complicated by the existence of errors and uncertainties in his or her analysis of the structural behaviour and estimation of the loads acting, and even in the structural requirements themselves. The designer usually simplifies this task by using a number of design criteria which allow him or her to relate the structural behaviour predicted by his or her analysis to the structural requirements. Thus the designer equates the satisfaction of these criteria by the predicted structural behaviour with satisfaction of the structural requirements by the actual structure.

In general, the various structural design requirements relate to corresponding *limit states*, and so the design of a structure to satisfy all the appropriate requirements is often referred to as a *limit states design*. The requirements are commonly presented in a deterministic fashion, by requiring that the structure shall not fail, or that its deflections shall not exceed prescribed limits. However, it is not possible to be completely certain about the structure and its loading, and so the structural requirements may also be presented in probabilistic forms, or in deterministic forms derived from probabilistic considerations. This may be done by defining an acceptably low risk of failure within the design life of the structure, after reaching some sort of balance between the initial cost of the structure and the economic and human losses resulting from failure. In many cases there will be a number of structural requirements which operate at different load levels, and it is not unusual to require a structure to suffer no damage at one load level, but to permit some minor damage to occur at a higher load level, provided there is no catastrophic failure.

The structural design criteria may be determined by the designer, or he or she may use those stated or implied in design codes. The stiffness design criteria adopted are usually related to the serviceability limit state of the structure under the service loads, and are concerned with ensuring that the structure has sufficient stiffness to prevent, excessive deflections such as sagging, distortion, and settlement, and excessive motions under dynamic load, including sway, bounce, and vibration.

The strength limit state design criteria are related to the possible methods of failure of the structure under overload and understrength conditions, and so these design criteria are concerned with yielding, buckling, brittle fracture, and fatigue. Also of importance is the ductility of the structure at and near failure: ductile structures give a warning of the impending failure and often redistribute the load effects away from the critical regions, while ductility provides a method of energy dissipation which will reduce the damage due to earthquake and blast loading. On the other hand, a brittle failure is more serious, as it occurs with no warning of failure, and in a catastrophic fashion with a consequent release of stored energy and increase in damage. Other design criteria may also be adopted, such as those related to corrosion and fire.

1.7.2   Errors and uncertainties

In determining the limitations prescribed by design criteria, account must be taken of the deliberate and accidental errors made by the designer, and of the uncertainties in his or her knowledge of the structure and its loads. Deliberate errors include those resulting from the assumptions made to simplify the analysis of the loading and of the structural behaviour. These assumptions are often made so that any errors involved are on the safe side, but in many cases the nature of the errors involved is not precisely known, and some possibility of danger exists.

Accidental errors include those due to a general lack of precision, either in the estimation of the loads and the analysis of the structural behaviour, or in the manufacture and erection of the structure. The designer usually attempts to control the magnitudes of these, by limiting them to what he or she judges to be suitably small values. Other accidental errors include what are usually termed blunders. These may be of a gross magnitude leading to failure or to uneconomic structures, or they may be less important. Attempts are usually made to eliminate blunders by using checking procedures, but often these are unreliable, and the logic of such a process is open to debate.

As well as the errors described above, there exist a number of uncertainties about the structure itself and its loads. The material properties of steel fluctuate, especially the yield stress and the residual stresses. The practice of using a minimum or characteristic yield stress for design purposes usually leads to oversafe designs, especially for redundant structures of reasonable size, for which an average yield stress would be more appropriate because of the redistribution of load which takes place after early yielding. Variations in the residual stress levels are not often accounted for in design codes, but there is a growing tendency to adjust design criteria in accordance with the method of manufacture so as to make some allowance for gross variations in the residual stresses. This is undertaken to some extent in EC3.

The cross-sectional dimensions of rolled-steel sections vary, and the values given in section handbooks are only nominal, especially for the thicknesses of universal sections. The fabricated lengths of a structural member will vary slightly from the nominal length, but this is usually of little importance, except where the variation induces additional stresses because of lack-of-fit problems, or where there is a cumulative geometrical effect. Of some significance to members subject to instability problems are the variations in their straightness which arise during manufacture, fabrication, and erection. Some allowances for these are usually made in design codes, while fabrication and erection tolerances are specified in EN1090 [47] to prevent excessive crookedness.

The loads acting on a structure vary significantly. Uncertainty exists in the designer's estimate of the magnitude of the dead load because of the variations in the densities of materials, and because of the minor modifications to the structure during or subsequent to its erection. Usually these variations are not very significant and a common practice is to err on the safe side by making conservative assumptions. Imposed loadings fluctuate significantly during the design usage of the structure, and may change dramatically with changes in usage. These fluctuations are usually accounted for by specifying what appear to be extreme values in loading codes, but there is often a finite chance that these values will be exceeded.

Wind loads vary greatly and the magnitudes specified in loading codes are usually obtained by probabilistic methods.

1.7.3   Strength design

1.7.3.1   Load and capacity factors, and factors of safety

The errors and uncertainties involved in the estimation of the loads on and the behaviour of a structure may be allowed for in strength design by using load factors to increase the nominal loads and capacity factors to decrease the structural strength. In the previous codes that employed the traditional working stress design, this was achieved by using factors of safety to reduce the failure stresses to permissible working stress values. The purpose of using various factors is to ensure that the probability of failure under the most adverse conditions of structural overload and understrength remains very small. The use of these factors is discussed in the following subsections.

1.7.3.2   Working stress design

The working stress methods of design given in previous codes and specifications required that the stresses calculated from the most adverse combination of loads must not exceed the specified permissible stresses. These specified stresses were obtained after making some allowances for the non-linear stability and material effects on the strength of isolated members, and in effect, were expressions of their ultimate strengths divided by the factors of safety SF. Thus

$$\text{Working stress} \leq \text{Permissible stress} \approx \frac{\text{Ultimate stress}}{\text{SF}} \tag{1.6}$$

It was traditional to use factors of safety of 1.7 approximately.

The working stress method of a previous steel design code [48] has been replaced by the limit states design method of EC3. Detailed discussions of the working stress method are available in the first edition of this book [49].

1.7.3.3   Ultimate load design

The ultimate load methods of designing steel structures required that the calculated ultimate load-carrying capacity of the complete structure must not exceed the most adverse combination of the loads obtained by multiplying the working loads by the appropriate load factors LF. Thus

$$\sum (\text{Working load} \times \text{LF}) \leq \text{Ultimate load} \tag{1.7}$$

These load factors allowed some margins for any deliberate and accidental errors, and for the uncertainties in the structure and its loads, and also provided the structure with a reserve of strength. The values of the factors should depend on the load type and combination, and also on the risk of failure that could be expected and the consequences of failure. A simplified approach often employed (perhaps illogically) was to use a single load factor on the most adverse combination of the working loads.

A previous code [48] allowed the use of the plastic method of ultimate load design when stability effects were unimportant. These have used load factors of 1.70 approximately. However, this ultimate load method has also been replaced by the limit states design method in EC3, and will not be discussed further.

1.7.3.4   Limit states design

It was pointed out in Section 1.5.6 that different types of load have different probabilities of occurrence and different degrees of variability, and that the probabilities associated with these loads change in different ways as the degree of overload considered increases. Because of this, different load factors should be used for the different load types.

Thus for limit states design, the structure is deemed to be satisfactory if its *design load effect* does not exceed its *design resistance*. The design load effect is an appropriate bending moment, torque, axial force, or shear force, and is calculated from the sum of the effects of the specified (or characteristic) loads F_k multiplied by partial factors $\gamma_{G,Q}$ which allow for the variabilities of the loads and the structural behaviour. The design resistance R_k/γ_M is calculated from the specified (or characteristic) resistance R_k divided by the partial factor γ_M which allows for the variability of the resistance. Thus

$$\text{Design load effect} \leq \text{Design resistance} \tag{1.8a}$$

or

$$\sum \gamma_{g,Q} \times (\text{effect of specified loads}) \leq (\text{specified resistance}/\gamma_M) \tag{1.8b}$$

Although the limit states design method is presented in deterministic form in equations 1.8, the partial factors involved are usually obtained by using probabilistic models based on statistical distributions of the loads and the capacities. Typical statistical distributions of the total load and the structural capacity are shown in Figure 1.16. The probability of failure p_F is indicated by the region for which the load distribution exceeds that for the structural capacity.

In the development of limit state codes, the probability of failure p_F is usually related to a parameter β, called the safety index, by the transformation

$$\Phi(-\beta) = p_F, \tag{1.9}$$

where the function Φ is the cumulative frequency distribution of a standard normal variate [35]. The relationship between β and p_F shown in Figure 1.17 indicates

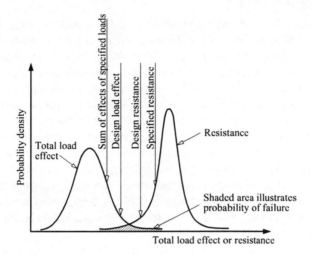

Figure 1.16 Limit states design.

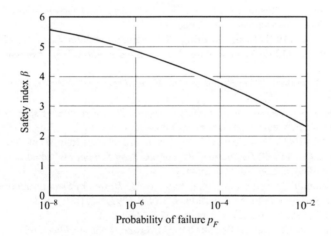

Figure 1.17 Relationship between safety index and probability of failure.

that an increase in β of 0.5 implies a decrease in the probability of failure by approximately an order of magnitude.

The concept of the safety index was used to derive the partial factors for EC3. This was done with reference to previous national codes such as BS 5950 [50] to obtain comparable values of the probability of failure p_F, although much of the detailed calibration treated the load and resistance sides of equations 1.8 separately.

1.7.4   Stiffness design

In the stiffness design of steel structures, the designer seeks to make the structure sufficiently stiff so that its deflections under the most adverse working load conditions will not impair its strength or serviceability. These deflections are usually calculated by a first-order linear elastic analysis, although the effects of geometrical non-linearities should be included when these are significant, as in structures which are susceptible to instability problems. The design criteria used in the stiffness design relate principally to the serviceability of the structure, in that the flexibility of the structure should not lead to damage of any non-structural components, while the deflections should not be unsightly, and the structure should not vibrate excessively. It is usually left to the designer to choose limiting values for use in these criteria which are appropriate to the structure, although a few values are suggested in some design codes. The stiffness design criteria which relate to the strength of the structure itself are automatically satisfied when the appropriate strength design criteria are satisfied.

References

1. British Standards Institution (2005) *BS4-1 Structural Steel Sections-Part 1: Specification for hot-rolled sections*, BSI, London.
2. O'Connor, C. (1971) *Design of Bridge Superstructures*, John Wiley, New York.
3. Chatterjee, S. (1991) *The Design of Modern Steel Bridges*, BSP Professional Books, Oxford.
4. Rhodes, J. (editor) (1991) *Design of Cold Formed Members*, Elsevier Applied Science, London.
5. Rhodes, J. and Lawson, R.M. (1992) *Design of Steel Structures Using Cold Formed Steel Sections*, Steel Construction Institute, Ascot.
6. Yu, W.-W. (2000) *Cold-Formed Steel Design*, 3rd edition, John Wiley, New York.
7. Hancock, G.J. (1998) *Design of Cold-Formed Steel Structures*, 3rd edition, Australian Institute of Steel Construction, Sydney.
8. British Standards Institution (2005) *Eurocode 3: Design of Steel Structures: Part 1.1 General Rules and Rules for Buildings, BS EN 1993-1-1*, BSI, London.
9. Nervi, P.L. (1956) *Structures*, McGraw-Hill, New York.
10. Salvadori, M.G. and Heller, R. (1963) *Structure in Architecture*, Prentice-Hall, Englewood Cliffs, New Jersey.
11. Torroja, E. (1958) *The Structures of Educardo Torroja*, F.W. Dodge Corp., New York.
12. Fraser, D.J. (1981) *Conceptual Design and Preliminary Analysis of Structures*, Pitman Publishing Inc., Marshfield, Massachusetts.
13. Krick, E.V. (1969) *An Introduction to Engineering and Engineering Design*, 2nd edition, John Wiley, New York.
14. British Standards Institution (2002) *Eurocode 1: Actions on Structures: Part 1.1 General Actions – Densities, Self-weight, Imposed Loads for Buildings, BS EN 1991-1-1*, BSI, London.
15. British Standards Institution (2005) *Eurocode 1: Actions on Structures: Part 1.4 General Actions – Wind Actions, BS EN 1991-1-4*, BSI, London.

16. British Standards Institution (2006) *Eurocode 3: Design of Steel Structures: Part 1.5 Plated Structural Elements, BS EN 1993-1-5*, BSI, London.
17. British Standards Institution (2006) *Eurocode 3: Design of Steel Structures: Part 1.8 Design of Joints, BS EN 1993-1-8*, BSI, London.
18. British Standards Institution (2006) *Eurocode 3: Design of Steel Structures: Part 1.10 Selection of Steel for Fracture Toughness and Through-Thickness Properties, BS EN 1993-1-10*, BSI, London.
19. British Standards Institution (2006) *Eurocode 3: Design of Steel Structures: Part 2 Steel Bridges, BS EN 1993-2*, BSI, London.
20. British Standards Institution (2005) *Eurocode 4: Design of Composite Steel and Concrete Structures: Part 1.1 General Rules and Rules for Buildings, BS EN 1994-1-1*, BSI, London.
21. Davison, B. and Owens, G.W. (eds) (2003) *Steel Designers' Manual*, 6th edition, Blackwell Publishing, Oxford.
22. CIMsteel (1997) *Design for Construction*, Steel Construction Institute, Ascot.
23. Antill, J.M. and Ryan, P.W.S. (1982) *Civil Engineering Construction*, 5th edition, McGraw-Hill, Sydney.
24. Australian Institute of Steel Construction (1991) *Economical Structural Steelwork*, AISC, Sydney.
25. Norris, C.H., Wilbur, J.B., and Utku, S. (1976) *Elementary Structural Analysis*, 3rd edition, McGraw-Hill, New York.
26. Coates, R.C., Coutie, M.G., and Kong, F.K. (1990) *Structural Analysis*, 3rd edition, Van Nostrand Reinhold (UK), Wokingham.
27. Ghali, A., Neville A.M., and Brown, T.G. (1997) *Structural Analysis – A Unified Classical and Matrix Approach*, 5th edition, Routledge, Oxford.
28. British Standards Institution (2006) *Eurocode 3: Design of Steel Structures: Part 1.9 Fatigue Strength of Steel Structures, BS EN 1993-1-9*, BSI, London. *Rules for Buildings*, ECS, Brussels.
29. Miner, M.A. (1945) Cumulative damage in fatigue, *Journal of Applied Mechanics, ASME*, **12**, No. 3, September, pp. A-159–A-164.
30. Gurney, T.R. (1979) *Fatigue of Welded Structures*, 2nd edition, Cambridge University Press.
31. Lay, M.G. (1982) *Structural Steel Fundamentals*, Australian Road Research Board, Melbourne.
32. Rolfe, S.T. and Barsoum, J.M. (1977) *Fracture and Fatigue Control in Steel Structures*, Prentice-Hall, Englewood Cliffs, New Jersey.
33. Grundy, P. (1985) Fatigue limit state for steel structures, *Civil Engineering Transactions*, Institution of Engineers, Australia, CE**27**, No. 1, February, pp. 143–8.
34. Navy Department Advisory Committee on Structural Steels (1970) *Brittle Fracture in Steel Structures* (ed. G.M. Boyd), Butterworth, London.
35. Walpole, R.E., Myers, R.H., Myers, S.L., and Ye, K. (2007) *Probability and Statistics for Engineers and Scientists*, 8th edition, Prentice-Hall, Englewood Cliffs.
36. Ravindra, M.K. and Galambos, T.V. (1978) Load and resistance factor design for steel, *Journal of the Structural Division, ASCE*, **104**, No. ST9, September, pp. 1337–54.
37. Clough, R.W. and Penzien, J. (2004) *Dynamics of Structures*, 2nd edition, CSI Books, Berkeley, California.
38. Irvine, H.M. (1986) *Structural Dynamics for the Practising Engineer*, Allen and Unwin, London.

39. Gulvanesian, H., Calgaro, J.-A., and Holicky, M. (2002) *Designers' Guide to EN 1990, Eurocode: Basis of Structural Design*, Thomas Telford Publishing, London.
40. Gardner, L. and Nethercot, D.A. (2005) *Designers' Guide to EN 1993-1-1*, Thomas Telford Publishing, London.
41. Gardner, L. and Grubb, P.J. (2008) *Guide to Eurocode Load Combinations for Steel Structures*, British Constructional Steelwork Association.
42. Harrison, H.B. (1973) *Computer Methods in Structural Analysis*, Prentice-Hall, Englewood Cliffs, New Jersey.
43. Harrison, H.B. (1990) *Structural Analysis and Design, Parts 1 and 2*, Pergamon Press, Oxford.
44. Computer Service Consultants (2006) *S-Frame 3D Structural Analysis Suite*, CSC (UK) Limited, Leeds.
45. Research Engineers Europe Limited (2006) *STAAD.Pro Space Frame Analysis*, REE, Bristol.
46. Clarke, M.J. (1994) Plastic-zone analysis of frames, Chapter 6 of *Advanced Analysis of Steel Frames: Theory, Software and Applications* (eds W.F. Chen and S. Toma), CRC Press, Inc., Boca Raton, Florida, pp. 259–319.
47. British Standards Institution (1998) *Execution of Steel Structures – General Rules and Rules for Buildings*, BS DD ENV 1090-1, BSI, London.
48. British Standards Institution (1969) *BS449: Part 2: Specification for the Use of Structural Steel in Building*, BSI, London.
49. Trahair, N.S. (1977) *The Behaviour and Design of Steel Structures*, 1st edition, Chapman and Hall, London.
50. British Standards Institution (2000) *BS5950. The Structural Use of Steelwork in Building: Part 1: Code of Practice for Design: Rolled and Welded Sections*, BSI, London.

Chapter 2

Tension members

2.1　Introduction

Concentrically loaded uniform tension members are perhaps the simplest structural elements, as they are nominally in a state of uniform axial stress. Because of this, their load-deformation behaviour very closely parallels the stress–strain behaviour of structural steel obtained from the results of tensile tests (see Section 1.3.1). Thus a member remains essentially linear and elastic until the general yield load is approached, even if it has residual stresses and initial crookedness.

However, in many cases a tension member is not loaded or connected concentrically or it has transverse loads acting, resulting in bending actions as well as an axial tension action. Simple design procedures are available which enable the bending actions in some members with eccentric connections to be ignored, but more generally special account must be taken of the bending action in design.

Tension members often have comparatively high average stresses, and in some cases the effects of local stress concentrations may be significant, especially when there is a possibility that the steel material may not act in a ductile fashion. In such cases, the causes of stress concentrations should be minimised, and the maximum local stresses should be estimated and accounted for.

In this chapter, the behaviour and design of steel tension members are discussed. The case of concentrically loaded members is dealt with first, and then a procedure which allows the simple design of some eccentrically connected tension members is presented. The design of tension members with eccentric or transverse loads is then considered, the effects of stress concentrations are discussed, and finally the design of tension members according to EC3 is dealt with.

2.2　Concentrically loaded tension members

2.2.1　Members without holes

The straight concentrically loaded steel tension member of length L and constant cross-sectional area A which is shown in Figure 2.1a has no holes and is free from residual stress. The axial extension e of the member varies with the load N in the

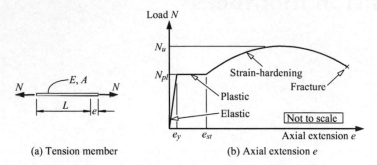

(a) Tension member　　　　　(b) Axial extension e

Figure 2.1 Load-extension behaviour of a perfect tension member.

same way as does the average strain $\varepsilon = e/L$ with the average stress $\sigma = N/A$, and so the load-extension relationship for the member shown in Figure 2.1b is similar to the material stress–strain relationship shown in Figure 1.6. Thus the extension at first increases linearly with the load and is equal to

$$e = \frac{NL}{EA},\tag{2.1}$$

where E is the Young's modulus of elasticity. This linear increase continues until the yield stress f_y of the steel is reached at the general yield load

$$N_{pl} = Af_y\tag{2.2}$$

when the extension increases with little or no increase in load until strain-hardening commences. After this, the load increases slowly until the maximum value

$$N_u = Af_u\tag{2.3}$$

is reached, in which f_u is the ultimate tensile strength of the steel. Beyond this, a local cross-section of the member necks down and the load N decreases until fracture occurs.

The behaviour of the tension member is described as ductile, in that it can reach and sustain the general yield load while significant extensions occur, before it fractures. The general yield load N_{pl} is often taken as the load capacity of the member.

If the tension member is not initially stress free, but has a set of residual stresses induced during its manufacture such as that shown in Figure 2.2b, then local yielding commences before the general yield load N_{pl} is reached (Figure 2.2c), and the range over which the load-extension behaviour is linear decreases. However, the general yield load N_{pl} at which the whole cross-section is yielded can still be reached because the early yielding causes a redistribution of the stresses. The

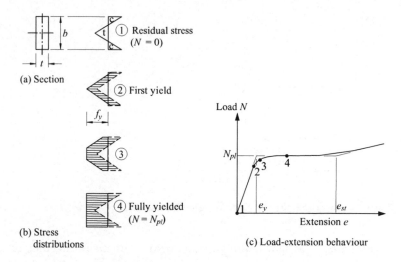

(a) Section

(b) Stress
    distributions

(c) Load-extension behaviour

Figure 2.2 Effect of residual stresses on load-extension behaviour.

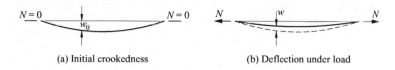

(a) Initial crookedness

(b) Deflection under load

Figure 2.3 Tension member with initial crookedness.

residual stresses also cause local early strain-hardening, and while the plastic range is shortened (Figure 2.2c), the member behaviour is still regarded as ductile.

If, however, the tension member has an initial crookedness w_0 (Figure 2.3a), the axial load causes it to bend and deflect laterally w (Figure 2.3b). These lateral deflections partially straighten the member so that the bending action in the central region is reduced. The bending induces additional axial stresses (see Section 5.3), which cause local early yielding and strain-hardening, in much the same way as do residual stresses (see Figure 2.2c). The resulting reductions in the linear and plastic ranges are comparatively small when the initial crookedness is small, as is normally the case, and the member behaviour is ductile.

2.2.2 Members with small holes

The presence of small local holes in a tension member (such as small bolt holes used for the connections of the member) causes early yielding around the holes, so that the load-deflection behaviour becomes non-linear. When the holes are small, the member may reach the gross yield load (see equation 2.2) calculated

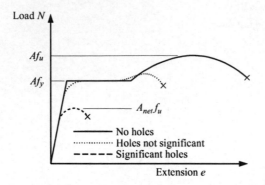

Figure 2.4 Effect of holes on load-extension behaviour.

on the gross area A, as shown in Figure 2.4, because of strain-hardening effects around the holes. In this case, the member behaviour is ductile, and the non-linear behaviour can be ignored because the axial extension of the member under load is not significantly increased, except when there are so many holes along the length of the member that the average cross-sectional area is significantly reduced. Thus the extension e can normally be calculated by using the gross cross-sectional area A in equation 2.1.

2.2.3   Members with significant holes

When the holes are large, the member may fail before the gross yield load N_{pl} is reached by fracturing at a hole, as shown in Figure 2.4. The local fracture load

$$N_u = A_{net} f_u \tag{2.4}$$

is calculated on the net area of the cross-section A_{net} measured perpendicular to the line of action of the load, and is given by

$$A_{net} = A - \sum d_0 t, \tag{2.5}$$

where d_0 is the diameter of a hole, t the thickness of the member at the hole, and the summation is carried out for all holes in the cross-section under consideration. The fracture load N_u is determined by the weakest cross-section, and therefore by the minimum net area A_{net}. A member which fails by fracture before the gross yield load can be reached is not ductile, and there is little warning of failure.

In many practical tension members with more than one row of holes, the reduction in the cross-sectional area may be reduced by staggering the rows of holes (Figure 2.5). In this case, the possibility must be considered of failure along a zig–zag path such as ABCDE in Figure 2.5, instead of across the section perpendicular

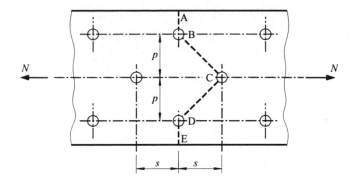

Figure 2.5 Possible failure path with staggered holes.

to the load. The minimum amount of stagger s_m for which a hole no longer reduces the area of the member depends on the diameter d_0 of the hole and the inclination p/s of the failure path, where p is the gauge distance between the rows of holes. An approximate expression for this minimum stagger is

$$s_m \approx (4pd_0)^{1/2}. \tag{2.6}$$

When the actual stagger s is less than s_m, some reduced part of the hole area A_h must be deducted from the gross area A, and this can be approximated by $A_h = d_0 t(1 - s^2/s_m^2)$, whence

$$A_h = d_0 t(1 - s^2/4pd_0), \tag{2.7}$$

$$A_{net} = A - \sum d_0 t + \sum s^2 t/4p, \tag{2.8}$$

where the summations are made for all the holes on the zig–zag path considered, and for all the staggers in the path. The use of equation 2.8 is allowed for in Clause 6.2.2.2 of EC3, and is illustrated in Section 2.7.1.

Holes in tension members also cause local stress increases at the hole boundaries, as well as the increased average stresses N/A_{net} discussed above. These local stress concentrations and their influence on member strength are discussed later in Section 2.5.

2.3 Eccentrically and locally connected tension members

In many cases the fabrication of tension members is simplified by making their end connections eccentric (i.e. the centroid of the connection does not coincide with the centroidal axis of the member) and by connecting to some but not all of the elements in the cross-section. It is common, for example, to make connections

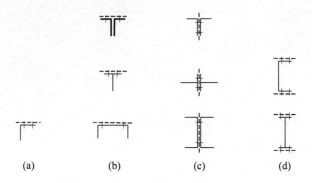

(a)   (b)   (c)   (d)

Figure 2.6 Eccentrically and locally connected tension members.

to an angle section through one leg only, to a tee-section through the flange (or table), or to a channel section through the web (Figure 2.6). The effect of eccentric connections is to induce bending moments in the member, whilst the effect of connecting to some but not all elements in the cross-section is to cause those regions most remote from the connection point(s) to carry less load. The latter is essentially a shear lag effect (see Section 5.4.5). Both of these effects are local to the connections, decreasing along the member length, and are reduced further by ductile stress redistribution after the onset of yielding. Members connected by some but not all of the elements in the cross-section (including those connected symmetrically as in Figure 2.6d) are also discussed in Section 2.5.

While tension members in bending can be designed rationally by using the procedure described in Section 2.4, simpler methods [1–3] also produce satisfactory results. In these simpler methods, the effects described above are approximated by reducing the cross-sectional area of the member (to an effective net area), and by designing it as if concentrically loaded. For a single angle in tension connected by a single row of bolts in one leg, the effective net section $A_{net,eff}$ to be used in place of the net area A_{net} in equation 2.4 is defined in EC3-1-8 [4]. It is dependent on the number of bolts and the pitch p_1, and for one bolt, is given by

$$A_{net,eff} = 2.0(e_2 - 0.5\,d_0)t,\qquad\qquad(2.9)$$

for two bolts by

$$A_{net,eff} = \beta_2 A_{net},\qquad\qquad(2.10)$$

and for three or more bolts by

$$A_{net,eff} = \beta_3 A_{net}.\qquad\qquad(2.11)$$

The symbols in equations 2.9–2.11 are defined below and in Figure 2.7, and A_{net} is the net area of the angle. For an unequal angle connected by its smaller leg, A_{net}

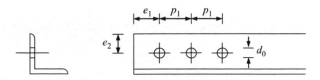

Figure 2.7 Definition of symbols for tension connections in single angles.

should be taken as the net section of an equivalent equal angle of leg length equal to the smaller leg of the unequal angle. In the case of welded end connections, for an equal angle, or an unequal angle connected by its larger leg, the eccentricity may be neglected, and the effective area A_{net} may be taken as equal to the gross area A (Clause 4.13(2) of EC3-1-8 [4]).

For a pair of bolts in a single row in an angle leg, β_2 in equation 2.10 is taken as 0.4 for closely spaced bolts ($p_1 \leq 2.5\ d_0$), as 0.7 if they are widely spaced ($p_1 \geq 5\ d_0$), or as a linear interpolation for intermediate spacings. For three or more bolts in a single row in an angle leg, β_3 in equation 2.11 is taken as 0.5 for closely spaced bolts ($p_1 \leq 2.5\ d_0$), as 0.7 if they are widely spaced ($p_1 \geq 5\ d_0$), or again as a linear interpolation for intermediate spacings.

2.4   Bending of tension members

Tension members often have bending actions caused by eccentric connections, eccentric loads, or transverse loads, including self-weight, as shown in Figure 2.8. These bending actions, which interact with the tensile loads, reduce the ultimate strengths of tension members, and must therefore be accounted for in design.

The axial tension N and transverse deflections w of a tension member caused by any bending action induce restoring moments Nw, which oppose the bending action. It is a common (and conservative) practice to ignore these restoring moments which are small when either the bending deflections w or the tensile load N are small. In this case the maximum stress σ_{max} in the member can be safely approximated by

$$\sigma_{max} = \sigma_{at} + \sigma_{bty} + \sigma_{btz}, \tag{2.12}$$

where $\sigma_{at} = N/A$ is the average tensile stress and σ_{bty} and σ_{btz} are the maximum tensile bending stresses caused by the major and minor axis bending actions M_y and M_z alone (see Section 5.3). The nominal first yield of the member therefore occurs when $\sigma_{max} = f_y$, whence

$$\sigma_{at} + \sigma_{bty} + \sigma_{btz} = f_y. \tag{2.13}$$

When either the tensile load N or the moments M_y and M_z are not small, the first yield prediction of equation 2.13 is inaccurate. A suggested interaction equation

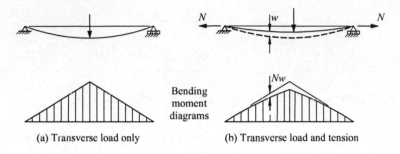

Figure 2.8 Bending of a tension member.

for failure is the linear inequality

$$\frac{M_{y,Ed}}{M_{ty,Rd}} + \frac{M_{z,Ed}}{M_{rz,Rd}} \le 1, \tag{2.14}$$

where $M_{rz,Rd}$ is the cross-section resistance for tension and bending about the minor principal (z) axis given by

$$M_{rz,Rd} = M_{cz,Rd}(1 - N_{t,Ed}/N_{t,Rd}) \tag{2.15}$$

and $M_{ty,Rd}$ is the lesser of the cross-section resistance $M_{ry,Rd}$ for tension and bending about the major principal (y) axis given by

$$M_{ry,Rd} = M_{cy,Rd}(1 - N_{t,Ed}/N_{t,Rd}) \tag{2.16}$$

and the out-of-plane member buckling resistance $M_{bt,Rd}$ for tension and bending about the major principal axis given by

$$M_{bt,Rd} = M_{b,Rd}(1 + N_{t,Ed}/N_{t,Rd}) \le M_{cy,Rd} \tag{2.17}$$

in which $M_{b,Rd}$ is the lateral buckling resistance when $N = 0$ (see Chapter 6).

In these equations, $N_{t,Rd}$ is the tensile resistance in the absence of bending (taken as the lesser of N_{pl} and N_u), while $M_{cy,Rd}$ and $M_{cz,Rd}$ are the cross-section resistances for bending alone about the y and z axes (see Sections 4.7.2 and 5.6.1.3). Equation 2.14 is similar to the first yield condition of equation 2.13, but includes a simple approximation for the possibility of lateral buckling under large values of $M_{y,Ed}$ through the use of equation 2.17.

2.5 Stress concentrations

High local stress concentrations are not usually important in ductile materials under static loading situations, because the local yielding resulting from these concentrations causes a favourable redistribution of stress. However, in situations for which it is doubtful whether the steel will behave in a ductile manner, as when there is a possibility of brittle fracture of a tension member under dynamic loads (see Section 1.3.3), or when repeated load applications may lead to fatigue failure (see Section 1.3.2), stress concentrations become very significant. It is usual to try to avoid or minimise stress concentrations by providing suitable joint and member details, but this is not always possible, in which case some estimate of the magnitudes of the local stresses must be made.

Stress concentrations in tension members occur at holes in the member, and where there are changes or very local reductions in the cross-section, and at points where concentrated forces act. The effects of a hole in a tension member (of net width b_{net} and thickness t) are shown by the uppermost curve in Figure 2.9, which is a plot of the variation of the stress concentration factor (the ratio of the maximum tensile stress to the nominal stress averaged over the reduced cross-section $b_{net}t$) with the ratio of the hole radius r to the net width b_{net}. The maximum stress approaches three times the nominal stress when the plate width is very large $(r/b_{net} \to 0)$.

This curve may also be used for plates with a series of holes spaced equally across the plate width, provided b_{net} is taken as the minimum width between adjacent holes. The effects of changes in the cross-section of a plate in tension are shown by the lower curves in Figure 2.9, while the effects of some notches or very local reductions in the cross-section are given in [5]. Methods of analysing these and other stress concentrations are discussed in [6].

Large concentrated forces are usually transmitted to structural members by local bearing, and while this produces high local compressive stresses, any subsequent

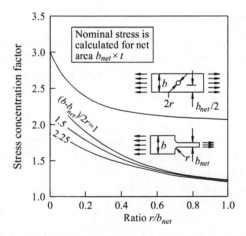

Figure 2.9 Stress concentrations in tension members.

plastic yielding is not usually serious. Of more importance are the increased tensile stresses which result when only some of the plate elements of a tension member are loaded (see Figure 2.6), as discussed in Section 2.3. In this case, conservative estimates can be made of these stresses by assuming that the resulting bending and axial actions are resisted solely by the loaded elements.

2.6   Design of tension members

2.6.1   General

In order to design or to check a tension member, the design tensile force $N_{t,Ed}$ at each cross-section in the member is obtained by a frame analysis (see Chapter 8) or by statics if the member is statically determinate, using the appropriate loads and partial load factors γ_G, γ_Q (see Section 1.7). If there is substantial bending present, the design moments $M_{y,Ed}$ and $M_{z,Ed}$ about the major and minor principal axes, respectively are also obtained.

The process of checking a specified tension member or of designing an unknown member is summarised in Figure 2.10 for the case where bending actions can be ignored. If a specified member is to be checked, then both the strength limit states of gross yielding and the net fracture are considered, so that both the yield strength f_y and the ultimate strength f_u are required. When the member size is not known, an approximate target area A can be established. The trial member selected may be checked for the fracture limit state once its holes, connection eccentricities, and net area A_{net} have been established, and modified if necessary.

The following sub-sections describe each of the EC3 check and design processes for a statically loaded tension member. Worked examples of their application are given in Sections 2.7.1–2.7.5.

2.6.2   Concentrically loaded tension members

The EC3 method of strength design of tension members which are loaded concentrically follows the philosophy of Section 2.2, with the two separate limit states of yield of the gross section and fracture of the net section represented by a single equation

$$N_{t,Ed} \leq N_{t,Rd} \tag{2.18}$$

where $N_{t,Rd}$ is the design tension resistance which is taken as the lesser of the yield (or plastic) resistance of the cross-section $N_{pl,Rd}$ and the ultimate (or fracture) resistance of the cross-section containing holes $N_{u,Rd}$.

The yield limit state of equation 2.2 is given in EC3 as

$$N_{pl,Rd} = Af_y/\gamma_{M0} \tag{2.19}$$

where A is the gross area of the cross-section and γ_{M0} is the partial factor for cross-section resistance, with a value of 1.0.

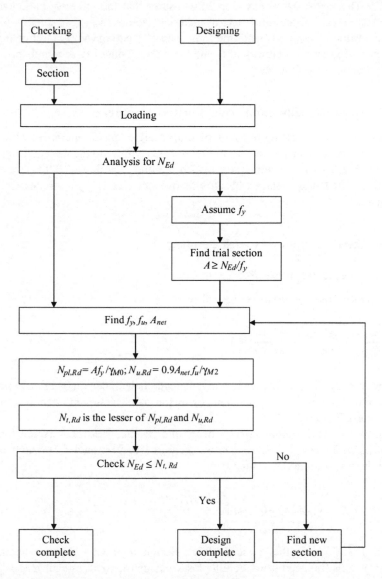

Figure 2.10 Flow chart for the design of tension members.

The fracture limit state of equation 2.4 is given in EC3 as

$$N_{u,Rd} = 0.9A_{net}f_u/\gamma_{M2}, \tag{2.20}$$

where A_{net} is the net area of the cross-section and γ_{M2} is the partial factor for resistance in tension to fracture, with a value of 1.1 given in the National Annex to EC3. The factor 0.9 in equation 2.20 ensures that the effective partial factor $\gamma_{M2}/0.9 (\approx 1.22)$ for the limit state of material fracture ($N_{u,Rd}$) is suitably higher than the value of $\gamma_{M0}(=1.0)$ for the limit state of yielding ($N_{pl,Rd}$), reflecting the influence of greater variability in f_u and the reduced ductility of members which fail by fracture at bolt holes.

2.6.3   Eccentrically connected tension members

The EC3 method of strength design of simple angle tension members which are connected eccentrically as shown in Figure 2.6 is similar to the method discussed in Section 2.6.2 for concentrically loaded members, with the '$0.9A_{net}$' used in equation 2.20 being replaced by an effective net area $A_{net,eff}$, as described in Section 2.3.

2.6.4   Tension members with bending

2.6.4.1   Cross-section resistance

EC3 provides the conservative inequality

$$\frac{N_{t,Ed}}{N_{t,Rd}} + \frac{M_{y,Ed}}{M_{y,Rd}} + \frac{M_{z,Ed}}{M_{z,Rd}} \leq 1 \tag{2.21}$$

for the cross-section resistance of tension members with design bending moments $M_{y,Ed}$ and $M_{z,Ed}$, where $M_{y,Rd}$ and $M_{z,Rd}$ are the cross-section moment resistances (Sections 4.7.2 and 5.6.1.3).

Equation 2.21 is rather conservative, and so EC3 allows I-section tension members with Class 1 or Class 2 cross-sections (see Section 4.7.2) with bending about the major (y) axis to satisfy

$$M_{y,Ed} \leq M_{N,y,Rd} = M_{pl,y,Rd} \left(\frac{1 - N_{t,Ed}/N_{pl,Rd}}{1 - 0.5a} \right) \leq M_{pl,y,Rd} \tag{2.22}$$

in which $M_{N,y,Rd}$ is the reduced plastic design moment resistance about the major (y) axis (reduced from the full plastic design moment resistance $M_{pl,y,Rd}$ to account for the axial force (see Section 7.2.4.1) and

$$a = (A - 2bt_f)/A \leq 0.5, \tag{2.23}$$

in which b and t_f are the flange width and thickness respectively. For I-section tension members with Class 1 or Class 2 cross-sections with bending about the minor (z) axis, EC3 allows

$$M_{z,Ed} \leq M_{N,z,Rd} = M_{pl,z,Rd} \left\{ 1 - \left(\frac{N_{t,Ed}/N_{pl,Rd} - a}{1 - a} \right)^2 \right\} \leq M_{pl,z,Rd}, \quad (2.24)$$

where $M_{N,z,Rd}$ is the reduced plastic design moment resistance about the minor (z) axis.

EC3 also provides an alternative to equation 2.21 for the section resistance to axial tension and bending about both principal axes of Class 1 or Class 2 cross-sections. This is the more accurate interaction equation (see Section 7.4.2)

$$\left(\frac{M_{y,Ed}}{M_{N,y,Rd}} \right)^\alpha + \left(\frac{M_{z,Ed}}{M_{N,z,Rd}} \right)^\beta \leq 1, \qquad (2.25)$$

where the values of α and β depend on the cross-section type (e.g. $\alpha = 2.0$ and $\beta = 5N_{t,Ed}/N_{pl,Rd}$ for I-sections and $\alpha = \beta = 2$ for circular hollow sections).

A worked example of a tension member with bending actions is given in Section 2.7.5.

2.6.4.2 Member resistance

EC3 does not provide any specific rules for determining the influence of lateral buckling on the member resistance of tension members with bending. This may be accounted for conservatively by using

$$M_{y,Ed} \leq M_{b,Rd} \qquad (2.26)$$

in equation 2.21, in which $M_{b,Rd}$ is the design buckling moment resistance of the member (see Section 6.5). This equation conservatively omits the strengthening effect of tension on lateral buckling. A less conservative method is provided by equations 2.14–2.17.

2.7 Worked examples

2.7.1 Example 1 – net area of a bolted universal column section member

Problem. Both flanges of a universal column section member have 22 mm diameter holes arranged as shown in Figure 2.11a. If the gross area of the section is 201×10^2 mm^2 and the flange thickness is 25 mm, determine the net area A_{net} of the member which is effective in tension.

Solution. Using equation 2.6, the minimum stagger is $s_m = \sqrt{(4 \times 60 \times 22)} = 72.7$ mm > 30 mm $= s$. The failure path through each flange is therefore staggered, and by inspection, it includes four holes and two staggers. The net area can

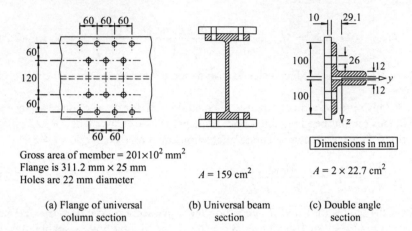

Gross area of member = 201×10^2 mm^2
Flange is 311.2 mm × 25 mm
Holes are 22 mm diameter

$A = 159$ cm^2

$A = 2 \times 22.7$ cm^2

Dimensions in mm

(a) Flange of universal
column section

(b) Universal beam
section

(c) Double angle
section

Figure 2.11 Examples 1–6.

therefore be calculated from equation 2.8 (or Clause 6.2.2.2(4) of EC3) as

$$A_{net} = [201 \times 10^2] - [2 \times 4 \times (22 \times 25)] + [2 \times 2 \times (30^2 \times 25)/(4 \times 60)]$$
$$= 161 \times 10^2 \text{ mm}^2.$$

2.7.2   Example 2 – checking a bolted universal column section member

Problem. Determine the tension resistance of the tension member of example 1 assuming it is of S355 steel.

Solution.

$t_f = 25$ mm, $f_y = 345$ N/mm^2, $f_u = 490$ N/mm^2 EN 10025-2

$N_{pl,Rd} = A f_y / \gamma_{M0} = 201 \times 10^2 \times 345/1.0 = 6935$ kN.   6.2.3(2)a

From Section 2.7.1,

$A_{net} = 161 \times 10^2$ mm^2

$N_{u,Rd} = 0.9 A_{net} f_u / \gamma_{M2} = 0.9 \times 161 \times 10^2 \times 490/1.1 = 6445$ kN.
     6.2.3(2)b

$N_{t,Rd} = 6445$ kN (the lesser of $N_{pl,Rd}$ and $N_{u,Rd}$)   6.2.3(2)

2.7.3   Example 3 – checking a bolted universal beam section member

Problem. A 610×229 UB 125 tension member of S355 steel is connected through both flanges by 20 mm bolts (in 22 mm diameter bolt holes) in four lines, two in each flange as shown in Figure 2.11b. Check the member for a design tension force of $N_{t,Ed} = 4000$ kN.

Solution.

$t_f = 19.6$ mm, $f_y = 345$ N/mm^2, $f_u = 490$ N/mm^2 EN 10025-2

$A = 15\,900$ mm^2

$N_{pl,Rd} = Af_y/\gamma_{M0} = 15900 \times 345/1.0 = 5486$ kN   6.2.3(2)a

$A_{net} = 15\,900 - (4 \times 22 \times 19.6) = 14175$ mm^2 6.2.2.2(3)

$N_{u,Rd} = 0.9A_{net}f_u/\gamma_{M2} = 0.9 \times 14175 \times 490/1.1 = 5683$ kN   6.2.3(2)b

$N_{t,Rd} = 5486$ kN (the lesser of $N_{pl,Rd}$ and $N_{u,Rd}$) > 4000 kN $= N_{t,Ed}$
                                                    6.2.3(2)

and so the member is satisfactory.

2.7.4   Example 4 – checking an eccentrically connected single (unequal) angle

Problem. A tension member consists of a $150 \times 75 \times 10$ single unequal angle whose ends are connected to gusset plates through the larger leg by a single row of four 22 mm bolts in 24 mm holes at 60 mm centres. Use the method of Section 2.3 to check the member for a design tension force of $N_{t,Ed} = 340$ kN, if the angle is of S355 steel and has a gross area of 21.7 cm^2.

Solution.

$t = 10$ mm, $f_y = 355$ N/mm^2, $f_u = 490$ N/mm^2 EN 10025-2

Gross area of cross-section, $A = 2170$ mm^2

$N_{pl,Rd} = Af_y/\gamma_{M0} = 2170 \times 355/1.0 = 770.4$ kN   6.2.3(2)a

$A_{net} = 2170 - (24 \times 10) = 1930$ mm^2 6.2.2.2(3)

$\beta_3 = 0.5$ (since the pitch $p_1 = 60$ mm $= 2.5d_0$)   EC3-1-8 3.10.3(2)

$N_{u,Rd} = \beta_3 A_{net}f_u/\gamma_{M2} = 0.5 \times 1930 \times 490/1.1 = 429.9$ kN   6.2.3(2)b

$N_{t,Rd} = 429.9$ kN (the lesser of $N_{pl,Rd}$ and $N_{u,Rd}$) > 340 kN $= N_{t,Ed}$
                                                    6.2.3(2)

and so the member is satisfactory.

2.7.5 Example 5 – checking a member under combined tension and bending

Problem. A tension member consists of two equal angles of S355 steel whose ends are connected to gusset plates as shown in Figure 2.11c. If the load eccentricity for the tension member is 39.1 mm and the tension and bending resistances are 1438 kN and 37.7 kNm, determine the resistance of the member by treating it as a member under combined tension and bending.

Solution.
Substituting into equation 2.21 leads to

$$N_{t,Ed}/1438 + N_{t,Ed} \times (39.1/1000)/37.7 \leq 1 \qquad\qquad 6.2.1(7)$$

so that $N_{t,Ed} \leq 577.2$ kN.

Because the load eccentricity causes the member to bend about its minor axis, there is no need to check for lateral buckling, which only occurs when there is major axis bending.

2.7.6 Example 6 – estimating the stress concentration factor

Problem. Estimate the maximum stress concentration factor for the tension member of Section 2.7.1.

Solution. For the inner line of holes, the net width is

$$b_{net} = 60 + 30 - 22 = 68 \text{ mm, and so}$$
$$r/b_{net} = (22/2)/68 = 0.16,$$

and so using Figure 2.9, the stress concentration factor is approximately 2.5.

However, the actual maximum stress is likely to be greater than 2.5 times the nominal average stress calculated from the effective area $A_{net} = 161 \times 10^2 \text{ mm}^2$ determined in Section 2.7.1, because the unconnected web is not completely effective. A safe estimate of the maximum stress can be determined on the basis of the flange areas only of

$$A_{net} = [2 \times 310.6 \times 25] - [2 \times 4 \times 22 \times 25] + [2 \times 2 \times (30^2 \times 25)/(4 \times 60)]$$
$$= 115.1 \times 10^2 \text{ mm}^2.$$

2.8 Unworked examples

2.8.1 Example 7 – bolting arrangement

A channel section tension member has an overall depth of 381 mm, width of 101.6 mm, flange and web thicknesses of 16.3 and 10.4 mm, respectively, and an yield strength of 345 N/mm^2. Determine an arrangement for bolting the web of the channel to a gusset plate so as to minimise the gusset plate length and maximise the tension member resistance.

2.8.2 Example 8 – checking a channel section member connected by bolts

Determine the design tensile resistance $N_{t,Rd}$ of the tension member of Section 2.8.1.

2.8.3 Example 9 – checking a channel section member connected by welds

If the tension member of Section 2.8.1 is fillet welded to the gusset plate instead of bolted, determine its design tensile resistance $N_{t,Rd}$.

2.8.4 Example 10 – checking a member under combined tension and bending

The beam of example 1 in Section 6.15.1 has a concentric axial tensile load $5Q$ in addition to the central transverse load Q. Determine the maximum value of Q.

References

1. Regan, P.E. and Salter, P.R. (1984) Tests on welded-angle tension members, *Structural Engineer*, **62B**(2), pp. 25–30.
2. Nelson, H.M. (1953) *Angles in Tension*. Publication No. 7, British Constructional Steel Association, pp. 9–18.
3. Bennetts, I.D., Thomas, I.R., and Hogan, T.J. (1986) Design of statically loaded tension members, *Civil Engineering Transactions*, Institution of Engineers, Australia, CE**28**, No. 4, November, pp. 318–27.
4. British Standards Institute (2006) *Eurocode 3: Design of Steel Structures – Part 1–8: Design of Joints*, BSI, London.
5. Roark, R.J. (1965) *Formulas for Stress and Strain*, 4th edition, McGraw-Hill, New York.
6. Timoshenko, S.P. and Goodier, J.N. (1970) *Theory of Elasticity*, 3rd edition, McGraw-Hill, New York.

Chapter 3

Compression members

3.1 Introduction

The compression member is the second type of axially loaded structural element, the first type being the tension member discussed in Chapter 2. Very stocky compression members behave in the same way as do tension members until after the material begins to flow plastically at the squash load $N_y = Af_y$. However, the resistance of a compression member decreases as its length increases, in contrast to the axially loaded tension member whose resistance is independent of its length. Thus, the compressive resistance of a very slender member may be much less than its tensile resistance, as shown in Figure 3.1.

This decrease in resistance is caused by the action of the applied compressive load N which causes bending in a member with initial curvature (see Figure 3.2a). In a tension member, the corresponding action decreases the initial curvature, and so this effect is usually ignored. However, the curvature and the lateral deflection of a compression member increase with the load, as shown in Figure 3.2b. The compressive stresses on the concave side of the member also increase until the member fails due to excessive yielding. This bending action is accentuated by the slenderness of the member, and so the resistance of a compression member decreases as its length increases.

For the hypothetical limiting case of a perfectly straight elastic member, there is no bending until the applied load reaches the elastic buckling value N_{cr} (EC3 refers to N_{cr} as the elastic critical force). At this load, the compression member begins to deflect laterally, as shown in Figure 3.2b, and these deflections grow until failure occurs at the beginning of compressive yielding. This action of suddenly deflecting laterally is called flexural buckling.

The elastic buckling load N_{cr} provides a measure of the slenderness of a compression member, while the squash load N_y gives an indication of its resistance to yielding. In this chapter, the influences of the elastic buckling load and the squash load on the behaviour of concentrically loaded compression members are discussed and related to their design according to EC3. Local buckling of thin-plate elements in compression members is treated in Chapter 4, while the behaviour and design of eccentrically loaded compression

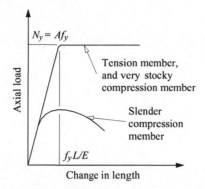

Figure 3.1 Resistances of axially loaded members.

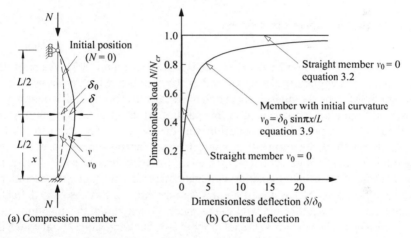

(a) Compression member   (b) Central deflection

Figure 3.2 Elastic behaviour of a compression member.

members are discussed in Chapter 7, and compression members in frames in Chapter 8.

3.2 Elastic compression members

3.2.1 Buckling of straight members

A perfectly straight member of a linear elastic material is shown in Figure 3.3a. The member has a frictionless hinge at each end, its lower end being fixed in position while its upper end is free to move vertically but is prevented from deflecting horizontally. It is assumed that the deflections of the member remain small.

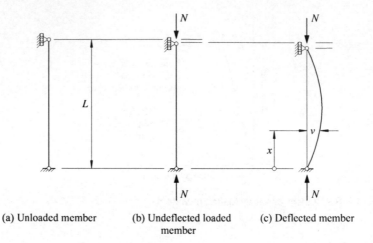

(a) Unloaded member   (b) Undeflected loaded   (c) Deflected member
                              member

Figure 3.3 Straight compression member.

The unloaded position of the member is shown in Figure 3.3a. A concentric axial load N is applied to the upper end of the member, which remains straight (Figure 3.3b). The member is then deflected laterally by a small amount v as shown in Figure 3.3c, and is held in this position. If the original straight position of Figure 3.3b is one of stable equilibrium, the member will return to it when released from the deflected position, while if the original position is one of unstable equilibrium, the member will collapse away from it when released. When the equilibrium of the original position is neither stable nor unstable, the member is in a condition described as one of neutral equilibrium. For this case, the deflected position is one of equilibrium, and the member will remain in this position when released. Thus, when the load N reaches the elastic buckling value N_{cr} at which the original straight position of the member is one of neutral equilibrium, the member may deflect laterally without any change in the load, as shown in Figure 3.2b.

The load N_{cr} at which a straight compression member buckles laterally can be determined by finding a deflected position which is one of equilibrium. It is shown in Section 3.8.1 that this position is given by

$$v = \delta \sin \pi x / L \tag{3.1}$$

in which δ is the undetermined magnitude of the central deflection, and that the elastic buckling load is

$$N_{cr} = \pi^2 EI / L^2 \tag{3.2}$$

in which EI is the flexural rigidity of the compression member.

The elastic buckling load N_{cr} and the elastic buckling stress

$$\sigma_{cr} = N_{cr}/A \tag{3.3}$$

can be expressed in terms of the geometrical slenderness ratio L/i by

$$N_{cr} = \sigma_{cr}A = \frac{\pi^2 EA}{(L/i)^2} \tag{3.4}$$

in which $i = \sqrt{(I/A)}$ is the radius of gyration (which can be determined for a number of sections by using Figure 5.6). The buckling load varies inversely as the square of the slenderness ratio L/i, as shown in Figure 3.4, in which the dimensionless buckling load N_{cr}/N_y is plotted against the generalised slenderness ratio

$$\bar{\lambda} = \sqrt{\frac{N_y}{N_{cr}}} = \sqrt{\frac{f_y}{\sigma_{cr}}} = \frac{L}{i}\sqrt{\frac{f_y}{\pi^2 E}} \tag{3.5}$$

in which

$$N_y = Af_y \tag{3.6}$$

is the squash load. If the material ceases to be linear elastic at the yield stress f_y, then the above analysis is only valid for $\bar{\lambda} = \sqrt{(N_y/N_{cr})} = \sqrt{(f_y/\sigma_{cr})} \geq 1$. This limit is equivalent to a slenderness ratio L/i of approximately 85 for a material with a yield stress f_y of 275 N/mm^2.

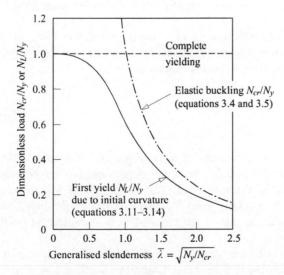

Figure 3.4 Buckling and yielding of compression members.

3.2.2   Bending of members with initial curvature

Real structural members are not perfectly straight, but have small initial curvatures, as shown in Figure 3.2a. The buckling behaviour of the hypothetical straight
members discussed in Section 3.2.1 must therefore be interpreted as the limiting behaviour of real members with infinitesimally small initial curvatures. The
initial curvature of the real member causes it to bend from the commencement
of application of the axial load, and this increases the maximum stress in the
member.

If the initial curvature is such that

$$v_0 = \delta_0 \sin \pi x/L, \tag{3.7}$$

then the deflection of member is given by

$$v = \delta \sin \pi x/L, \tag{3.8}$$

where

$$\frac{\delta}{\delta_0} = \frac{N/N_{cr}}{1 - N/N_{cr}}, \tag{3.9}$$

as shown in Section 3.8.2. The variation of the dimensionless central deflection
δ/δ_0 is shown in Figure 3.2b, and it can be seen that deflection begins at the
commencement of loading and increases rapidly as the elastic buckling load N_{cr}
is approached.

The simple load-deflection relationship of equation 3.9 is the basis of the
Southwell plot technique for extrapolating the elastic buckling load from experimental measurements. If equation 3.9 is rearranged as

$$\frac{\delta}{N} = \frac{1}{N_{cr}}\delta + \frac{\delta_0}{N_{cr}}, \tag{3.10}$$

then the linear relation between δ/N and δ shown in Figure 3.5 is obtained. Thus,
if a straight line is drawn which best fits the points determined from experimental measurements of N and δ, the reciprocal of the slope of this line gives an
experimental estimate of the buckling load N_{cr}. An estimate of the magnitude
δ_0 of the initial crookedness can also be determined from the intercept on the
horizontal axis.

As the deflections v increase with the load N, so also do the bending moments
and the stresses. It is shown in Section 3.8.2 that the limiting axial load N_L at which
the compression member first yields (due to a combination of axial plus bending

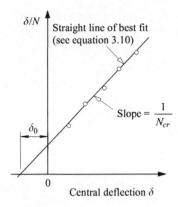

Figure 3.5 Southwell plot.

stresses) is given by

$$\frac{N_L}{N_y} = \frac{1}{\varPhi + \sqrt{\varPhi^2 - \bar{\lambda}^2}}, \tag{3.11}$$

in which

$$\varPhi = \frac{1 + \eta + \bar{\lambda}^2}{2}, \tag{3.12}$$

$$\eta = \frac{\delta_0 b}{2i^2}, \tag{3.13}$$

and b is the width of the member. The variation of the dimensionless limiting axial load N_L/N_y with the generalised slenderness ratio $\bar{\lambda}$ is shown in Figure 3.4 for the case when

$$\eta = \frac{1}{4}\frac{N_y}{N_{cr}}. \tag{3.14}$$

For stocky members, the limiting load N_L approaches the squash load N_y, while for slender members the limiting load approaches the elastic buckling load N_{cr}. Equations 3.11 and 3.12 are the basis for the member buckling resistance in EC3.

3.3   Inelastic compression members

3.3.1   Tangent modulus theory of buckling

The analysis of a perfectly straight elastic compression member given in Section 3.2.1 applies only to a material whose stress–strain relationship remains

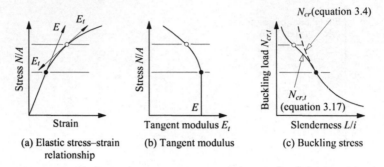

Figure 3.6 Tangent modulus theory of buckling.

linear. However, the buckling of an elastic member of a non-linear material, such as that whose stress–strain relationship is shown in Figure 3.6a, can be analysed by a simple modification of the linear elastic treatment. It is only necessary to note that the small bending stresses and strains, which occur during buckling, are related by the tangent modulus of elasticity E_t corresponding to the average compressive stress N/A (Figure 3.6a and b), instead of the initial modulus E. Thus the flexural rigidity is reduced from EI to $E_t I$, and the tangent modulus buckling load $N_{cr,t}$ is obtained from equation 3.4 by substituting E_t for E, whence

$$N_{cr,t} = \frac{\pi^2 E_t A}{(L/i)^2}. \tag{3.15}$$

The deviation of this tangent modulus buckling load $N_{cr,t}$ from the elastic buckling load N_{cr} is shown in Figure 3.6c. It can be seen that the deviation of $N_{cr,t}$ from N_{cr} increases as the slenderness ratio L/i decreases.

3.3.2   Reduced modulus theory of buckling

The tangent modulus theory of buckling is only valid for elastic materials. For inelastic nonlinear materials, the changes in the stresses and strains are related by the initial modulus E when the total strain is decreasing, and the tangent modulus E_t only applies when the total strain is increasing, as shown in Figure 3.7. The flexural rigidity $E_r I$ of an inelastic member during buckling therefore depends on both E and E_t. As a direct consequence of this, the effective (or reduced) modulus of the section E_r depends on the geometry of the cross-section as well. It is shown in Section 3.9.1 that the reduced modulus of a rectangular section is given by

$$E_r = \frac{4EE_t}{\left(\sqrt{E} + \sqrt{E_t}\right)^2}. \tag{3.16}$$

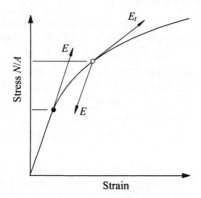

Figure 3.7 Inelastic stress–strain relationship.

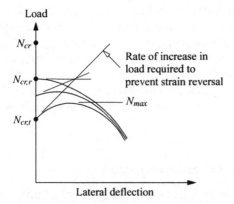

Figure 3.8 Shanley's theory of inelastic buckling.

The reduced modulus buckling load $N_{cr,r}$ can be obtained by substituting E_r for E in equation 3.4, whence

$$N_{cr,r} = \frac{\pi^2 E_r A}{(L/i)^2}.$$

(3.17)

Since the tangent modulus E_t is less than the initial modulus E, it follows that $E_t < E_r < E$, and so the reduced modulus buckling load $N_{cr,r}$ lies between the tangent modulus buckling load $N_{cr,t}$ and the elastic buckling load N_{cr}, as indicated in Figure 3.8.

3.3.3 Shanley's theory of inelastic buckling

Although the tangent modulus theory appears to be invalid for inelastic materials, careful experiments have shown that it leads to more accurate predictions than the apparently rigorous reduced modulus theory. This paradox was resolved by Shanley [1], who reasoned that the tangent modulus theory is valid when buckling is accompanied by a simultaneous increase in the applied load (see Figure 3.8) of sufficient magnitude to prevent strain reversal in the member. When this happens, all the bending stresses and strains are related by the tangent modulus of elasticity E_t, the initial modulus E does not feature, and so the buckling load is equal to the tangent modulus value $N_{cr,t}$.

As the lateral deflection of the member increases as shown in Figure 3.8, the tangent modulus E_t decreases (see Figure 3.6b) because of the increased axial and bending strains, and the post-buckling curve approaches a maximum load N_{max} which defines the ultimate resistance of the member. Also shown in Figure 3.8 is a post-buckling curve which commences at the reduced modulus load $N_{cr,r}$ (at which buckling can take place without any increase in the load). The tangent modulus load $N_{cr,t}$ is the lowest load at which buckling can begin, and the reduced modulus load $N_{cr,r}$ is the highest load for which the member can remain straight. It is theoretically possible for buckling to begin at any load between $N_{cr,t}$ and $N_{cr,r}$.

It can be seen that not only is the tangent modulus load more easily calculated, but it also provides a conservative estimate of the member resistance, and is in closer agreement with experimental results than the reduced modulus load. For these reasons, the tangent modulus theory of inelastic buckling has gained wide acceptance.

3.3.4 Buckling of members with residual stresses

The presence of residual stresses in an intermediate length steel compression member may cause a significant reduction in its buckling resistance. Residual stresses are established during the cooling of a hot-rolled or welded steel member (and during plastic deformation such as cold-rolling). The shrinking of the late-cooling regions of the member induces residual compressive stresses in the early-cooling regions, and these are balanced by equilibrating tensile stresses in the late-cooling regions. In hot-rolled I-section members, the flange – web junctions are the least exposed to cooling influences, and so these are the regions of residual tensile stress, as shown in Figure 3.9, while the more exposed flange tips are regions of residual compressive stress. In a straight intermediate length compression member, the residual compressive stresses cause premature yielding under reduced axial loads, as shown in Figure 3.10, and the member buckles inelastically at a load which is less than the elastic buckling load N_{cr}.

In applying the tangent modulus concept of inelastic buckling to a steel which has the stress–strain relationships shown in Figure 1.6 (see Chapter 1), the strain-hardening modulus E_{st} is sometimes used in the yielded regions as well as in

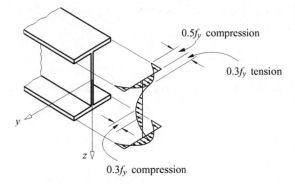

Figure 3.9 Idealised residual stress pattern.

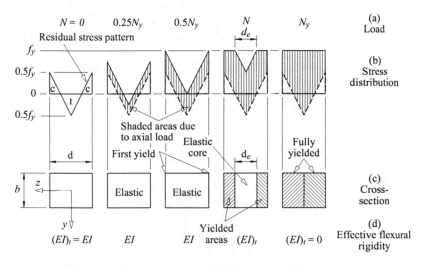

Figure 3.10 Effective section of a member with residual stresses.

the strain-hardened regions. This use is based on the slip theory of dislocation, in which yielding is represented by a series of dynamic jumps instead of by a smooth quasi-static flow. Thus the material in the yielded region is either elastic or strain-hardened, and its subsequent behaviour may be estimated conservatively by using the strain-hardening modulus. However, the even more conservative assumption that the tangent modulus E_t is zero in both the yielded and strain-hardened regions is frequently used because of its simplicity. According to this assumption, these regions of the member are ineffective during buckling, and the moment of resistance is entirely due to the elastic core of the section.

Thus, for a rectangular section member which has the simplified residual stress distribution shown in Figure 3.10, the effective flexural rigidity $(EI)_t$ about the z axis is (see Section 3.9.2)

$$(EI)_t = EI[2(1 - N/N_y)]^{1/2}, \tag{3.18}$$

when the axial load N is greater than $0.5N_y$ at which first yield occurs, and the axial load at buckling $N_{cr,t}$ is given by

$$\frac{N_{cr,t}}{N_y} = \left(\frac{N_{cr}}{N_y}\right)^2 \{[1 + 2(N_y/N_{cr})^2]^{1/2} - 1\}. \tag{3.19}$$

The variation of this dimensionless tangent modulus buckling load $N_{cr,t}/N_y$ with the generalised slenderness ratio $\bar{\lambda} = \sqrt{(N_y/N_{cr})}$ is shown in Figure 3.11. For stocky members, the buckling load $N_{cr,t}$ approaches the squash load N_y, while for intermediate length members it approaches the elastic buckling load N_{cr} as $\bar{\lambda} = \sqrt{(N_y/N_{cr})}$ approaches $\sqrt{2}$. For more slender members, premature yielding does not occur, and these members buckle at the elastic buckling load N_{cr}.

Also shown in Figure 3.11 is the dimensionless tangent modulus buckling load $N_{cr,t}/N_y$ given by

$$\frac{N_{cr,t}}{N_y} = 1 - \frac{1}{4}\frac{N_y}{N_{cr}}, \tag{3.20}$$

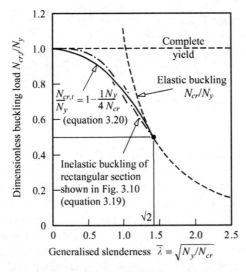

Figure 3.11 Inelastic buckling of compression members with residual stresses.

which was developed as a compromise between major and minor axis buckling of hot-rolled I-section members. It can be seen that this simple compromise is very similar to the relationship given by equation 3.19 for the rectangular section member.

3.4 Real compression members

The conditions under which real members act differ in many ways from the idealised conditions assumed in Section 3.2.1 for the analysis of the elastic buckling of a perfect member. Real members are not perfectly straight, and their loads are applied eccentrically, while accidental transverse loads may act. The effects of small imperfections of these types are qualitatively the same as those of initial curvature, which were described in Section 3.2.2. These imperfections can therefore be represented by an increased equivalent initial curvature which has a similar effect on the behaviour of the member as the combined effect of all of these imperfections. The resulting behaviour is shown by curve A in Figure 3.12.

A real member also has residual stresses, and its elastic modulus E and yield stress f_y may vary throughout the member. The effects of these material variations are qualitatively the same as those of the residual stresses, which were described in Section 3.3.4, and so they can be represented by an equivalent set of the residual stresses. The resulting behaviour of the member is shown by curve B in Figure 3.12.

Since real members have both kinds of imperfections, their behaviour is as shown by curve C in Figure 3.12, which is a combination of curves A and B. Thus the real member behaves as a member with equivalent initial curvature (curve A) until the elastic limit is reached. It then follows a path which is similar to and approaches that of a member with equivalent residual stresses (curve B).

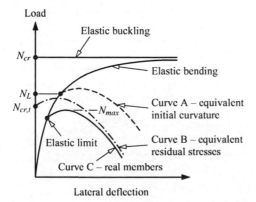

Figure 3.12 Behaviour of real compression members.

3.5   Design of compression members

3.5.1   EC3 design buckling resistance

Real compression members may be analysed using a model in which the initial crookedness and load eccentricity are specified. Residual stresses may be included, and the realistic stress–strain behaviour may be incorporated in the prediction of the load versus deflection relationship. Thus curve C in Figure 3.12 may be generated, and the maximum load N_{max} ascertained.

Rational computer analyses based on the above modelling are rarely used except in research, and are inappropriate for the routine design of real compression members because of the uncertainties and variations that exist in the initial crookedness and residual stresses. Instead, simplified design predictions for the maximum load (or resistance) are used in EC3 that have been developed from the results of computer analyses and correlations with available test data.

A close prediction of numerical solutions and test results may be obtained by using equations 3.11 and 3.12 which are based on the first yield of a geometrically imperfect member. This is achieved by writing the imperfection parameter η of equation 3.12 as

$$\eta = \alpha(\overline{\lambda} - 0.2) \geq 0 \tag{3.21}$$

in which α is a constant (imperfection factor) which shifts the resistance curve as shown in Figure 3.13 for different cross-section types, proportions, thicknesses, buckling axes, and material strengths (Table 6.2 of EC3). The advantage of this approach is that the resistances (referred to as buckling resistances in EC3) of a particular group of sections can be determined by assigning an appropriate value

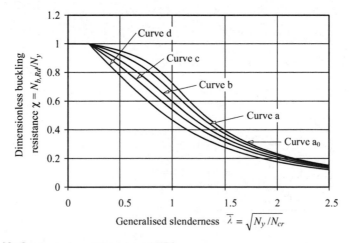

Figure 3.13 Compression resistances of EC3.

of α to them. The design buckling resistance $N_{b,Rd}$ is then given by

$$N_{b,Rd} = \chi N_y / \gamma_{M1} \tag{3.22}$$

in which γ_{M1} is the partial factor for member instability, which has a recommended value of 1.0 in EC3.

The dimensionless compressive design buckling resistances $N_{b,Rd}/N_y$ for $\alpha = 0.13, 0.21, 0.34, 0.49$, and 0.76, which represent the EC3 compression member curves (a_0), (a), (b), (c), and (d), respectively, are shown in Figure 3.13. Members with higher initial crookedness have lower strengths due to premature yielding, and these are associated with higher values of α. On the other hand, members with lower initial crookedness are not as greatly affected by premature yielding, and have lower values of α assigned to them.

3.5.2 Elastic buckling load

Although the design buckling resistance $N_{b,Rd}$ given by equation 3.22 was derived by using the expression for the elastic buckling load N_{cr} given by equation 3.2 for a pin-ended compression member, equation 3.22 is also used for other compression members by generalising equation 3.2 to

$$N_{cr} = \pi^2 EI / L_{cr}^2 \tag{3.23}$$

in which L_{cr} is the effective length (referred to as the buckling length in EC3).

The elastic buckling load N_{cr} varies with the member geometry, loading, and restraints, but there is no guidance given in EC3 for determining either N_{cr} or L_{cr}. Section 3.6 following summarises methods of determining the effects of restraints on the effective length L_{cr}, while Section 3.7 discusses the design of compression members with variable sections or loading.

3.5.3 Effects of local buckling

Compression members containing thin-plate elements are likely to be affected by local buckling of the cross-section (see Chapter 4). Local buckling reduces the resistances of short compression members below their squash loads N_y and the resistances of longer members which fail by flexural buckling.

Local buckling of a short compression member is accounted for by using a reduced effective area A_{eff} instead of the gross area A, as discussed in Section 4.7.1. Local buckling of a longer compression member is accounted for by using A_{eff} instead of A and by reducing the generalised slenderness to

$$\bar{\lambda} = \sqrt{A_{eff} f_y / N_{cr}} \tag{3.24}$$

but using the gross cross-sectional properties to determine N_{cr}.

3.5.4 Design procedures

For the design of a compression member, the design axial force N_{Ed} is determined by a rational frame analysis, as in Chapter 8, or by statics for a statically determinate structure. The design loads (factored loads) F_{Ed} are for the ultimate limit state, and are determined by summing up the specified loads multiplied by the appropriate partial load factors γ_F (see Section 1.5.6).

To check the compression resistance of a member, both the cross-section resistance $N_{c,Rd}$ and the member buckling resistance $N_{b,Rd}$ should generally be considered, though it is usually the case for practical members that the buckling resistance will govern. The cross-section resistance of a compression member $N_{c,Rd}$ must satisfy

$$N_{Ed} \leq N_{c,Rd} \tag{3.25}$$

in which

$$N_{c,Rd} = A f_y / \gamma_{M0} \tag{3.26}$$

for a fully effective section, in which γ_{M0} is the partial factor for cross-section resistance (with a recommended value of 1.0 in EC3), or

$$N_{c,Rd} = A_{eff} f_y / \gamma_{M0} \tag{3.27}$$

for the cross-sections that are susceptible to local buckling prior to yielding.

The member buckling resistance $N_{b,Rd}$ must satisfy

$$N_{Ed} \leq N_{b,Rd} \tag{3.28}$$

where the member buckling resistance $N_{b,Rd}$ is given by equation 3.22.

The procedure for checking a specified compression member is summarised in Figure 3.14. The cross-section is checked to determine if it is fully effective ($A_{eff} = A$, where A is the gross area) or slender (in which case the reduced effective area A_{eff} is used), and the design buckling resistance $N_{b,Rd}$ is then found and compared with the design compression force N_{Ed}.

An iterative series of calculations is required when designing a compression member, as indicated in Figure 3.14. An initial trial section is first selected, either by using tabulations, formulations or graphs of design buckling resistance $N_{b,Rd}$ versus effective length L_{cr}, or by making initial guesses for f_y, A_{eff}/A, and χ (say 0.5) and calculating a target area A. A trial section is then selected and checked. If the section is not satisfactory, then a new section is selected using the latest values of f_y, A_{eff}/A, and χ, and the checking process is repeated. The iterations usually converge within a few cycles, but convergence can be hastened by using the mean of the previous and current values of χ in the calculation of the target area A.

A worked example of checking the resistance of a compression member is given in Section 3.12.1, while worked examples of the design of compression members are given in Sections 3.12.2 and 3.12.3.

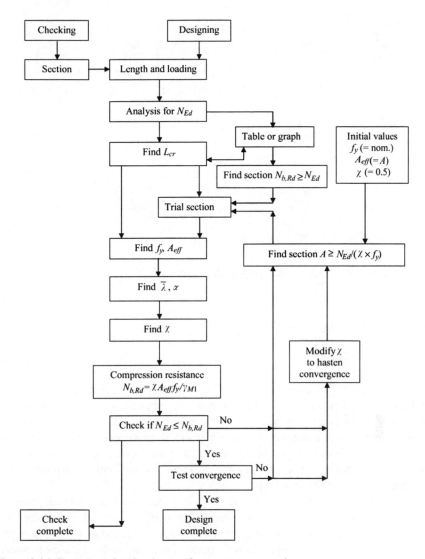

Figure 3.14 Flow chart for the design of compression members.

3.6 Restrained compression members

3.6.1 Simple supports and rigid restraints

In the previous sections it was assumed that the compression member was supported only at its ends, as shown in Figure 3.15a. If the member has an additional lateral support which prevents it from deflecting at its centre so that $(v)_{L/2} = 0$,

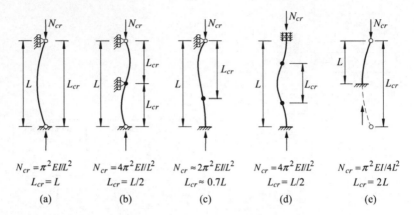

$$N_{cr} = \pi^2 EI/L^2 \qquad N_{cr} = 4\pi^2 EI/L^2 \qquad N_{cr} \approx 2\pi^2 EI/L^2 \qquad N_{cr} = 4\pi^2 EI/L^2 \qquad N_{cr} = \pi^2 EI/4L^2$$
$$L_{cr} = L \qquad\qquad L_{cr} = L/2 \qquad\qquad L_{cr} \approx 0.7L \qquad\qquad L_{cr} = L/2 \qquad\qquad L_{cr} = 2L$$
$$\quad\text{(a)} \qquad\qquad\qquad \text{(b)} \qquad\qquad\qquad \text{(c)} \qquad\qquad\qquad \text{(d)} \qquad\qquad\qquad \text{(e)}$$

Figure 3.15 Effective lengths of columns.

as shown in Figure 3.15b, then its buckled shape v is given by

$$v = \delta \sin 2\pi x/L, \tag{3.29}$$

and its elastic buckling load N_{cr} is given by

$$N_{cr} = \frac{4\pi^2 EI}{L^2}. \tag{3.30}$$

The end supports of a compression member may also differ from the simple supports shown in Figure 3.15a which allow the member ends to rotate but prevent them from deflecting laterally. For example, one or both ends may be rigidly built-in so as to prevent end rotation (Figure 3.15c and d), or one end may be completely free (Figure 3.15e). In each case the elastic buckling load of the member may be obtained by finding the solution of the differential equilibrium equation which satisfies the boundary conditions.

All of these buckling loads can be expressed by the generalisation of equation 3.23 which replaces the member length L of equation 3.2 by the effective length L_{cr}. Expressions for L_{cr} are shown in Figure 3.15, and in each case it can be seen that the effective length of the member is equal to the distance between the inflexion points of its buckled shape.

The effects of the variations in the support and restraint conditions on the compression member buckling resistance $N_{b,Rd}$ may be accounted for by replacing the actual length L used in equation 3.2 by the effective length L_{cr} in the calculation of the elastic buckling load N_{cr} and hence the generalised slenderness $\bar{\lambda}$, and using this modified generalised slenderness throughout the resistance equations. It should be noted that it is often necessary to consider the member behaviour in each principal plane, since the effective lengths $L_{cr,y}$ and $L_{cr,z}$ may also differ, as well as the radii of gyration i_y and i_z.

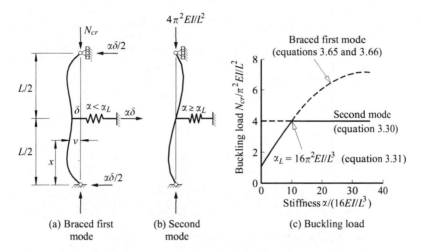

Figure 3.16 Compression member with an elastic intermediate restraint.

3.6.2 Intermediate restraints

In Section 3.6.1 it was shown that the elastic buckling load of a simply supported member is increased by a factor of 4 to $N_{cr} = 4\pi^2 EI/L^2$ when an additional lateral restraint is provided which prevents it from deflecting at its centre, as shown in Figure 3.15b. This restraint need not be completely rigid, but may be elastic (Figure 3.16), provided its stiffness exceeds a certain minimum value. If the stiffness α of the restraint is defined by the force $\alpha\delta$ acting on the restraint which causes its length to change by δ, then the minimum stiffness α_L is determined in Section 3.10.1 as

$$\alpha_L = 16\pi^2 EI/L^3. \tag{3.31}$$

The limiting stiffness α_L can be expressed in terms of the buckling load $N_{cr} = 4\pi^2 EI/L^2$ as

$$\alpha_L = 4N_{cr}/L. \tag{3.32}$$

Compression member restraints are generally required to be able to transmit 1% (Clause 5.3.3(3)) of EC3) of the force in the member restrained. This is a little less than the value of 1.5% suggested [2] as leading to the braces which are sufficiently stiff.

3.6.3 Elastic end restraints

When the ends of a compression member are rigidly connected to its adjacent elastic members, as shown in Figure 3.17a, then at buckling the adjacent members

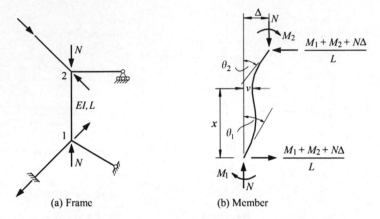

(a) Frame   (b) Member

Figure 3.17 Buckling of a member of a rigid frame.

exert total elastic restraining moments M_1, M_2 which oppose buckling and which are proportional to the end rotations θ_1, θ_2 of the compression member. Thus

$$\left.\begin{array}{l} M_1 = -\alpha_1 \theta_1 \\ M_2 = -\alpha_2 \theta_2 \end{array}\right\} \tag{3.33}$$

in which

$$\alpha_1 = \sum_1 \alpha, \tag{3.34}$$

where α is the stiffness of any adjacent member connected to the end 1 of the compression member, and α_2 is similarly defined. The stiffness α of an adjacent member depends not only on its length L and flexural rigidity EI but also on its support conditions and on the magnitude of any axial load transmitted by it.

The particular case of a braced restraining member, which acts as if simply supported at both ends as shown in Figure 3.18a, and which provides equal and opposite end moments M and has an axial force N, is analysed in Section 3.10.2, where it is shown that when the axial load N is compressive, the stiffness α is given by

$$\alpha = \frac{2EI}{L} \frac{(\pi/2)\sqrt{N/N_{cr,L}}}{\tan(\pi/2)\sqrt{N/N_{cr,L}}} \tag{3.35}$$

where

$$N_{cr,L} = \pi^2 EI/L^2, \tag{3.36}$$

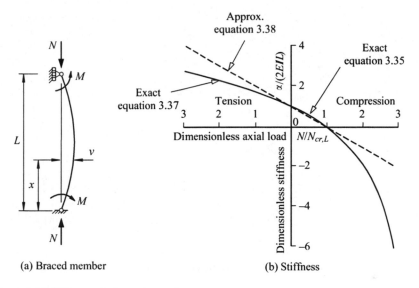

(a) Braced member   (b) Stiffness

Figure 3.18 Stiffness of a braced member.

and by

$$\alpha = \frac{2EI}{L} \frac{(\pi/2)\sqrt{N/N_{cr,L}}}{\tanh(\pi/2)\sqrt{N/N_{cr,L}}} \tag{3.37}$$

when the axial load N is tensile. These relationships are shown in Figure 3.18b, and it can be seen that the stiffness decreases almost linearly from $2EI/L$ to zero as the compressive axial load increases from zero to $N_{cr,L}$, and that the stiffness is negative when the axial load exceeds $N_{cr,L}$. In this case the adjacent member no longer restrains the buckling member, but disturbs it. When the axial load causes tension, the stiffness is increased above the value $2EI/L$.

Also shown in Figure 3.18b is the simple approximation

$$\alpha = \frac{2EI}{L}\left(1 - \frac{N}{N_{cr,L}}\right). \tag{3.38}$$

The term $(1 - N/N_{cr,L})$ in this equation is the reciprocal of the amplification factor, which expresses the fact that the first-order rotations $ML/2EI$ associated with the end moments M are amplified by the compressive axial load N to $(ML/2EI)/(1 - N/N_{cr,L})$. The approximation provided by equation 3.38 is close and conservative in the range $0 < N/N_{cr,L} < 1$, but errs on the unsafe side and with increasing error as the axial load N increases away from this range.

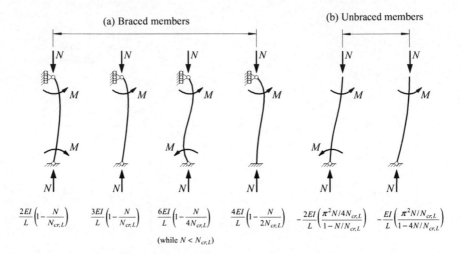

Figure 3.19 Approximate stiffnesses of restraining members.

Similar analyses may be made of braced restraining members under other end conditions, and some approximate solutions for the stiffnesses of these are summarised in Figure 3.19. These approximations have accuracies comparable with that of equation 3.38.

When a restraining member is unbraced, its ends sway Δ as shown in Figure 3.20a, and restoring end moments M are required to maintain equilibrium. Such a member which receives equal end moments is analysed in Section 3.10.3, where it is shown that its stiffness is given by

$$\alpha = -\left(\frac{2EI}{L}\right)\frac{\pi}{2}\sqrt{\frac{N}{N_{cr,L}}}\,\tan\frac{\pi}{2}\sqrt{\frac{N}{N_{cr,L}}} \qquad (3.39)$$

when the axial load N is compressive, and by

$$\alpha = \left(\frac{2EI}{L}\right)\frac{\pi}{2}\sqrt{\frac{N}{N_{cr,L}}}\,\tanh\frac{\pi}{2}\sqrt{\frac{N}{N_{cr,L}}} \qquad (3.40)$$

when the axial load N is tensile. These relationships are shown in Figure 3.20b, and it can be seen that the stiffness α decreases as the tensile load N decreases, and becomes negative as the load changes to compressive. Also shown in Figure 3.20b is the approximation

$$\alpha = -\frac{2EI}{L}\frac{(\pi^{2}/4)(N/N_{cr,L})}{(1-N/N_{cr,L})}, \qquad (3.41)$$

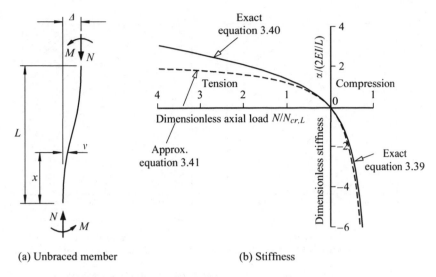

(a) Unbraced member   (b) Stiffness

Figure 3.20 Stiffness of an unbraced member.

which is close and conservative when the axial load N is compressive, and which errs on the safe side when the axial load N is tensile. It can be seen from Figure 3.20b that the stiffness of the member is negative when it is subjected to compression, so that it disturbs the buckling member, rather than restraining it.

Similar analyses may be made of unbraced restraining members with other end conditions, and another approximate solution is given in Figure 3.19. This has an accuracy comparable with that of equation 3.41.

3.6.4 Buckling of braced members with end restraints

The buckling of an end-restrained compression member 1–2 is analysed in Section 3.10.4, where it is shown that if the member is braced so that it cannot sway ($\Delta = 0$), then its elastic buckling load N_{cr} can be expressed in the general form of equation 3.23 when the effective length ratio $k_{cr} = L_{cr}/L$ is the solution of

$$\frac{\gamma_1 \gamma_2}{4} \left(\frac{\pi}{k_{cr}} \right)^2 + \left(\frac{\gamma_1 + \gamma_2}{2} \right) \left(1 - \frac{\pi}{k_{cr}} \cot \frac{\pi}{k_{cr}} \right) + \frac{\tan(\pi/2k_{cr})}{\pi/2k_{cr}} = 1, \quad (3.42)$$

where the relative stiffness of the braced member at its end 1 is

$$\gamma_1 = \frac{(2EI/L)_{12}}{\sum_1 \alpha}, \qquad (3.43)$$

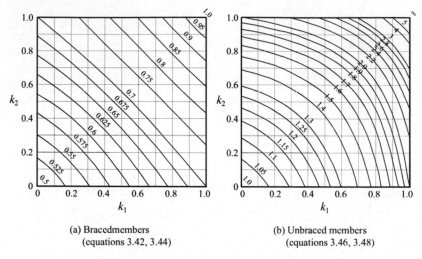

(a) Braced members
(equations 3.42, 3.44)

(b) Unbraced members
(equations 3.46, 3.48)

Figure 3.21 Effective length ratios L_{cr}/L for members in rigid-jointed frames.

in which the summation $\sum_1 \alpha$ is for all the other members at end 1, γ_2 is similarly defined, and $0 \leq \gamma \leq \infty$. The relative stiffnesses may also be expressed as

$$k_1 = \frac{(2EI/L)_{12}}{0.5 \sum_1 \alpha + (2EI/L)_{12}} = \frac{2\gamma_1}{1 + 2\gamma_1} \tag{3.44}$$

and a similar definition of k_2 ($0 \leq k \leq 1$). Values of the effective length ratio L_{cr}/L which satisfy equations 3.42–3.44 are presented in chart form in Figure 3.21a.

In using the chart of Figure 3.21a, the stiffness factors k may be calculated from equation 3.44 by using the stiffness approximations of Figure 3.19. Thus for braced restraining members of the type shown in Figure 3.18a, the factor k_1 can be approximated by

$$k_1 = \frac{(I/L)_{12}}{0.5 \sum_1 (I/L)(1 - N/N_{cr,L}) + (I/L)_{12}} \tag{3.45}$$

in which N is the compression force in the restraining member and $N_{cr,L}$ its elastic buckling load.

3.6.5 Buckling of unbraced members with end restraints

The buckling of an end-restrained compression member 1–2 is analysed in Section 3.10.5, where it is shown that if the member is unbraced against sidesway, then its elastic buckling load N_{cr} can be expressed in the general form of equation 3.23 when the effective length ratio $k_{cr} = L_{cr}/L$ is the solution of

$$\frac{\gamma_1\gamma_2(\pi/k_{cr})^2 - 36}{6(\gamma_1 + \gamma_2)} = \frac{\pi}{k_{cr}} \cot \frac{\pi}{k_{cr}}, \tag{3.46}$$

where the relative stiffness of the unbraced member at its end 1 is

$$\gamma_1 = \frac{(6EI/L)_{12}}{\sum_1 \alpha} \tag{3.47}$$

in which the summation $\sum_1 \alpha$ is for all the other members at end 1, and γ_2 is similarly defined. The relative stiffnesses may also be expressed as

$$k_1 = \frac{(6EI/L)_{12}}{1.5 \sum_1 \alpha + (6EI/L)_{12}} = \frac{\gamma_1}{1.5 + \gamma_1} \tag{3.48}$$

and a similar definition of k_2. Values of the effective length ratio L_{cr}/L which satisfy equation 3.46 are presented in chart form in Figure 3.21b.

In using the chart of Figure 3.21b, the stiffness factors k may be calculated from equation 3.48 by using the stiffness approximations of Figure 3.19. Thus for braced restraining members of the third type shown in Figure 3.19a,

$$k_1 = \frac{(I/L)_{12}}{1.5 \sum_1 (I/L)(1 - N/4N_{cr,L}) + (I/L)_{12}}. \tag{3.49}$$

3.6.6   Bracing stiffness required for a braced member

The elastic buckling load of an unbraced compression member may be increased substantially by providing a translational bracing system which effectively prevents sway. The bracing system need not be completely rigid, but may be elastic as shown in Figure 3.22, provided its stiffness α exceeds a certain minimum value α_L. It is shown in Section 3.10.6 that the minimum value for a pin-ended compression member is

$$\alpha_L = \pi^2 EI/L^3. \tag{3.50}$$

This conclusion can be extended to compression members with rotational end restraints, and it can be shown that if the sway bracing stiffness α is greater than

$$\alpha_L = N_{cr}/L, \tag{3.51}$$

where N_{cr} is the elastic buckling load for the braced mode, then the member is effectively braced against sway.

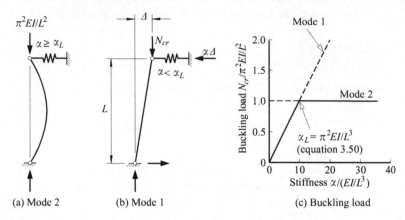

(a) Mode 2   (b) Mode 1   (c) Buckling load

Figure 3.22 Compression member with an elastic sway brace.

3.7   Other compression members

3.7.1   General

The design method outlined in Section 3.5, together with the effective length concept developed in Section 3.6, are examples of a general approach to the analysis and design of the compression members whose ultimate resistances are governed by the interaction between yielding and buckling. This approach, often termed design by buckling analysis, originates from the dependence of the design compression resistance $N_{b,Rd}$ of a simply supported uniform compression member on its squash load N_y and its elastic buckling load N_{cr}, as shown in Figure 3.23, which is adapted from Figure 3.13. The generalisation of this relationship to other compression members allows the design buckling resistance $N_{b,Rd}$ of any member to be determined from its yield load N_y and its elastic buckling load N_{cr} by using Figure 3.23.

In the following subsections, this design by buckling analysis method is extended to the in-plane behaviour of rigid-jointed frames with joint loading, and also to the design of compression members which either are non-uniform, or have intermediate loads, or twist during buckling. This method has also been used for compression members with oblique restraints [3]. Similar and related methods may be used for the flexural–torsional buckling of beams (Sections 6.6–6.9).

3.7.2   Rigid-jointed frames with joint loads only

The application of the method of design by buckling analysis to rigid-jointed frames which only have joint loads is a simple extrapolation of the design

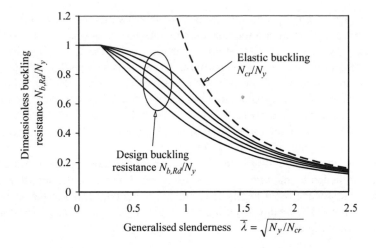

Figure 3.23 Compression resistance of structures designed by buckling analysis.

method for simply supported compression members. For this extrapolation, it is assumed that the resistance $N_{b,Rd}$ of a compression member in a frame is related to its squash load N_y and to the axial force N_{cr} carried by it when the frame buckles elastically, and that this relationship (Figure 3.23) is the same as that used for simply supported compression members of the same type (Figure 3.13).

This method of design by buckling analysis is virtually the same as the effective length method which uses the design charts of Figure 3.21, because these charts were obtained from the elastic buckling analyses of restrained members. The only difference is that the elastic buckling load N_{cr} may be calculated directly for the method of design by buckling analysis, as well as by determining the approximate effective length from the design charts.

The elastic buckling load of the member may often be determined approximately by the effective length method discussed in Section 3.6. In cases where this is not satisfactory, a more accurate method must be used. Many such methods have been developed, and some of these are discussed in Sections 8.3.5.3 and 8.3.5.4 and in [4–13], while other methods are referred to in the literature [14–16]. The buckling loads of many frames have already been determined, and extensive tabulations of approximations for some of these are available [14, 16–20].

While the method of design by buckling analysis might also be applied to rigid frames whose loads act between joints, the bending actions present in those frames make this less rational. Because of this, consideration of the effects of buckling on the design of frames with bending actions will be deferred until Chapter 8.

3.7.3 Non-uniform members

It was assumed in the preceding discussions that the members are uniform, but some practical compression members are of variable cross-section. Non-uniform members may be stepped or tapered, but in either case the elastic buckling load N_{cr} can be determined by solving a differential equilibrium equation similar to that governing the buckling of uniform members (see equation 3.58), but which has variable values of EI. In general, this can best be done using numerical techniques [4–6], and the tedium of these can be relieved by making use of a suitable computer program. Many particular cases have been solved, and tabulations and graphs of solutions are available [5, 6, 16, 21–23].

Once the elastic buckling load N_{cr} of the member has been determined, the method of design by buckling analysis can be used. For this, the squash load N_y is calculated for the most highly stressed cross-section, which is the section of minimum area. The design buckling resistance $N_{b,Rd} = \chi N_y/\gamma_{M1}$ can then be obtained from Figure 3.23.

3.7.4 Members with intermediate axial loads

The elastic buckling of members with intermediate as well as end loads is also best analysed by numerical methods [4–6], while solutions of many particular cases are available [5, 16, 21]. Once again, the method of design by buckling analysis can be used, and for this the squash load will be determined by the most heavily stressed section. If the elastic buckling load N_{cr} is calculated for the same member, then the design buckling resistance $N_{b,Rd}$ can be determined from Figure 3.23.

3.7.5 Flexural–torsional buckling

In the previous sections, attention was confined to compression members which buckle by deflecting laterally, either perpendicular to the section minor axis at an elastic buckling load

$$N_{cr,z} = \pi^2 EI_z/L_{cr,z}^2,$$ (3.52)

or perpendicular to the major axis at

$$N_{cr,y} = \pi^2 EI_y/L_{cr,y}^2.$$ (3.53)

However, thin-walled open section compression members may also buckle by twisting about a longitudinal axis, as shown in Figure 3.24 for a cruciform section, or by combined bending and twisting.

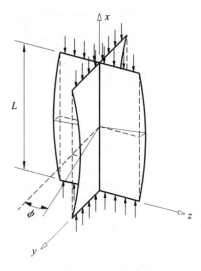

Figure 3.24 Torsional buckling of a cruciform section.

A compression member of doubly symmetric cross-section may buckle elastically by twisting at a torsional buckling load (see Section 3.11) given by

$$N_{cr,T} = \frac{1}{i_0^2} \left(GI_t + \frac{\pi^2 EI_w}{L_{cr,T}^2} \right) \tag{3.54}$$

in which GI_t and EI_w are the torsional and warping rigidities (see Chapter 10), $L_{cr,T}$ is the distance between inflexion points of the twisted shape, and

$$i_0^2 = i_p^2 + y_0^2 + z_0^2 \tag{3.55}$$

in which y_0, z_0 are the shear centre coordinates (which are zero for doubly symmetric sections, see Section 5.4.3), and

$$i_p = \sqrt{\{(I_y + I_z)/A\}} \tag{3.56}$$

is the polar radius of gyration. For most rolled steel sections, the minor axis buckling load $N_{cr,z}$ is less than $N_{cr,T}$, and the possibility of torsional buckling can be ignored. However, short members which have low torsional and warping rigidities (such as thin-walled cruciforms) should be checked. Such members can be designed by using Figure 3.23 with the value of $N_{cr,T}$ substituted for the elastic buckling load N_{cr}.

Monosymmetric and asymmetric section members (such as thin-walled tees and angles) may buckle in a combined mode by twisting and deflecting. This action takes place because the axis of twist through the shear centre does not

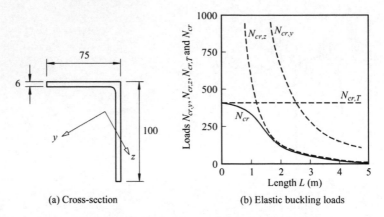

(a) Cross-section   (b) Elastic buckling loads

Figure 3.25 Elastic buckling of a simply supported angle section column.

coincide with the loading axis through the centroid, and any twisting which occurs causes the centroidal axis to deflect. For simply supported members, it is shown in [5, 24, 25] that the (lowest) elastic buckling load N_{cr} is the lowest root of the cubic equation

$$N_{cr}^3 \left\{ i_0^2 - y_0^2 - z_0^2 \right\} - N_{cr}^2 \left\{ (N_{cr,y} + N_{cr,z} + N_{cr,T})i_0^2 - N_{cr,z}y_0^2 - N_{cr,y}z_0^2 \right\}$$
$$+ N_{cr}i_0^2 \{N_{cr,y}N_{cr,z} + N_{cr,z}N_{cr,T} + N_{cr,T}N_{cr,y}\} - N_{cr,y}N_{cr,z}N_{cr,T}i_0^2 = 0. \tag{3.57}$$

For example, the elastic buckling load N_{cr} for a pin-ended unequal angle is shown in Figure 3.25, where it can be seen that N_{cr} is less than any of $N_{cr,y}$, $N_{cr,z}$, or $N_{cr,T}$.

These and other cases of flexural–torsional buckling (referred to as torsional–flexural buckling in EC3) are treated in a number of textbooks and papers [4–6, 24–26], and tabulations of solutions are also available [16, 27]. Once the buckling load N_{cr} has been determined, the compression resistance $N_{b,Rd}$ can be found from Figure 3.23.

3.8   Appendix – elastic compression members

3.8.1   Buckling of straight members

The elastic buckling load N_{cr} of the compression member shown in Figure 3.3 can be determined by finding a deflected position which is one of equilibrium. The differential equilibrium equation of bending of the member is

$$EI\frac{d^2v}{dx^2} = -N_{cr}v. \tag{3.58}$$

This equation states that for equilibrium, the internal moment of resistance $EI(\mathrm{d}^2v/\mathrm{d}x^2)$ must exactly balance the external disturbing moment $-N_{cr}v$ at any point along the length of the member. When this equation is satisfied at all points, the displaced position is one of equilibrium.

The solution of equation 3.58 which satisfies the boundary condition at the lower end that $(v)_0 = 0$ is

$$v = \delta \sin \frac{\pi x}{k_{cr}L},$$

where

$$\frac{1}{k_{cr}^2} = \frac{N_{cr}}{\pi^2 EI / L^2},$$

and δ is an undetermined constant. The boundary condition at the upper end that $(v)_L = 0$ is satisfied when either

$$\left.\begin{array}{c} \delta = 0 \\ v = 0 \end{array}\right\} \tag{3.59}$$

or $k_{cr} = 1/n$ in which n is an integer, so that

$$N_{cr} = n^2 \pi^2 EI / L^2, \tag{3.60}$$

$$v = \delta \sin n\pi x / L. \tag{3.61}$$

The first solution (equations 3.59) defines the straight stable equilibrium position which is valid for all loads N less than the lowest value of N_{cr}, as shown in Figure 3.2b. The second solution (equations 3.60 and 3.61) defines the buckling loads N_{cr} at which displaced equilibrium positions can exist. This solution does not determine the magnitude δ of the central deflection, as indicated in Figure 3.2b. The lowest buckling load is the most important, and this occurs when $n = 1$, so that

$$N_{cr} = \pi^2 EI / L^2, \tag{3.2}$$

$$v = \delta \sin \pi x / L. \tag{3.1}$$

3.8.2   Bending of members with initial curvature

The bending of the compression member with initial curvature shown in Figure 3.2a can be analysed by considering the differential equilibrium equation

$$EI \frac{\mathrm{d}^2v}{\mathrm{d}x^2} = -N(v + v_0), \tag{3.62}$$

which is obtained from equation 3.58 for a straight member by adding the additional bending moment $-Nv_0$ induced by the initial curvature.

If the initial curvature of the member is such that

$$v_0 = \delta_0 \sin \pi x/L, \tag{3.7}$$

then the solution of equation 3.62 which satisfies the boundary conditions $(v)_{0,L} = 0$ is the deflected shape

$$v = \delta \sin \pi x/L, \tag{3.8}$$

where

$$\frac{\delta}{\delta_0} = \frac{N/N_{cr}}{1 - N/N_{cr}}. \tag{3.9}$$

The maximum moment in the compression member is $N(\delta + \delta_0)$, and so the maximum bending stress is $N(\delta + \delta_0)/W_{el}$, where W_{el} is the elastic section modulus. Thus the maximum total stress is

$$\sigma_{max} = \frac{N}{A} + \frac{N(\delta + \delta_0)}{W_{el}}.$$

If the elastic limit is taken as the yield stress f_y, then the limiting axial load N_L for which the above elastic analysis is valid is given by

$$N_L = N_y - \frac{N_L(\delta + \delta_0)A}{W_{el}}, \tag{3.63}$$

where $N_y = Af_y$ is the squash load. By writing $W_{el} = 2I/b$, in which b is the member width, equation 3.63 becomes

$$N_L = N_y - \frac{\delta_0 b}{2i^2} \frac{N_L}{(1 - N_L/N_{cr})},$$

which can be solved for the dimensionless limiting load N_L/N_y as

$$\frac{N_L}{N_y} = \left[\frac{1 + (1 + \eta)N_{cr}/N_y}{2}\right] - \left\{\left[\frac{1 + (1 + \eta)N_{cr}/N_y}{2}\right]^2 - \frac{N_{cr}}{N_y}\right\}^{1/2}, \tag{3.64}$$

where

$$\eta = \frac{\delta_0 b}{2i^2}. \tag{3.13}$$

Alternatively, equation 3.64 can be rearranged to give the dimensionless limiting load N_L/N_y as

$$\frac{N_L}{N_y} = \frac{1}{\Phi + \sqrt{\Phi^2 - \bar{\lambda}^2}} \tag{3.11}$$

in which

$$\Phi = \frac{1 + \eta + \overline{\lambda}^2}{2} \tag{3.12}$$

and

$$\overline{\lambda} = \sqrt{N_y / N_{cr}}. \tag{3.5}$$

3.9 Appendix – inelastic compression members

3.9.1 Reduced modulus theory of buckling

The reduced modulus buckling load $N_{cr,r}$ of a rectangular section compression member which buckles in the y direction (see Figure 3.26a) can be determined from the bending strain and stress distributions, which are related to the curvature $\kappa (= -\mathrm{d}^2 v / \mathrm{d}x^2)$ and the moduli E and E_t as shown in Figure 3.26b and c. The position of the line of zero bending stress can be found by using the condition that the axial force remains constant during buckling, from which it follows that the force resultant of the bending stresses must be zero, so that

$$\frac{1}{2} d b_c E_t b_c \kappa = \frac{1}{2} d b_t E b_t \kappa,$$

or

$$\frac{b_c}{b} = \frac{\sqrt{E}}{\sqrt{E} + \sqrt{E_t}}.$$

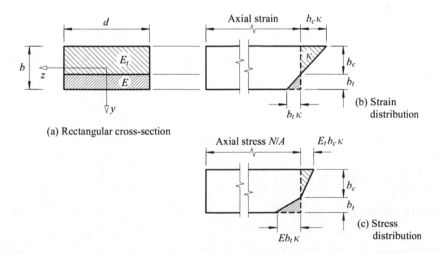

Figure 3.26 Reduced modulus buckling of a rectangular section member.

The moment of resistance of the section $E_r I (d^2 v/dx^2)$ is equal to the moment resultant of the bending stresses, so that

$$E_r I \frac{d^2 v}{dx^2} = -\frac{db_c}{2} E_t b_c \kappa \frac{2b_c}{3} - \frac{db_t}{2} E b_t \kappa \frac{2b_t}{3}$$

or

$$E_r I \frac{d^2 v}{dx^2} = -\frac{4b_c^2 E_t}{b^2} \frac{db^3}{12} \kappa.$$

The reduced modulus of elasticity E_r is therefore given by

$$E_r = \frac{4EE_t}{(\sqrt{E} + \sqrt{E_t})^2}. \tag{3.16}$$

3.9.2 Buckling of members with residual stresses

A rectangular section member with a simplified residual stress distribution is shown in Figure 3.10. The section first yields at the edges $z = \pm d/2$ at a load $N = 0.5N_y$, and yielding then spreads through the section as the load approaches the squash load N_y. If the depth of the elastic core is d_e, then the flexural rigidity for bending about the z axis is

$$(EI)_t = EI \left(\frac{d_e}{d} \right),$$

where $I = b^3 d/12$, and the axial force is

$$N = N_y \left(1 - \frac{1}{2} \frac{d_e^2}{d^2} \right).$$

By combining these two relationships, the effective flexural rigidity of the partially yielded section can be written as

$$(EI)_t = EI[2(1 - N/N_y)]^{1/2}. \tag{3.18}$$

Thus the axial load at buckling N_t is given by

$$\frac{N_{cr,t}}{N_{cr}} = [2(1 - N_{cr,t}/N_y)]^{1/2},$$

which can be rearranged as

$$\frac{N_{cr,t}}{N_y} = \left(\frac{N_{cr}}{N_y} \right)^2 \{[1 + 2(N_y/N_{cr})^2]^{1/2} - 1\}. \tag{3.19}$$

3.10 Appendix – effective lengths of compression members

3.10.1 Intermediate restraints

A straight pin-ended compression member with a central elastic restraint is shown in Figure 3.16a. It is assumed that when the buckling load N_{cr} is applied to the member, it buckles symmetrically as shown in Figure 3.16a with a central deflection δ, and the restraint exerts a restoring force $\alpha\delta$. The equilibrium equation for this buckled position is

$$EI\frac{d^2v}{dx^2} = -N_{cr}v + \frac{\alpha\delta}{2}x$$

for $0 \le x \le L/2$.

The solution of this equation which satisfies the boundary conditions $(v)_0 = (dv/dx)_{L/2} = 0$ is given by

$$v = \frac{\alpha\delta L}{2N_{cr}}\left(\frac{x}{L} - \frac{\sin \pi x/k_{cr}L}{2(\pi/2k_{cr})\cos \pi/2k_{cr}}\right),$$

where $k_{cr} = L_{cr}/L$ and L_{cr} is given by equation 3.23. Since $\delta = (v)_{L/2}$, it follows that

$$\frac{\left(\dfrac{\pi}{2k_{cr}}\right)^3 \cot \dfrac{\pi}{2k_{cr}}}{\left(\dfrac{\pi}{2k_{cr}} \cot \dfrac{\pi}{2k_{cr}} - 1\right)} = \frac{\alpha L^3}{16EI}. \tag{3.65}$$

The variation with the dimensionless restraint stiffness $\alpha L^3/16EI$ of the dimensionless buckling load

$$\frac{N_{cr}}{\pi^2 EI/L^2} = \frac{4}{\pi^2}\left(\frac{\pi}{2k_{cr}}\right)^2 \tag{3.66}$$

which satisfies equation 3.65 is shown in Figure 3.16c. It can be seen that the buckling load for this symmetrical mode varies from $\pi^2 EI/L^2$ when the restraint is of zero stiffness to approximately $8\pi^2 EI/L^2$ when the restraint is rigid. When the restraint stiffness exceeds

$$\alpha_L = 16\pi^2 EI/L^3, \tag{3.31}$$

the buckling load obtained from equations 3.65 and 3.66 exceeds the value of $4\pi^2 EI/L^2$ for which the member buckles in the antisymmetrical second mode shown in Figure 3.16b. Since buckling always takes place at the lowest possible load, it follows that the member buckles at $4\pi^2 EI/L^2$ in the second mode shown in Figure 3.16b for all restraint stiffnesses α which exceed α_L.

3.10.2 Stiffness of a braced member

When a structural member is braced so that its ends act as if simply supported as shown in Figure 3.18a, then its response to equal and opposite disturbing end moments M can be obtained by considering the differential equilibrium equation

$$EI\frac{d^2v}{dx^2} = -Nv - M. \tag{3.67}$$

The solution of this which satisfies the boundary conditions $(v)_0 = (v)_L = 0$ when N is compressive is

$$v = \frac{M}{N}\left\{\frac{1 - \cos\pi\sqrt{N/N_{cr,L}}}{\sin\pi\sqrt{N/N_{cr,L}}}\sin\left[\pi\sqrt{\frac{N}{N_{cr,L}}}\frac{x}{L}\right] + \cos\left[\pi\sqrt{\frac{N}{N_{cr,L}}}\frac{x}{L}\right] - 1\right\},$$

where

$$N_{cr,L} = \pi^2 EI/L^2. \tag{3.36}$$

The end rotation $\theta = (dv/dx)_0$ is

$$\theta = \frac{2M}{NL}\frac{\pi}{2}\sqrt{\frac{N}{N_{cr,L}}}\tan\frac{\pi}{2}\sqrt{\frac{N}{N_{cr,L}}},$$

whence

$$\alpha = \frac{M}{\theta} = \frac{2EI}{L}\frac{(\pi/2)\sqrt{N/N_{cr,L}}}{\tan(\pi/2)\sqrt{N/N_{cr,L}}}. \tag{3.35}$$

When the axial load N is tensile, the solution of equation 3.67 is

$$v = \frac{M}{N}\left\{\frac{1 - \cosh\pi\sqrt{N/N_{cr,L}}}{\sinh\pi\sqrt{N/N_{cr,L}}}\sinh\left[\pi\sqrt{\frac{N}{N_{cr,L}}}\frac{x}{L}\right]\right.$$
$$\left. + \cosh\left[\pi\sqrt{\frac{N}{N_{cr,L}}}\frac{x}{L}\right] - 1\right\},$$

whence

$$\alpha = \frac{2EI}{L}\frac{(\pi/2)\sqrt{N/N_{cr,L}}}{\tanh(\pi/2)\sqrt{N/N_{cr,L}}}. \tag{3.37}$$

3.10.3   Stiffness of an unbraced member

When a structural member is unbraced so that its ends sway Δ as shown in Figure 3.20a, restoring end moments M are required to maintain equilibrium. The differential equation for equilibrium of a member with equal end moments $M = N\Delta/2$ is

$$EI\frac{\mathrm{d}^2 v}{\mathrm{d}x^2} = -Nv + M, \tag{3.68}$$

and its solution which satisfies the boundary conditions $(v)_0 = 0$, $(v)_L = \Delta$ is

$$v = \frac{M}{N}\left\{\frac{1 + \cos\pi\sqrt{N/N_{cr,L}}}{\sin\pi\sqrt{N/N_{cr,L}}}\sin\left[\pi\sqrt{\frac{N}{N_{cr,L}}}\frac{x}{L}\right] - \cos\left[\pi\sqrt{\frac{N}{N_{cr,L}}}\frac{x}{L}\right] + 1\right\}.$$

The end rotation $\theta = (\mathrm{d}v/\mathrm{d}x)_0$ is

$$\theta = \frac{2M}{NL}\frac{\pi}{2}\sqrt{\frac{N}{N_{cr,L}}}\cot\frac{\pi}{2}\sqrt{\frac{N}{N_{cr,L}}},$$

whence

$$\alpha = -\left(\frac{2EI}{L}\right)\frac{\pi}{2}\sqrt{\frac{N}{N_{cr,L}}}\tan\frac{\pi}{2}\sqrt{\frac{N}{N_{cr,L}}}, \tag{3.39}$$

where the negative sign occurs because the end moments M are restoring instead of disturbing moments.

When the axial load N is tensile, the end moments M are disturbing moments, and the solution of equation 3.68 is

$$v = \frac{M}{N}\left\{\frac{1 + \cosh\pi\sqrt{N/N_{cr,L}}}{\sinh\pi\sqrt{N/N_{cr,L}}}\sinh\left[\pi\sqrt{\frac{N}{N_{cr,L}}}\frac{x}{L}\right]\right.$$

$$\left. - \cosh\left[\pi\sqrt{\frac{N}{N_{cr,L}}}\frac{x}{L}\right] + 1\right\},$$

whence

$$\alpha = \left(\frac{2EI}{L}\right)\frac{\pi}{2}\sqrt{\frac{N}{N_{cr,L}}}\tanh\frac{\pi}{2}\sqrt{\frac{N}{N_{cr,L}}}. \tag{3.40}$$

3.10.4   Buckling of a braced member

A typical compression member 1–2 in a rigid-jointed frame is shown in Figure 3.17b. When the frame buckles, the member sways Δ, and its ends rotate by

θ_1, θ_2 as shown. Because of these actions, there are end shears $(M_1 + M_2 + N\varDelta)/L$ and end moments M_1, M_2. The equilibrium equation for the member is

$$EI \frac{d^2 v}{dx^2} = -Nv - M_1 + (M_1 + M_2)\frac{x}{L} + N\varDelta \frac{x}{L},$$

and the solution of this is

$$v = -\left(\frac{M_1 \cos \pi/k_{cr} + M_2}{N_{cr} \sin \pi/k_{cr}}\right) \sin \frac{\pi x}{k_{cr}L} + \frac{M_1}{N_{cr}}\left(\cos \frac{\pi x}{k_{cr}L} - 1\right)$$
$$+ \left(\frac{M_1 + M_2}{N_{cr}}\right)\frac{x}{L} + \varDelta \frac{x}{L},$$

where N_{cr} is the elastic buckling load transmitted by the member (see equation 3.23) and $k_{cr} = L_{cr}/L$. By using this to obtain expressions for the end rotations θ_1, θ_2, and by substituting equation 3.33, the following equations can be obtained:

$$\left. \begin{array}{l} M_1\left(\dfrac{N_{cr}L}{\alpha_1} + 1 - \dfrac{\pi}{k_{cr}} \cot \dfrac{\pi}{k_{cr}}\right) + M_2\left(1 - \dfrac{\pi}{k_{cr}}\mathrm{cosec}\,\dfrac{\pi}{k_{cr}}\right) + N_{cr}\varDelta = 0 \\[3mm] M_1\left(1 - \dfrac{\pi}{k_{cr}}\mathrm{cosec}\,\dfrac{\pi}{k_{cr}}\right) + M_2\left(\dfrac{N_{cr}L}{\alpha_2} + 1 - \dfrac{\pi}{k_{cr}} \cot \dfrac{\pi}{k_{cr}}\right) + N_{cr}\varDelta = 0 \end{array} \right\}.$$

$$(3.69)$$

If the compression member is braced so that joint translation is effectively prevented, the sway terms $N_{cr}\varDelta$ disappear from equations 3.69. The moments M_1, M_2 can then be eliminated, whence

$$\frac{(EI/L)^2}{\alpha_1 \alpha_2}\left(\frac{\pi}{k_{cr}}\right)^2 + \frac{EI}{L}\left(\frac{1}{\alpha_1} + \frac{1}{\alpha_2}\right)\left(1 - \frac{\pi}{k_{cr}} \cot \frac{\pi}{k_{cr}}\right) + \frac{\tan \pi/2k_{cr}}{\pi/2k_{cr}} = 1.$$

$$(3.70)$$

By writing the relative stiffness of the braced member at the end 1 as

$$\gamma_1 = \frac{2EI/L}{\alpha_1} = \frac{(2EI/L)_{12}}{\sum_1 \alpha}, \qquad (3.43)$$

with a similar definition for γ_2, equation 3.70 becomes

$$\frac{\gamma_1 \gamma_2}{4}\left(\frac{\pi}{k_{cr}}\right)^2 + \left(\frac{\gamma_1 + \gamma_2}{2}\right)\left(1 - \frac{\pi}{k_{cr}} \cot \frac{\pi}{k_{cr}}\right) + \frac{\tan(\pi/2k_{cr})}{\pi/2k_{cr}} = 1. \quad (3.42)$$

3.10.5 Buckling of an unbraced member

If there are no reaction points or braces to supply the member end shears $(M_1 + M_2 + N\varDelta)/L$, the member will sway as shown in Figure 3.17b, and the sway moment $N\Delta$ will be completely resisted by the end moments M_1 and M_2, so that

$$N\Delta = -(M_1 + M_2).$$

If this is substituted into equation 3.69 and if the moments M_1 and M_2 are eliminated, then

$$\frac{(EI/L)^2}{\alpha_1\alpha_2}\left(\frac{\pi}{k_{cr}}\right)^2 - 1 = \frac{EI}{L}\left(\frac{1}{\alpha_1} + \frac{1}{\alpha_2}\right)\frac{\pi}{k_{cr}}\cot\frac{\pi}{k_{cr}} \tag{3.71}$$

where $k_{cr} = L_{cr}/L$. By writing the relative stiffnesses of the unbraced member at end 1 as

$$\gamma_1 = \frac{6EI/L}{\alpha_1} = \frac{(6EI/L)_{12}}{\sum\limits_1 \alpha} \tag{3.47}$$

with a similar definition of γ_2, equation 3.71 becomes

$$\frac{\gamma_1\gamma_2(\pi/k_{cr})^2 - 36}{6(\gamma_1 + \gamma_2)} = \frac{\pi}{k_{cr}}\cot\frac{\pi}{k_{cr}}. \tag{3.46}$$

3.10.6 Stiffness of bracing for a rigid frame

If the simply supported member shown in Figure 3.22 has a sufficiently stiff sway brace acting at one end, then it will buckle as if rigidly braced, as shown in Figure 3.22a, at a load $\pi^2 EI/L^2$. If the stiffness α of the brace is reduced below a minimum value α_L, the member will buckle in the rigid body sway mode shown in Figure 3.22b. The elastic buckling load N_{cr} for this mode can be obtained from the equilibrium condition that

$$N_{cr}\Delta = \alpha\Delta L,$$

where $\alpha\Delta$ is the restraining force in the brace, so that

$$N_{cr} = \alpha L.$$

The variation of N_{cr} with α is compared in Figure 3.22c with the braced mode buckling load of $\pi^2 EI/L^2$. It can be seen that when the brace stiffness α is less than

$$\alpha_L = \pi^2 EI/L^3, \tag{3.50}$$

the member buckles in the sway mode, and that when α is greater than α_L, it buckles in the braced mode.

3.11   Appendix – torsional buckling

Thin-walled open-section compression members may buckle by twisting, as shown in Figure 3.24 for a cruciform section, or by combined bending and twisting. When this type of buckling takes place, the twisting of the member causes the axial compressive stresses N/A to exert a disturbing torque which is opposed by the torsional resistance of the section. For the typical longitudinal element of cross-sectional area δA shown in Figure 3.27, which rotates $a_0(d\phi/dx)$ (where a_0 is the distance to the axis of twist) when the section rotates ϕ, the axial force $(N/A)\delta A$ has a component $(N/A)\delta A a_0(d\phi/dx)$ which exerts a torque $(N/A)\delta A a_0(d\phi/dx)a_0$ about the axis of twist (which passes through the shear centre y_0, z_0 as shown in Chapter 5). The total disturbing torque T_P exerted is therefore

$$T_P = \frac{N}{A}\frac{d\phi}{dx}\int_A a_0^2 dA,$$

where

$$a_0^2 = (y - y_0)^2 + (z - z_0)^2.$$

This torque can also be written as

$$T_P = Ni_0^2\frac{d\phi}{dx},$$

where

$$i_0^2 = i_p^2 + y_0^2 + z_0^2, \tag{3.55}$$

$$i_p = \sqrt{\{(I_y + I_z)/A\}}. \tag{3.56}$$

For members of doubly symmetric cross-section ($y_0 = z_0 = 0$), a twisted equilibrium position is possible when the disturbing torque T_P exactly balances

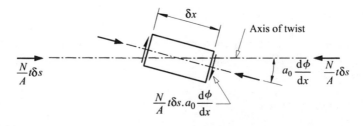

Figure 3.27 Torque exerted by axial load during twisting.

the internal resisting torque

$$M_x = GI_t \frac{d\phi}{dx} - EI_w \frac{d^3\phi}{dx^3}$$

in which GI_t and EI_w are the torsional and warping rigidities (see Chapter 10). Thus, at torsional buckling

$$\frac{N}{A} i_0^2 \frac{d\phi}{dx} = GI_t \frac{d\phi}{dx} - EI_w \frac{d^3\phi}{dx^3}. \tag{3.72}$$

The solution of this which satisfies the boundary conditions of end twisting prevented $((\phi)_{0,L} = 0)$ and ends free to warp $((d^2\phi/dx^2)_{0,L} = 0)$ (see Chapter 10) is $\phi = (\phi)_{L/2} \sin \pi x/L$ in which $(\phi)_{L/2}$ is the undetermined magnitude of the angle of twist rotation at the centre of the member, and the buckling load $N_{cr,T}$ is

$$N_{cr,T} = \frac{1}{i_0^2} \left(GI_t + \frac{\pi^2 EI_w}{L^2} \right).$$

This solution may be generalised for compression members with other end conditions by writing it in the form

$$N_{cr,T} = \frac{1}{i_0^2} \left(GI_t + \frac{\pi^2 EI_w}{L_{cr,T}^2} \right) \tag{3.54}$$

in which the torsional buckling effective length $L_{cr,T}$ is the distance between inflexion points in the twisted shape.

3.12 Worked examples

3.12.1 Example 1 – checking a UB compression member

Problem. The 457×191 UB 82 compression member of S275 steel of Figure 3.28a is simply supported about both principal axes at each end ($L_{cr,y} = 12.0$ m), and has a central brace which prevents lateral deflections in the minor principal plane ($L_{cr,z} = 6.0$ m). Check the adequacy of the member for a factored axial compressive load corresponding to a nominal dead load of 160 kN and a nominal imposed load of 230 kN.

Factored axial load. $N_{Ed} = (1.35 \times 160) + (1.5 \times 230) = 561$ kN

Classifying the section.
For S275 steel with $t_f = 16$ mm, $f_y = 275$ N/mm^2 EN 10025-2

$$\varepsilon = (235/275)^{0.5} = 0.924$$

$$c_f/(t_f\varepsilon) = [(191.3 - 9.9 - 2 \times 10.2)/2]/(16.0 \times 0.924) = 5.44 < 14$$
           T5.2

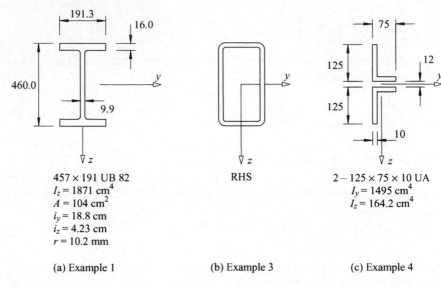

(a) Example 1   (b) Example 3   (c) Example 4

Figure 3.28 Examples 1, 3, and 4.

$$c_w = (460.0 - 2 \times 16.0 - 2 \times 10.2) = 407.6 \text{ mm} \qquad \text{T5.2}$$

$$c_w/(t_w \varepsilon) = 407.6/(9.9 \times 0.924) = 44.5 > 42 \qquad \text{T5.2}$$

and so the web is Class 4 (slender).   T5.2

Effective area.

$$\bar{\lambda}_p = \sqrt{\frac{f_y}{\sigma_{cr}}} = \frac{\bar{b}/t}{28.4\varepsilon\sqrt{k_\sigma}} = \frac{407.6/9.9}{28.4 \times 0.924 \times \sqrt{4.0}} = 0.784 \qquad \text{EC3-1-5 4.4(2)}$$

$$\rho = \frac{\bar{\lambda}_p - 0.055(3 + \psi)}{\bar{\lambda}_p^2} = \frac{0.784 - 0.055(3 + 1)}{0.784^2} = 0.918 \qquad \text{EC3-1-5 4.4(2)}$$

$$d - d_{eff} = (1 - 0.918) \times 407.6 = 33.6 \text{ mm}$$

$$A_{eff} = 104 \times 10^2 - 33.6 \times 9.9 = 10\,067 \text{ mm}^2$$

Cross-section compression resistance.

$$N_{c,Rd} = \frac{A_{eff}\,f_y}{\gamma_{M0}} = \frac{10\,067 \times 275}{1.0} = 2768 \text{ kN} > 561 \text{ kN} = N_{Ed} \qquad \text{6.2.4(2)}$$

Member buckling resistance.

$$\bar{\lambda}_y = \sqrt{\frac{A_{\it eff}\, f_y}{N_{\it cr,y}}} = \frac{L_{\it cr,y}}{i_y} \frac{\sqrt{A_{\it eff}/A}}{\lambda_1} = \frac{12\,000}{(18.8 \times 10)} \frac{\sqrt{10\,067/10\,400}}{93.9 \times 0.924} = 0.724$$

6.3.1.3(1)

$$\bar{\lambda}_z = \sqrt{\frac{A_{\it eff}\, f_y}{N_{\it cr,z}}} = \frac{L_{\it cr,z}}{i_z} \frac{\sqrt{A_{\it eff}/A}}{\lambda_1} = \frac{6000}{(4.23 \times 10)} \frac{\sqrt{10\,067/10\,400}}{93.9 \times 0.924}$$

$$= 1.608 > 0.724$$

6.3.1.3(1)

Buckling will occur about the minor (z) axis. For a rolled UB section (with $h/b >$ 1.2 and $t_f \leq 40$ mm), buckling about the z-axis, use buckling curve (b) with $\alpha = 0.34$ T6.2, T6.1

$$\varPhi_z = 0.5[1 + 0.34(1.608 - 0.2) + 1.608^2] = 2.032$$

6.3.1.2(1)

$$\chi_z = \frac{1}{2.032 + \sqrt{2.032^2 - 1.608^2}} = 0.305$$

6.3.1.2(1)

$$N_{b,z,Rd} = \frac{\chi A_{\it eff}\, f_y}{\gamma_{M1}} = \frac{0.305 \times 10\,067 \times 275}{1.0} = 844 \text{ kN} > 561 \text{ kN} = N_{Ed}$$

6.3.1.1(3)

and so the member is satisfactory.

3.12.2   Example 2 – designing a UC compression member

Problem. Design a suitable UC of S355 steel to resist the loading of example 1 in Section 3.12.1.

Design axial load. $N_{Ed} = 561$ kN, as in Section 3.12.1.

Target area and first section choice.
Assume $f_y = 355$ N/mm^2 and $\chi = 0.5$

$$A \geq 561 \times 10^3/(0.5 \times 355) = 3161 \text{ mm}^2$$

Try a 152×152 UC 30 with $A = 38.3$ cm^2, $i_y = 6.76$ cm, $i_z = 3.83$ cm, $t_f = 9.4$ mm.

$$\varepsilon = (235/355)^{0.5} = 0.814$$

T5.2

$$\bar{\lambda}_y = \sqrt{\frac{A\, f_y}{N_{\it cr,y}}} = \frac{L_{\it cr,y}}{i_y} \frac{1}{\lambda_1} = \frac{12\,000}{(6.76 \times 10)} \frac{1}{93.9 \times 0.814} = 2.322 \quad \text{6.3.1.3(1)}$$

$$\bar{\lambda}_z = \sqrt{\frac{A\, f_y}{N_{\it cr,z}}} = \frac{L_{\it cr,z}}{i_z} \frac{1}{\lambda_1} = \frac{6000}{(3.83 \times 10)} \frac{1}{93.9 \times 0.814} = 2.050 < 2.322$$

6.3.1.3(1)

Buckling will occur about the major (y) axis. For a rolled UC section (with $h/b \leq$ 1.2 and $t_f \leq 100$ mm), buckling about the y-axis, use buckling curve (b) with $\alpha = 0.34$ T6.2, T6.1

$$\Phi_y = 0.5[1 + 0.34(2.322 - 0.2) + 2.322^2] = 3.558 \qquad\qquad 6.3.1.2(1)$$

$$\chi_y = \frac{1}{3.558 + \sqrt{3.558^2 - 2.322^2}} = 0.160 \qquad\qquad 6.3.1.2(1)$$

which is much less than the guessed value of 0.5.

Second section choice.

Guess $\chi = (0.5 + 0.160)/2 = 0.33$

$A \geq 561 \times 10^3/(0.33 \times 355) = 4789$ mm^2

Try a 203 $\times$ 203 UC 52, with $A = 66.3$ cm^2, $i_y = 8.91$ cm, $t_f = 12.5$ mm.
For S355 steel with $t_f = 12.5$ mm, $f_y = 355$ N/mm^2 EN 10025-2

$\varepsilon = (235/355)^{0.5} = 0.814$ T5.2

$c_f/(t_f\varepsilon) = [(204.3 - 7.9 - 2 \times 10.2)/2](12.5 \times 0.814) = 8.65 < 14$ T5.2

$c_w/(t_w\varepsilon) = (206.2 - 2 \times 12.5 - 2 \times 10.2)/(7.9 \times 0.814) = 25.0 < 42$
                                                                                                                          T5.2

and so the cross-section is fully effective.

$$\bar{\lambda}_y = \sqrt{\frac{Af_y}{N_{cr,y}}} = \frac{L_{cr,y}}{i_y}\frac{1}{\lambda_1} = \frac{12\,000}{(8.91 \times 10)}\frac{1}{93.9 \times 0.814} = 1.763 \quad 6.3.1.3(1)$$

For a rolled UC section (with $h/b \leq 1.2$ and $t_f \leq 100$ mm), buckling about the y-axis, use buckling curve (b) with $\alpha = 0.34$ T6.2, T6.1

$$\Phi_y = 0.5[1 + 0.34(1.763 - 0.2) + 1.763^2] = 2.320 \qquad\qquad 6.3.1.2(1)$$

$$\chi_y = \frac{1}{2.320 + \sqrt{2.320^2 - 1.763^2}} = 0.261 \qquad\qquad 6.3.1.2(1)$$

$$N_{b,y,Rd} = \frac{\chi Af_y}{\gamma_{M1}} = \frac{0.261 \times 66.3 \times 10^2 \times 355}{1.0} = 615 \text{ kN} > 561 \text{ kN} = N_{Ed}$$
                                                                                                                     6.3.1.1(3)

and so the 203 $\times$ 203 UC 52 is satisfactory.

3.12.3   Example 3 – designing an RHS compression member

Problem. Design a suitable hot-finished RHS of S355 steel to resist the loading of example 1 in Section 3.12.1.

Design axial load. $N_{Ed} = 561$ kN, as in Section 3.12.1.
Solution.
Guess $\chi = 0.3$

$$A \geq 561 \times 10^3 / (0.3 \times 355) = 5268 \text{ mm}^2$$

Try a $250 \times 150 \times 8$ RHS, with $A = 60.8$ cm^2, $i_y = 9.17$ cm, $i_z = 6.15$ cm, $t = 8.0$ mm.
For S355 steel with $t = 8$ mm, $f_y = 355$ N/mm^2 EN 10025-2

$$\varepsilon = (235/355)^{0.5} = 0.814$$
$$c_w/(t\varepsilon) = (250.0 - 2 \times 8.0 - 2 \times 4.0)/(8.0 \times 0.814) = 34.7 < 42 \quad \text{T5.2}$$

and so the cross-section is fully effective.

$$\bar{\lambda}_y = \sqrt{\frac{Af_y}{N_{cr,y}}} = \frac{L_{cr,y}}{i_y}\frac{1}{\lambda_1} = \frac{12\,000}{(9.18 \times 10)}\frac{1}{93.9 \times 0.814} = 1.710 \quad 6.3.1.3(1)$$

$$\bar{\lambda}_z = \sqrt{\frac{Af_y}{N_{cr,z}}} = \frac{L_{cr,z}}{i_z}\frac{1}{\lambda_1} = \frac{6000}{(6.15 \times 10)}\frac{1}{93.9 \times 0.814} = 1.276 < 1.710$$
$$6.3.1.3(1)$$

Buckling will occur about the major (y) axis. For a hot-finished RHS, use buckling curve (a) with $\alpha = 0.21$ T6.2, T6.1

$$\Phi_y = 0.5[1 + 0.21(1.710 - 0.2) + 1.710^2] = 2.121 \qquad 6.3.1.2(1)$$

$$\chi_y = \frac{1}{2.121 + \sqrt{2.121^2 - 1.710^2}} = 0.296 \qquad 6.3.1.2(1)$$

$$N_{b,y,Rd} = \frac{\chi A f_y}{\gamma_{M1}} = \frac{0.296 \times 60.8 \times 10^2 \times 355}{1.0} = 640 \text{ kN} > 561 \text{ kN} = N_{Ed}$$
$$6.3.1.1(3)$$

and so the $250 \times 150 \times 8$ RHS is satisfactory.

3.12.4   Example 4 – buckling of double angles

Problem. Two steel $125 \times 75 \times 10$ UA are connected together at 1.5 m intervals to form the long compression member whose properties are given in Figure 3.28c. The minimum second moment of area of each angle is 49.9 cm^4. The member is simply supported about its major axis at 4.5 m intervals and about its minor axis at 1.5 m intervals. Determine the elastic buckling load of the member.

Member buckling about the major axis.

$$N_{cr,y} = \pi^2 \times 210\,000 \times 1495 \times 10^4/4500^2 \text{ N} = 1530 \text{ kN.}$$

Member buckling about the minor axis.

$$N_{cr,z} = \pi^2 \times 210\,000 \times 164.2 \times 10^4/1500^2 \text{ N} = 1513 \text{ kN.}$$

Single angle buckling.

$$N_{cr,min} = \pi^2 \times 210\,000 \times 49.9 \times 10^4/1500^2 \text{ N} = 459.7 \text{ kN}$$

and so for both angles $2N_{cr,min} = 2 \times 459.7 = 919 \text{ kN} < 1520 \text{ kN}$.

It can be seen that the lowest buckling load of 919 kN corresponds to the case where each unequal angle buckles about its own minimum axis.

3.12.5 Example 5 – effective length factor in an unbraced frame

Problem. Determine the effective length factor of member 1–2 of the unbraced frame shown in Figure 3.29a.

Solution. Using equation 3.48,

$$k_1 = \frac{6EI/L}{1.5 \times 0 + 6EI/L} = 1.0$$

$$k_2 = \frac{6EI/L}{1.5 \times 6(2EI)/(2L) + 6EI/L} = 0.4$$

Using Figure 3.21b, $L_{cr}/L = 2.3$

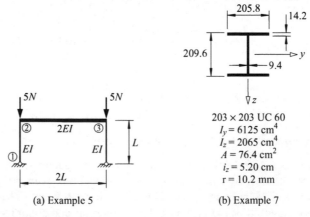

(a) Example 5

203 × 203 UC 60
$I_y = 6125 \text{ cm}^4$
$I_z = 2065 \text{ cm}^4$
$A = 76.4 \text{ cm}^2$
$i_z = 5.20 \text{ cm}$
$r = 10.2 \text{ mm}$

(b) Example 7

Figure 3.29 Examples 5 and 7.

3.12.6 Example 6 – effective length factor in a braced frame

Problem. Determine the effective length factor of member 1–2 of the frame shown in Figure 3.29a if bracing is provided which prevents sway buckling.

Solution.
Using equation 3.44,

$$k_1 = \frac{2EI/L}{0.5 \times 0 + 2EI/L} = 1.0$$

$$k_2 = \frac{2EI/L}{0.5 \times 2(2EI)/(2L) + 2EI/L} = 0.667$$

Using Figure 3.21a, $L_{cr}/L = 0.87$.

3.12.7 Example 7 – checking a non-uniform member

Problem. The 5.0 m long simply supported 203 × 203 UC 60 member shown in Figure 3.29b has two steel plates 250 mm × 12 mm × 2 m long welded to it, one to the central length of each flange. This increases the elastic buckling load about the minor axis by 70%. If the yield strengths of the plates and the UC are 355 N/mm², determine the compression resistance.

Solution.

$$\varepsilon = (235/355)^{0.5} = 0.814 \qquad\qquad \text{T5.2}$$

$$c_f/(t_f\varepsilon) = [(205.8 - 9.3 - 2 \times 10.2)/2]/(14.2 \times 0.814) = 7.62 < 14$$
$$\text{T5.2}$$

$$c_w/(t_w\varepsilon) = (209.6 - 2 \times 14.2 - 2 \times 10.2)/(9.3 \times 0.814) = 21.24 < 42$$
$$\text{T5.2}$$

and so the cross-section is fully effective.

$$N_{cr,z} = 1.7 \times \pi^2 \times 210\,000 \times 2065 \times 10^4/5000^2 = 2910 \text{ kN}.$$

$$\bar{\lambda}_z = \sqrt{\frac{Af_y}{N_{cr,z}}} = \sqrt{\frac{76.4 \times 10^2 \times 355}{2910 \times 10^3}} = 0.965$$

For a rolled UC section with welded flanges ($t_f \leq 40$ mm), buckling about the z-axis, use buckling curve (c) with $\alpha = 0.49$ \qquad\qquad T6.2, T6.1

$$\Phi_z = 0.5[1 + 0.49(0.965 - 0.2) + 0.965^2] = 1.153 \qquad\qquad 6.3.1.2(1)$$

$$\chi_z = \frac{1}{1.153 + \sqrt{1.153^2 - 0.965^2}} = 0.560 \qquad\qquad 6.3.1.2(1)$$

$$N_{b,z,Rd} = \frac{\chi Af_y}{\gamma_{M1}} = \frac{0.560 \times 76.4 \times 10^2 \times 355}{1.0} = 1520 \text{ kN} \qquad 6.3.1.1(3)$$

3.12.8   Example 8 – checking a member with intermediate axial loads

Problem. A 5 m long simply supported 457×191 UB 82 compression member of S355 steel whose properties are given in Figure 3.28a has concentric axial loads of N_{Ed} and $2N_{Ed}$ at its ends, and a concentric axial load N_{Ed} at its midpoint. At elastic buckling the maximum compression force in the member is 2000 kN. Determine the compression resistance.

Solution.
From Section 3.12.1, the cross-section is Class 4 slender and $A_{eff} = 10\,067$ mm^2.

$$N_{cr,z} = 2000 \text{ kN}.$$

$$\overline{\lambda}_z = \sqrt{\frac{A_{eff}f_y}{N_{cr,z}}} = \sqrt{\frac{10\,067 \times 355}{2000 \times 10^3}} = 1.337$$

For a rolled UB section (with $h/b > 1.2$ and $t_f \leq 40$ mm), buckling about the z-axis, use buckling curve (b) with $\alpha = 0.34$ \hfill T6.2, T6.1

$$\Phi_z = 0.5[1 + 0.34(1.337 - 0.2) + 1.337^2] = 1.587 \hfill 6.3.1.2(1)$$

$$\chi_z = \frac{1}{1.587 + \sqrt{1.587^2 - 1.337^2}} = 0.410 \hfill 6.3.1.2(1)$$

$$N_{b,z,Rd} = \frac{\chi A_{eff}f_y}{\gamma_{M1}} = \frac{0.410 \times 10\,067 \times 355}{1.0} = 1465 \text{ kN} \hfill 6.3.1.1(3)$$

3.13   Unworked examples

3.13.1   Example 9 – checking a welded column section

The 14.0 m long welded column section compression member of S355 steel shown in Figure 3.30(a) is simply supported about both principal axes at each end ($L_{cr,y} = 14.0$ m), and has a central brace that prevents lateral deflections in the minor principal plane ($L_{cr,z} = 7.0$ m). Check the adequacy of the member for a factored axial compressive force corresponding to a nominal dead load of 420 kN and a nominal imposed load of 640 kN.

3.13.2   Example 10 – designing a UC section

Design a suitable UC of S275 steel to resist the loading of example 9 in Section 3.13.1.

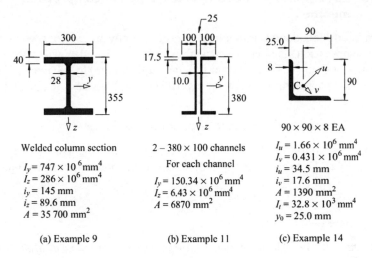

Welded column section

$I_y = 747 \times 10^6\,\text{mm}^4$
$I_z = 286 \times 10^6\,\text{mm}^4$
$i_y = 145\,\text{mm}$
$i_z = 89.6\,\text{mm}$
$A = 35\,700\,\text{mm}^2$

(a) Example 9

$2 - 380 \times 100$ channels
For each channel

$I_y = 150.34 \times 10^6\,\text{mm}^4$
$I_z = 6.43 \times 10^6\,\text{mm}^4$
$A = 6870\,\text{mm}^2$

(b) Example 11

$90 \times 90 \times 8$ EA
$I_u = 1.66 \times 10^6\,\text{mm}^4$
$I_v = 0.431 \times 10^6\,\text{mm}^4$
$i_u = 34.5\,\text{mm}$
$i_v = 17.6\,\text{mm}$
$A = 1390\,\text{mm}^2$
$I_t = 32.8 \times 10^3\,\text{mm}^4$
$y_0 = 25.0\,\text{mm}$

(c) Example 14

Figure 3.30 Examples 9, 11, and 14.

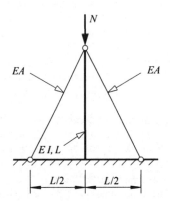

Figure 3.31 Example 12.

3.13.3 Example 11 – checking a compound section

Determine the minimum elastic buckling load of the two laced 380×100 channels shown in Figure 3.30(b) if the effective length about each axis is $L_{cr} = 3.0$ m.

3.13.4 Example 12 – elastic buckling

Determine the elastic buckling load of the guyed column shown in Figure 3.31. It should be assumed that the guys have negligible flexural stiffness, that $(EA)_{guy} = 2(EI)_{column}/L^2$, and that the column is built-in at its base.

3.13.5   Example 13 – checking a non-uniform compression member

Determine the maximum design load N_{Ed} of a lightly welded box section cantilever compression member 1–2 made from plates of S355 steel which taper from 300 mm × 300 mm × 12 mm at end 1 (at the cantilever tip) to 500 mm × 500 mm × 12 mm at end 2 (at the cantilever support). The cantilever has a length of 10.0 m.

3.13.6   Example 14 – flexural–torsional buckling

A simply supported $90 \times 90 \times 8$ EA compression member is shown in Figure 3.30(c). Determine the variation of the elastic buckling load with member length.

References

1. Shanley, F.R. (1947) Inelastic column theory, *Journal of the Aeronautical Sciences*, **14**, No. 5, May, pp. 261–7.
2. Trahair, N.S. (2000) Column bracing forces, *Australian Journal of Structural Engineering*, Institution of Engineers, Australia, **3**, Nos 2 & 3, pp. 163–8.
3. Trahair, N.S. and Rasmussen, K.J.R. (2005) Finite-element analysis of the flexural buckling of columns with oblique restraints, *Journal of Structural Engineering, ASCE*, **131**, No. 3, pp. 481–7.
4. Structural Stability Research Council (1998) *Guide to Stability Design Criteria for Metal Structures*, 5th edition (ed. T.V. Galambos), John Wiley, New York.
5. Timoshenko, S.P. and Gere, J.M. (1961) *Theory of Elastic Stability*, 2nd edition, McGraw-Hill, New York.
6. Bleich, F. (1952) *Buckling Strength of Metal Structures*, McGraw-Hill, New York.
7. Livesley, R.K. (1964) *Matrix Methods of Structural Analysis*, Pergamon Press, Oxford.
8. Horne, M.R. and Merchant, W. (1965) *The Stability of Frames*, Pergamon Press, Oxford.
9. Gregory, M.S. (1967) *Elastic Instability*, E. & F.N. Spon, London.
10. McMinn, S.J. (1961) The determination of the critical loads of plane frames, *The Structural Engineer*, **39**, No. 7, July, pp. 221–7.
11. Stevens, L.K. and Schmidt, L.C. (1963) Determination of elastic critical loads, *Journal of the Structural Division, ASCE*, **89**, No. ST6, pp. 137–58.
12. Harrison, H.B. (1967) The analysis of triangulated plane and space structures accounting for temperature and geometrical changes, *Space Structures*, (ed. R.M. Davies), Blackwell Scientific Publications, Oxford, pp. 231–43.
13. Harrison, H.B. (1973) *Computer Methods in Structural Analysis*, Prentice-Hall, Englewood Cliffs, New Jersey.
14. Lu, L.W. (1962) A survey of literature on the stability of frames, *Welding Research Council Bulletin*, No. 81, pp. 1–11.
15. Archer, J.S. *et al.* (1963) Bibliography on the use of digital computers in structural engineering, *Journal of the Structural Division, ASCE*, **89**, No. ST6, pp. 461–91.
16. Column Research Committee of Japan (1971) *Handbook of Structural Stability*, Corona, Tokyo.

17. Brotton, D.M. (1960) Elastic critical loads of multibay pitched roof portal frames with rigid external stanchions, *The Structural Engineer*, **38**, No. 3, March, pp. 88–99.
18. Switzky, H. and Wang, P.C. (1969) Design and analysis of frames for stability, *Journal of the Structural Division, ASCE*, **95**, No. ST4, pp. 695–713.
19. Davies, J.M. (1990) In-plane stability of portal frames, *The Structural Engineer*, **68**, No. 8, April, pp. 141–7.
20. Davies, J.M. (1991) The stability of multi-bay portal frames, *The Structural Engineer*, **69**, No. 12, June, pp. 223–9.
21. Bresler, B., Lin, T.Y., and Scalzi, J.B. (1968) *Design of Steel Structures*, 2nd edition, John Wiley, New York.
22. Gere, J.M. and Carter, W.O. (1962) Critical buckling loads for tapered columns, *Journal of the Structural Division, ASCE*, **88**, No. ST1, pp. 1–11.
23. Bradford, M.A. and Abdoli Yazdi, N. (1999) A Newmark-based method for the stability of columns, *Computers and Structures*, **71**, No. 6, pp. 689–700.
24. Trahair, N.S. (1993) *Flexural–Torsional Buckling of Structures*, E. & F.N. Spon, London.
25. Galambos, T.V. (1968) *Structural Members and Frames*, Prentice-Hall, Englewood Cliffs, New Jersey.
26. Trahair, N.S. and Rasmussen, K.J.R. (2005) Flexural–torsional buckling of columns with oblique eccentric restraints, *Journal of Structural Engineering, ASCE*, **131**, No. 11, pp. 1731–7.
27. Chajes, A. and Winter, G. (1965) Torsional–flexural buckling of thin-walled members, *Journal of the Structural Division, ASCE*, **91**, No. ST4, pp. 103–24.

Chapter 4

Local buckling of thin-plate elements

4.1 Introduction

The behaviour of compression members was discussed in Chapter 3, where it was assumed that no local distortion of the cross-section took place, so that failure was only due to overall buckling and yielding. This treatment is appropriate for solid section members, and for members whose cross-sections are composed of comparatively thick-plate elements, including many hot-rolled steel sections.

However, in some cases the member cross-section is composed of more slender-plate elements, as for example in some built-up members and in most light-gauge, cold-formed members. These slender-plate elements may buckle locally as shown in Figure 4.1, and the member may fail prematurely, as indicated by the reduction from the full line to the dashed line in Figure 4.2. A slender-plate element does not fail by elastic buckling, but exhibits significant post-buckling behaviour, as indicated in Figure 4.3. Because of this, the plate's resistance to local failure depends not only on its slenderness, but also on its yield strength and residual stresses, as shown in Figure 4.4. The resistance of a plate element of intermediate slenderness is also influenced significantly by its geometrical imperfections, while the resistance of a stocky-plate element depends primarily on its yield stress and strain-hardening moduli of elasticity, as indicated in Figure 4.4.

In this chapter, the behaviour of thin rectangular plates subjected to in-plane compression, shear, bending, or bearing is discussed. The behaviour under compression is applied to the design of plate elements in compression members and compression flanges in beams. The design of beam webs is also discussed, and the influence of the behaviour of thin plates on the design of plate girders to EC3 is treated in detail.

Figure 4.1 Local buckling of an I-section column.

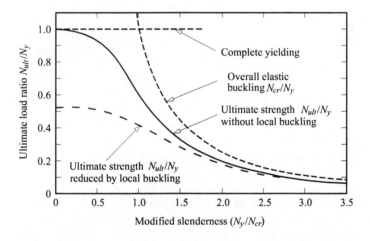

Figure 4.2 Effects of local buckling on the resistances of compression members.

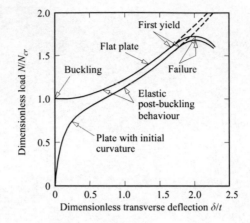

Figure 4.3 Post-buckling behaviour of thin plates.

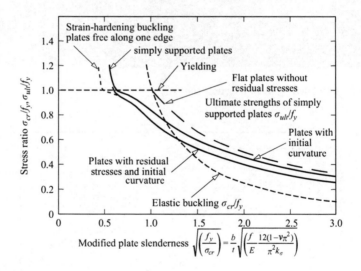

Figure 4.4 Ultimate strengths of plates in compression.

4.2 Plate elements in compression

4.2.1 Elastic buckling

4.2.1.1 Simply supported plates

The thin flat plate element of length L, width b, and thickness t shown in Figure 4.5b is simply supported along all four edges. The applied compressive loads N are

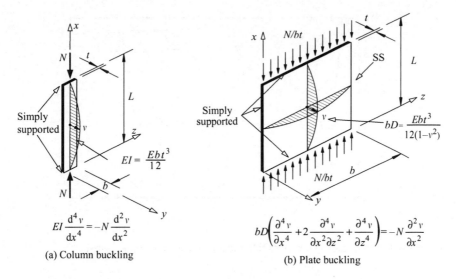

Figure 4.5 Comparison of column and plate buckling.

uniformly distributed over each end of the plate. When the applied loads are equal to the elastic buckling loads, the plate can buckle by deflecting v laterally out of its original plane into an adjacent position [1–4].

It is shown in Section 4.8.1 that this equilibrium position is given by

$$v = \delta \sin \frac{m\pi x}{L} \sin \frac{n\pi z}{b} \tag{4.1}$$

with $n = 1$, where δ is the undetermined magnitude of the central deflection. The elastic buckling load N_{cr} at which the plate buckles corresponds to the buckling stress

$$\sigma_{cr} = N_{cr}/bt, \tag{4.2}$$

which is given by

$$\sigma_{cr} = \frac{\pi^2 E}{12(1 - v^2)} \frac{k_\sigma}{(b/t)^2}. \tag{4.3}$$

The lowest value of the buckling coefficient k_σ (see Figure 4.6) is

$$k_\sigma = 4, \tag{4.4}$$

which is appropriate for the high aspect ratios L/b of most structural steel members (see Figure 4.7).

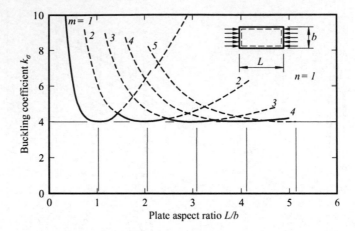

Figure 4.6 Buckling coefficients of simply supported plates in compression.

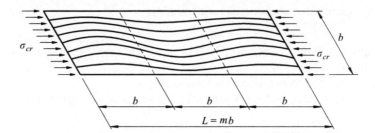

Figure 4.7 Buckled pattern of a long simply supported plate in compression.

The buckling stress σ_{cr} varies inversely as the square of the plate slenderness or width–thickness ratio b/t, as shown in Figure 4.4, in which the dimensionless buckling stress σ_{cr}/f_y is plotted against a modified plate slenderness ratio

$$\sqrt{\frac{f_y}{\sigma_{cr}}} = \frac{b}{t}\sqrt{\frac{f_y}{E}\frac{12(1-\nu^2)}{\pi^2 k_\sigma}}. \tag{4.5}$$

If the material ceases to be linear elastic at the yield stress f_y, the above analysis is only valid for $\sqrt{(f_y/\sigma_{cr})} \geq 1$. This limit is equivalent to a width–thickness ratio b/t given by

$$\frac{b}{t}\sqrt{\frac{f_y}{235}} = 56.8 \tag{4.6}$$

for a steel with $E = 210\,000\,\text{N/mm}^2$ and $\nu = 0.3$. (The yield stress f_y in equation 4.6 and in all of the similar equations which appear later in this chapter must be expressed in N/mm².)

The values of the buckling coefficient k_σ shown in Figure 4.6 indicate that the use of intermediate transverse stiffeners to increase the elastic buckling stress of a plate in compression is not effective, except when their spacing is significantly less than the plate width. Because of this, it is more economical to use intermediate longitudinal stiffeners which cause the plate to buckle in a number of half waves across its width. When such stiffeners are used, equation 4.3 still holds, provided b is taken as the stiffener spacing. Longitudinal stiffeners have an additional advantage when they are able to transfer compressive stresses, in which case their cross-sectional areas may be added to that of the plate.

An intermediate longitudinal stiffener must be sufficiently stiff flexurally to prevent the plate from deflecting at the stiffener. An approximate value for the minimum second moment of area I_{st} of a longitudinal stiffener at the centre line of a simply supported plate is given by

$$I_{st} = 4.5bt^3 \left[1 + 2.3\frac{A_{st}}{bt} \left(1 + 0.5\frac{A_{st}}{bt} \right) \right] \tag{4.7}$$

in which b is now the half width of the plate and A_{st} is the area of the stiffener. This minimum increases with the stiffener area because the load transmitted by the stiffener also increases with its area, and the additional second moment of area is required to resist the buckling action of the stiffener load.

Stiffeners should also be proportioned to resist local buckling. A stiffener is usually fixed to one side of the plate rather than placed symmetrically about the mid-plane, and in this case its effective second moment of area is greater than the value calculated for its centroid. It is often suggested [2, 3] that the value of I_{st} can be approximated by the value calculated for the mid-plane of the plate, but this may provide an overestimate in some cases [4–6].

The behaviour of an edge stiffener is different from that of an intermediate stiffener, in that theoretically it must be of infinite stiffness before it can provide an effective simple support to the plate. However, if a minimum value of the second moment of area of an edge stiffener of

$$I_{st} = 2.25bt^3 \left[1 + 4.6\frac{A_{st}}{bt} \left(1 + \frac{A_{st}}{bt} \right) \right] \tag{4.8}$$

is provided, the resulting reduction in the plate buckling stress is only a few percent [5]. Equation 4.8 is derived from equation 4.7 on the basis that an edge stiffener only has to support a plate on one side of the stiffener, while an intermediate stiffener has to support plates on both sides.

The effectiveness of longitudinal stiffeners decreases as their number increases, and the minimum stiffness required for them to provide effective lateral support increases. Some guidance on this is given in [3, 5].

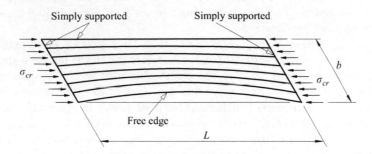

Figure 4.8 Buckled pattern of a plate free along one edge.

4.2.1.2   Plates free along one longitudinal edge

The thin flat plate shown in Figure 4.8 is simply supported along both transverse edges and one longitudinal edge, and is free along the other. The differential equation of equilibrium of the plate in a buckled position is the same as equation 4.105 (see Section 4.8.1). The buckled shape which satisfies this equation differs, however, from the approximately square buckles of the simply supported plate shown in Figure 4.7. The different boundary conditions along the free edge cause the plate to buckle with a single half wave along its length, as shown in Figure 4.8. Despite this, the solution for the elastic buckling stress σ_{cr} can still be expressed in the general form of equation 4.3 in which the buckling coefficient k_σ is now approximated by

$$k_\sigma = 0.425 + \left(\frac{b}{L}\right)^2,$$ (4.9)

as shown in Figure 4.9.

For the long-plate elements which are used as flange outstands in many structural steel members, the buckling coefficient k_σ is close to the minimum value of 0.425. In this case the elastic buckling stress (for a steel for which $E = 210\,000$ N/mm^2 and $v = 0.3$) is equal to the yield stress f_y when

$$\frac{b}{t}\sqrt{\frac{f_y}{235}} = 18.5.$$ (4.10)

Once again it is more economical to use longitudinal stiffeners to increase the elastic buckling stress than transverse stiffeners. The stiffness requirements of intermediate and edge longitudinal stiffeners are discussed in Section 4.2.1.1.

4.2.1.3   Plates with other support conditions

The edges of flat-plate elements may be fixed or elastically restrained, instead of being simply supported or free. The elastic buckling loads of flat plates with

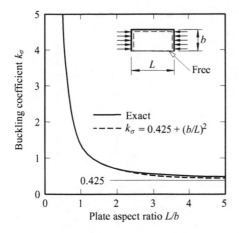

Figure 4.9 Buckling coefficients of plate free along one edge.

various support conditions have been determined, and many values of the buckling coefficient k_σ to be used in equation 4.3 are given in [2–6].

4.2.1.4 Plate assemblies

Many structural steel compression members are assemblies of flat-plate elements which are rigidly connected together along their common boundaries. The local buckling of such an assembly can be analysed approximately by assuming that the plate elements are hinged along their common boundaries, so that each plate acts as if simply supported along its connected boundary or boundaries and free along any unconnected boundary. The buckling stress of each plate element can then be determined from equation 4.3 with $k_\sigma = 4$ or 0.425 as appropriate, and the lowest of these can be used as an approximation for determining the buckling load of the member.

This approximation is conservative because the rigidity of the joints between the plate elements causes all plates to buckle simultaneously at a stress intermediate between the lowest and the highest of the buckling stresses of the individual plate elements. A number of analyses have been made of the stress at which simultaneous buckling takes place [4–6].

For example, values of the elastic buckling coefficient k_σ for an I-section in uniform compression are shown in Figure 4.10, and for a box section in uniform compression in Figure 4.11. The buckling stress can be obtained from these figures by using equation 4.3 with the plate thickness t replaced by the flange thickness t_f. The use of these stresses and other results [4–6] leads to economic thin-walled compression members.

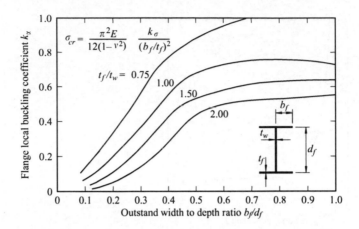

Figure 4.10 Local buckling coefficients for I-section compression members.

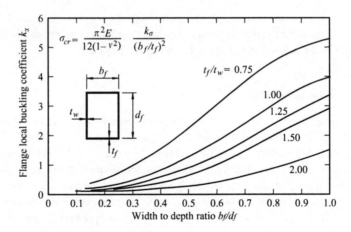

Figure 4.11 Local buckling coefficients for box section compression members.

4.2.2   Ultimate strength

4.2.2.1   Inelastic buckling of thick plates

The discussion given in Section 4.2.1 on the elastic buckling of rectangular plates applies only to materials whose stress–strain relationships remain linear. Thus, for stocky steel plates for which the calculated elastic buckling stress exceeds the yield stress f_y, the elastic analysis must be modified accordingly.

One particularly simple modification which can be applied to strain-hardened steel plates is to use the strain-hardening modulus E_{st} and $\sqrt{(EE_{st})}$ instead of E

in the terms of equation 4.105 (Section 4.8.1) which represent the longitudinal bending and the twisting resistances to buckling, while still using E in the term for the lateral bending resistance. With these modifications, the strain-hardening buckling stress of a long simply supported plate is equal to the yield stress f_y when

$$\frac{b}{t}\sqrt{\frac{f_y}{235}} = 23.7. \tag{4.11}$$

More accurate investigations [7] of the strain-hardening buckling of steel plates have shown that this simple approach is too conservative, and that the strain-hardening buckling stress is equal to the yield stress when

$$\frac{b}{t}\sqrt{\frac{f_y}{235}} = 32.1 \tag{4.12}$$

for simply supported plates, and when

$$\frac{b}{t}\sqrt{\frac{f_y}{235}} = 8.2 \tag{4.13}$$

for plates which are free along one longitudinal edge. If these are compared with equations 4.6 and 4.10, then it can be seen that these limits can be expressed in terms of the elastic buckling stress σ_{cr} by

$$\sqrt{\frac{f_y}{\sigma_{cr}}} = 0.57 \tag{4.14}$$

for simply supported plates, and by

$$\sqrt{\frac{f_y}{\sigma_{cr}}} = 0.46 \tag{4.15}$$

for plates which are free along one longitudinal edge. These limits are shown in Figure 4.4.

4.2.2.2 Post-buckling strength of thin plates

A thin elastic plate does not fail soon after it buckles, but can support loads significantly greater than its elastic buckling load without deflecting excessively. This is in contrast to the behaviour of an elastic compression member which can only carry very slightly increased loads before its deflections become excessive, as indicated in Figure 4.12. This post-buckling behaviour of a thin plate is due to a number of causes, but the main reason is that the deflected shape of the buckled plate cannot be developed from the pre-buckled configuration without some redistribution of the in-plane stresses within the plate. This redistribution, which is ignored in the

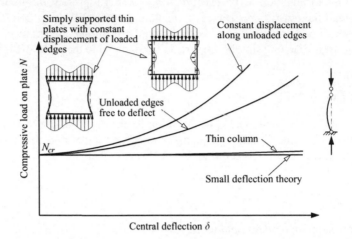

Figure 4.12 Post-buckling behaviour of thin elastic plates.

small deflection theory of elastic buckling, usually favours the less stiff portions of the plate, and causes an increase in the efficiency of the plate.

One of the most common causes of this redistribution is associated with the in-plane boundary conditions at the loaded edges of the plate. In long structural members, the continuity conditions along the transverse lines dividing consecutive buckled panels require that each of these boundary lines deflects a constant amount longitudinally. However, the longitudinal shortening of the panel due to its transverse deflections varies across the panel from a maximum at the centre to a minimum at the supported edges. This variation must, therefore, be compensated for by a corresponding variation in the longitudinal shortening due to axial strain, from a minimum at the centre to a maximum at the edges. The longitudinal stress distribution must be similar to the shortening due to strain, and so the stress at the centre of the panel is reduced below the average stress while the stress at the supported edges is increased above the average. This redistribution, which is equivalent to a transfer of stress from the more flexible central region of the panel to the regions near the supported edges, leads to a reduction in the transverse deflection, as indicated in Figure 4.12.

A further redistribution of the in-plane stresses takes places when the longitudinal edges of the panel are supported by very stiff elements which ensure that these edges deflect a constant amount laterally, in the plane of the panel. In this case the variation along the panel of the lateral shortening due to the transverse deflections induces a self-equilibrating set of lateral in-plane stresses which are tensile at the centre of the panel and compressive at the loaded edges. The tensile stresses help support the less stiff central region of the panel, and lead to further reductions in the transverse deflections, as indicated in Figure 4.12. However, this action is not usually fully developed, because the elements supporting the panels of most structural members are not very stiff.

The post-buckling effect is greater in plates supported along both longitudinal edges than it is in plates which are free along one longitudinal edge. This is because the deflected shapes of the latter have much less curvature than the former and the redistributions of the in-plane stresses are not as pronounced. In addition, it is not possible to develop any lateral in-plane stresses along free edges, and so it is not uncommon to ignore any post-buckling reserves of slender flange outstands.

The redistribution of the in-plane stresses after buckling continues with increasing load until the yield stress f_y is reached at the supported edges. Yielding then spreads rapidly and the plate fails soon after, as indicated in Figure 4.3. The occurrence of the first yield in an initially flat plate depends on its slenderness, and a thick plate yields before its elastic buckling stress σ_{cr} is reached. As the slenderness increases and the elastic buckling stress decreases below the yield stress, the ratio of the ultimate stress f_{ult} to the elastic buckling stress increases, as shown in Figure 4.4.

Although the analytical determination of the ultimate strength of a thin flat plate is difficult, it has been found that the use of an effective width concept can lead to satisfactory approximations. According to this concept, the actual ultimate stress distribution in a simply supported plate (see Figure 4.13) is replaced by a simplified distribution for which the central portion of the plate is ignored and the remaining effective width b_{eff} carries the yield stress f_y. It was proposed that this effective width should be approximated by

$$\frac{b_{eff}}{b} = \sqrt{\frac{\sigma_{cr}}{f_y}}, \tag{4.16}$$

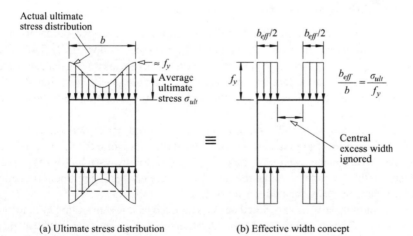

(a) Ultimate stress distribution   (b) Effective width concept

Figure 4.13 Effective width concept for simply supported plates.

which is equivalent to supposing that the ultimate load carrying capacity of the plate $f_y b_{eff} t$ is equal to the elastic buckling load of a plate of width b_{eff}. Alternatively, this proposal can be regarded as determining an effective average ultimate stress f_{ult} which acts on the full width b of the plate. This average ultimate stress, which is given by

$$\frac{f_{ult}}{f_y} = \sqrt{\frac{\sigma_{cr}}{f_y}} \qquad (4.17)$$

is shown in Figure 4.4.

Experiments on real plates with initial curvatures and residual stresses have confirmed the qualitative validity of this effective width approach, but suggest that the quantitative values of the effective width for hot-rolled and welded plates should be obtained from equations of the type

$$\frac{b_{eff}}{b} = \alpha \sqrt{\frac{\sigma_{cr}}{f_y}}, \qquad (4.18)$$

where α reflects the influence of the initial curvatures and residual stresses. For example, test results for the ultimate stresses $f_{ult} (= f_y b_{eff}/b)$ of hot-rolled simply supported plates with residual stresses and initial curvatures are shown in Figure 4.14, and suggest a value of α equal to 0.65. Other values of α are given in [8]. Tests on cold-formed members also support the effective width concept, with quantitative values of the effective width being obtained from

$$\frac{b_{eff}}{b} = \sqrt{\frac{\sigma_{cr}}{f_y}} \left(1 - 0.22\sqrt{\frac{\sigma_{cr}}{f_y}}\right). \qquad (4.19)$$

4.2.2.3   Effects of initial curvature and residual stresses

Real plates are not perfectly flat, but have small initial curvatures similar to those in real columns (see Section 3.2.2) and real beams (see Section 6.2.2). The initial curvature of a plate causes it to deflect transversely as soon as it is loaded, as shown in Figure 4.3. These deflections increase rapidly as the elastic buckling stress is approached, but slow down beyond the buckling stress and approach those of an initially flat plate.

In a thin plate with initial curvature, the first yield and failure occur only slightly before they do in a flat plate, as indicated in Figure 4.3, while the initial curvature has little effect on the resistance of a thick plate (see Figure 4.4). It is only in a plate of intermediate slenderness that the initial curvature causes a significant reduction in the resistance, as indicated in Figure 4.4.

Real plates usually have residual stresses induced by uneven cooling after rolling or welding. These stresses are generally tensile at the junctions between plate elements, and compressive in the regions away from the junctions. Residual compressive stresses in the central region of a simply supported thin plate cause it

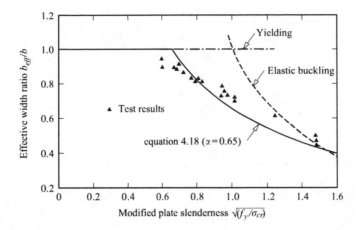

Figure 4.14 Effective widths of simply supported plates.

to buckle prematurely and reduce its ultimate strength, as shown in Figure 4.4. Residual stresses also cause premature yielding in plates of intermediate slenderness, as indicated in Figure 4.4, but have a negligible effect on the strain-hardening buckling of stocky plates.

Some typical test results for thin supported plates with initial curvatures and residual stresses are shown in Figure 4.14.

4.3 Plate elements in shear

4.3.1 Elastic buckling

The thin flat plate of length L, depth d, and thickness t shown in Figure 4.15 is simply supported along all four edges. The plate is loaded by shear stresses distributed uniformly along its edges. When these stresses are equal to the elastic buckling value τ_{cr}, then the plate can buckle by deflecting v laterally out of its original plane into an adjacent position. For this adjacent position to be one of equilibrium, the differential equilibrium equation [1–4]

$$\left(\frac{\partial^4 v}{\partial x^4} + 2\frac{\partial^4 v}{\partial x^2 \partial z^2} + \frac{\partial^4 v}{\partial z^4}\right) = -\frac{2\tau_{cr}t}{D}\frac{\partial^2 v}{\partial x \partial z} \tag{4.20}$$

must be satisfied (this may be compared with the corresponding equation 4.105 of Section 4.8.1, for a plate in compression).

Closed form solutions of this equation are not available, but numerical solutions have been obtained. These indicate that the plate tends to buckle along compression diagonals, as shown in Figure 4.15. The shape of the buckle is influenced by the tensile forces acting along the other diagonal, while the number of buckles

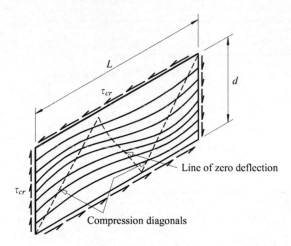

Figure 4.15 Buckling pattern of a simply supported plate in shear.

increases with the aspect ratio L/d. The numerical solutions for the elastic buckling stress τ_{cr} can be expressed in the form

$$\tau_{cr} = \frac{\pi^2 E}{12(1 - v^2)} \frac{k_\tau}{(d/t)^2} \qquad (4.21)$$

in which the buckling coefficient k_τ is approximated by

$$k_\tau = 5.34 + 4 \left(\frac{d}{L}\right)^2 \qquad (4.22)$$

when $L \geq d$, and by

$$k_\tau = 5.34 \left(\frac{d}{L}\right)^2 + 4 \qquad (4.23)$$

when $L \leq d$, as shown in Figure 4.16.

The shear stresses in many structural members are transmitted by unstiffened webs, for which the aspect ratio L/d is large. In this case the buckling coefficient k_τ approaches 5.34, and the buckling stress can be closely approximated by

$$\tau_{cr} = \frac{5.34\pi^2 E/12(1 - v^2)}{(d/t)^2}. \qquad (4.24)$$

This elastic buckling stress is equal to the yield stress in shear $\tau_y = f_y/\sqrt{3}$ (see Section 1.3.1) of a steel for which $E = 210\,000 \text{ N/mm}^2$ and $v = 0.3$ when

$$\frac{d}{t}\sqrt{\frac{f_y}{235}} = 86.4. \qquad (4.25)$$

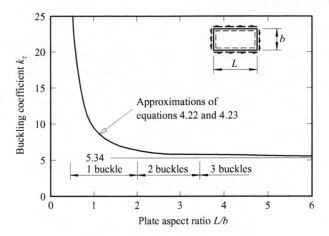

Figure 4.16 Buckling coefficients of plates in shear.

The values of the buckling coefficient k_τ shown in Figure 4.16 and the form of equation 4.21 indicate that the elastic buckling stress may be significantly increased either by using intermediate transverse stiffeners to decrease the aspect ratio L/d and to increase the buckling coefficient k_τ, or by using longitudinal stiffeners to decrease the depth–thickness ratio d/t. It is apparent from Figure 4.16 that such stiffeners are likely to be most efficient when the stiffener spacing a is such that the aspect ratio of each panel lies between 0.5 and 2 so that only one buckle can form in each panel.

Any intermediate stiffeners used must be sufficiently stiff to ensure that the elastic buckling stress τ_{cr} is increased to the value calculated for the stiffened panel. Some values of the required stiffener second moment of area I_{st} are given in [3, 5] for various panel aspect ratios a/d. These can be conservatively approximated by using

$$\frac{I_{st}}{at^3/12(1-v^2)} = \frac{6}{a/d} \tag{4.26}$$

when $a/d \geq 1$, and by using

$$\frac{I_{st}}{at^3/12(1-v^2)} = \frac{6}{(a/d)^4} \tag{4.27}$$

when $a/d \leq 1$.

4.3.2 Ultimate strength

4.3.2.1 Unstiffened plates

A stocky unstiffened web in an I-section beam in pure shear is shown in Figure 4.17a. The elastic shear stress distribution in such a section is analysed in Section 5.10.2. The web behaves elastically in shear until first yield occurs at $\tau_y = f_y/\sqrt{3}$, and then undergoes increasing plastification until the web is fully yielded in shear (Figure 4.17b). Because the shear stress distribution at first yield is nearly uniform, the nominal first yield and fully plastic loads are nearly equal, and the shear shape factor is usually very close to 1.0. Stocky unstiffened webs in steel beams reach first yield before they buckle elastically, so that their resistances are determined by the shear stress τ_y, as indicated in Figure 4.18.

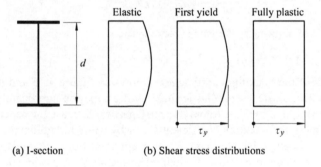

(a) I-section   (b) Shear stress distributions

Figure 4.17 Plastification of an I-section web in shear.

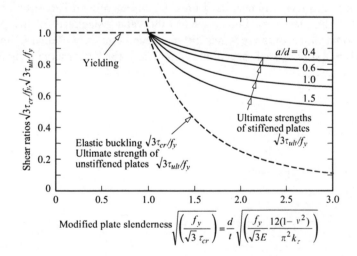

Figure 4.18 Ultimate strengths of plates in shear.

Thus the resistance of a stocky web in a flanged section for which the shear shape factor is close to unity is closely approximated by the web plastic shear resistance

$$V_{pl} = dt\tau_y. \tag{4.28}$$

When the depth–thickness ratio d/t of a long unstiffened steel web exceeds $86.4/\sqrt{(f_y/235)}$, its elastic buckling shear stress τ_{cr} is less than the shear yield stress (see equation 4.25). The post-buckling reserve of shear strength of such an unstiffened web is not great, and its ultimate stress τ_{ult} can be approximated with reasonable accuracy by the elastic buckling stress τ_{cr} given by equation 4.24 and shown by the dashed line in Figure 4.18.

4.3.2.2   Stiffened plates

Stocky stiffened webs in steel beams yield in shear before they buckle, and so the stiffeners do not contribute to the ultimate strength. Because of this, stocky webs are usually unstiffened.

In a slender web with transverse stiffeners which buckles elastically before it yields, there is a significant reserve of strength after buckling. This is caused by a redistribution of stress, in which the diagonal tension stresses in the web panel continue to increase with the applied shear, while the diagonal compressive stresses remain substantially unchanged. The increased diagonal tension stresses form a tension field, which combines with the transverse stiffeners and the flanges to transfer the additional shear in a truss-type action [7, 9], as shown in Figure 4.19. The ultimate shear stresses τ_{ult} of the web is reached soon after yield occurs in the tension field, and this can be approximated by

$$\tau_{ult} = \tau_{cr} + \tau_{tf} \tag{4.29}$$

in which τ_{cr} is the elastic buckling stress given by equations 4.21–4.23 with L equal to the stiffener spacing a, and τ_{tf} is the tension field contribution at yield which can be approximated by

$$\tau_{tf} = \frac{f_y}{2} \frac{(1 - \sqrt{3}\tau_{cr}/f_y)}{\sqrt{1 + (a/d)^2}}. \tag{4.30}$$

Thus the approximate ultimate strength can be obtained from

$$\frac{\sqrt{3}\tau_{ult}}{f_y} = \frac{\sqrt{3}\tau_{cr}}{f_y} + \frac{\sqrt{3}}{2} \frac{(1 - \sqrt{3}\tau_{cr}/f_y)}{\sqrt{1 + (a/d)^2}}, \tag{4.31}$$

which is shown in Figure 4.18. This equation is conservative, as it ignores any contributions made by the bending resistance of the flanges to the ultimate strength of the web [9, 10].

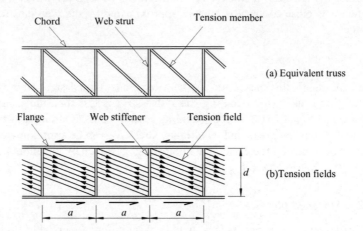

Figure 4.19 Tension fields in a stiffened web.

Not only must the intermediate transverse stiffeners be sufficiently stiff to ensure that the elastic buckling stress τ_{cr} is reached, as explained in Section 4.3.1, but they must also be strong enough to transmit the stiffener force

$$N_{st} = \frac{f_y d t}{2}\left(1 - \frac{\sqrt{3}\tau_{cr}}{f_y}\right)\left[\frac{a}{d} - \frac{(a/d)^2}{\sqrt{1 + (a/d)^2}}\right] \tag{4.32}$$

required by the tension field action. Intermediate transverse stiffeners are often so stocky that their ultimate strengths can be approximated by their squash loads, in which case the required area A_{st} of a symmetrical pair of stiffeners is given by

$$A_{st} = N_{st}/f_y. \tag{4.33}$$

However, the stability of very slender stiffeners should be checked, in which case the second moment of area I_{st} required to resist the stiffener force can be conservatively estimated from

$$I_{st} = \frac{N_{st}d^2}{\pi^2 E} \tag{4.34}$$

which is based on the elastic buckling of a pin-ended column of length d.

4.4   Plate elements in bending

4.4.1   Elastic buckling

The thin flat plate of length L, width d, and thickness t shown in Figure 4.20 is simply supported along all four edges. The plate is loaded by bending stress

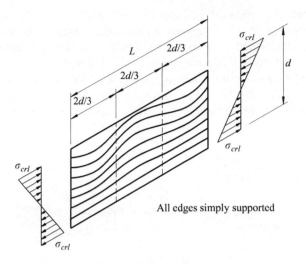

Figure 4.20 Buckled pattern of a plate in bending.

distributions which vary linearly across its width. When the maximum stress reaches the elastic buckling value σ_{crl}, the plate can buckle out of its original plane as shown in Figure 4.20. The elastic buckling stress can be expressed in the form

$$\sigma_{crl} = \frac{\pi^2 E}{12(1 - v^2)} \frac{k_\sigma}{(d/t)^2} \tag{4.35}$$

where the buckling coefficient k_σ depends on the aspect ratio L/d of the plate and the number of buckles along the plate. For long plates, the value of k_σ is close to its minimum value of

$$k_\sigma = 23.9 \tag{4.36}$$

for which the length of each buckle is approximately $2d/3$. By using this value of k_σ, it can be shown that the elastic buckling stress for a steel for which $E = 210\,000$ N/mm^2 and $v = 0.3$ is equal to the yield stress f_y when

$$\frac{d}{t} \sqrt{\frac{f_y}{235}} = 138.9. \tag{4.37}$$

The buckling coefficient k_σ is not significantly greater than 23.9 except when the buckle length is reduced below $2d/3$. For this reason, transverse stiffeners are ineffective unless more closely spaced than $2d/3$. On the other hand, longitudinal stiffeners may be quite effective in changing the buckled shape, and therefore the value of k_σ. Such a stiffener is most efficient when it is placed in the compression region at about $d/5$ from the compression edge. This is close to the position of the crests in the buckles of an unstiffened plate. Such a stiffener may increase the buckling coefficient k_σ significantly, the maximum value of 129.4 being achieved when the stiffener acts as if rigid. The values given in [5, 6] indicate that a hypothetical stiffener of zero area A_{st} will produce this effect if its second moment of area I_{st} is equal to $4dt^3$. This value should be increased to allow for the compressive load in the real stiffener, and it is suggested that a suitable value might be given by

$$I_{st} = 4dt^3 \left[1 + 4\frac{A_{st}}{dt} \left(1 + \frac{A_{st}}{dt} \right) \right], \tag{4.38}$$

which is of a similar form to that given by equation 4.7 for the longitudinal stiffeners of plates in uniform compression.

4.4.2   Plate assemblies

The local buckling of a plate assembly in bending, such as an I-beam bent about its major axis, can be analysed approximately by assuming that the plate elements are hinged along their common boundaries in much the same fashion as that described in Section 4.2.1.4 for compression members. The elastic buckling moment can then be approximated by using $k_\sigma = 0.425$ or 4 for the flanges, as appropriate, and $k_\sigma = 23.9$ for the web, and determining the element which is closest to buckling. The results of more accurate analyses of the simultaneous buckling of the flange and web elements in I-beams and box section beams in bending are given in Figures 4.21 and 4.22, respectively.

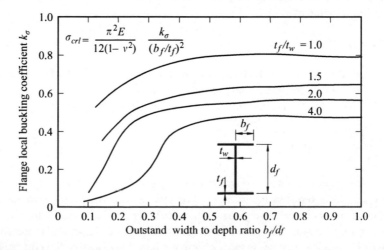

Figure 4.21 Local buckling coefficients for I-beams.

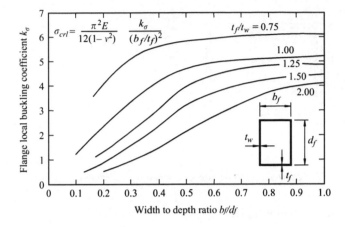

Figure 4.22 Local buckling coefficients for box section beams.

4.4.3   Ultimate strength

The ultimate strength of a thick plate in bending is governed by its yield stress f_y and by its plastic shape factor which is equal to 1.5 for a constant thickness plate, as discussed in Section 5.5.2.

When the width to thickness ratio d/t exceeds $138.9/\sqrt{(f_y/235)}$, the elastic buckling stress σ_{crl} of a simply supported plate is less than the yield stress f_y (see equation 4.37). A long slender plate such as this has a significant reserve of strength after buckling, because it is able to redistribute the compressive stresses from the buckled region to the area close to the supported compression edge, in the same way as a plate in uniform compression (see Section 4.2.2.2). Solutions for the post-buckling reserve of strength of a plate in bending given in [6, 11–13] show that an effective width treatment can be used, with the effective width being taken over part of the compressive portion of the plate. In the more general case of the slender plate shown in Figure 4.20 being loaded by combined bending and axial force, the post-buckling strength reserve of the plate can be substantial. For such plates, the effective width is taken over a part of the compressive portion of the plate. Effective widths can be obtained in this way from the local buckling stress σ_{cr} under combined bending and axial force using the modified plate slenderness $\sqrt{(f_y/\sigma_{cr})}$ in a similar fashion to that depicted in Figure 4.14, although the procedure is more complex.

4.5   Plate elements in bending and shear

4.5.1   Elastic buckling

The elastic buckling of the simply supported thin flat plate shown in Figure 4.23 which is subjected to combined shear and bending can be predicted by using the

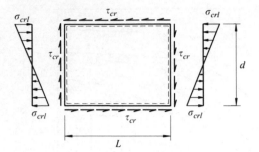

Figure 4.23 Simply supported plate under shear and bending.

approximate interaction equation [2, 4, 5, 13]

$$\left(\frac{\tau_{cr}}{\tau_{cro}}\right)^2 + \left(\frac{\sigma_{crl}}{\sigma_{crlo}}\right)^2 = 1 \qquad (4.39)$$

in which τ_{cro} is the elastic buckling stress when the plate is in pure shear (see equations 4.21–4.24) and σ_{crlo} is the elastic buckling stress for pure bending (see equations 4.35 and 4.36). If the elastic buckling stresses τ_{cr}, σ_{crl} are used in the Hencky–Von Mises yield criterion (see Section 1.3.1)

$$3\tau_{cr}^2 + \sigma_{crl}^2 = f_y^2, \qquad (4.40)$$

it can shown from equations 4.39 and 4.40 that the most severe loading condition for which elastic buckling and yielding occur simultaneously is that of pure shear. Thus, in an unstiffened web of a steel for which $E = 210\,000\,\text{N/mm}^2$ and $\nu = 0.3$, yielding will occur before buckling while (see equation 4.25)

$$\frac{d}{t}\sqrt{\frac{f_y}{235}} \le 86.4. \qquad (4.41)$$

4.5.2   Ultimate strength

4.5.2.1   Unstiffened plates

Stocky unstiffened webs in steel beams yield before they buckle, and their design resistances can be estimated approximately by using the Hencky–Von Mises yield criterion (see Section 1.3.1), so that the shear force V_w and moment M_w in the web satisfy

$$3\left(\frac{V_w}{dt}\right)^2 + \left(\frac{6M_w}{d^2 t}\right)^2 = f_y^2. \qquad (4.42)$$

This approximation is conservative, as it assumes that the plastic shape factor for web bending is 1.0 instead of 1.5.

While the elastic buckling of slender unstiffened webs under combined shear and bending has been studied, the ultimate strengths of such webs have not been fully investigated. However, it seems possible that there is only a small reserve of strength after elastic buckling, so that the ultimate strength can be approximated conservatively by the elastic buckling stress combinations which satisfy equation 4.39.

4.5.2.2   Stiffened plates

The ultimate strength of a stiffened web under combined shear and bending can be discussed in terms of the interaction diagram shown in Figure 4.24 for the ultimate shear force V and bending moment M acting on a plate girder. When there is no shear, the ultimate moment capacity is equal to the full plastic moment M_p, providing the flanges are Class 1 or Class 2 (see Section 4.7.2), while the ultimate shear capacity in the absence of bending moment is

$$V_{ult} = dt\tau_{ult}, \tag{4.43}$$

where the ultimate stress τ_{ult} is given approximately by equation 4.31 (which ignores any contributions made by the bending resistance of the flanges). This shear capacity remains unchanged as the bending moment increases to the value M_{fp} which is sufficient to fully yield the flanges if they alone resist the moment. This fact forms the basis for the widely used proportioning procedure for which it is assumed that the web resists only the shear and that the flanges are required to

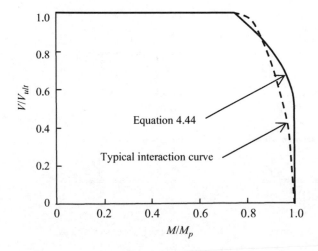

Figure 4.24 Ultimate strengths of stiffened webs in shear and bending.

resist the full moment. As the bending moment increases beyond M_{fp}, the ultimate shear capacity falls off rapidly as shown. A simple approximation for the reduced shear capacity is given by

$$\frac{V}{V_{ult}} = 0.5 \left(1 + \sqrt{\frac{1 - M/M_p}{1 - M_{fp}/M_p}} \right) \tag{4.44}$$

while $M_{fp}/M_p \leq M/M_p \leq 1$. This approximation ignores the influence of the bending stresses on the tension field and the bending resistance of the flanges. A more accurate method of analysis is discussed in [14].

4.6   Plate elements in bearing

4.6.1   Elastic buckling

Plate elements are subjected to bearing stresses by concentrated or locally distributed edge loads. For example, a concentrated load applied to the top flange of a plate girder induces local bearing stresses in the web immediately beneath the load. In a slender girder with transverse web stiffeners, the load is resisted by vertical shear stresses acting at the stiffeners, as shown in Figure 4.25a. Bearing commonly occurs in conjunction with shear (as at an end support, see Figure 4.25b), bending (as at mid-span, see Figure 4.25c), and with combined shear and bending (as at an interior support, see Figure 4.25d).

In the case of a panel of a stiffened web, the edges of the panel may be regarded as simply supported. When the bearing load is distributed along the full length a of the panel and there is no shear or bending, the elastic bearing buckling stress σ_{crp}

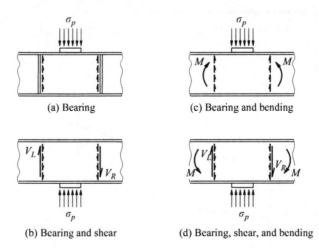

(a) Bearing

(c) Bearing and bending

(b) Bearing and shear

(d) Bearing, shear, and bending

Figure 4.25 Plate girder webs in bearing.

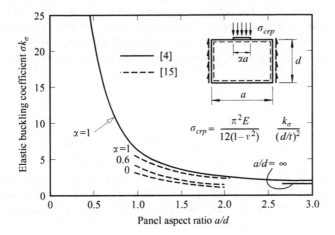

Figure 4.26 Buckling coefficients of plates in bearing.

can be expressed as

$$\sigma_{crp} = \frac{\pi^2 E}{12(1 - v^2)} \frac{k_\sigma}{(d/t)^2} \qquad (4.45)$$

in which the buckling coefficient k_σ varies with the panel aspect ratio a/d as shown in Figure 4.26.

When a patch-bearing load acts along a reduced length αa of the top edge of the panel, the value of αk_σ is decreased. This effect has been investigated [15], and some values of αk_σ are shown in Figure 4.26. These values suggest a limiting value of $\alpha k_\sigma = 0.8$ approximately for the case of unstiffened webs ($a/d \rightarrow \infty$) with concentrated loads ($\alpha \rightarrow 0$).

For the case of a panel in combined bearing, shear, and bending, it has been suggested [6, 15] that the elastic buckling stresses can be determined from the interaction equation

$$\frac{\sigma_{crp}}{\sigma_{crpo}} + \left(\frac{\tau_{cr}}{\tau_{cro}}\right)^2 + \left(\frac{\sigma_{crl}}{\sigma_{crlo}}\right)^2 = 1 \qquad (4.46)$$

in which the final subscripts o indicate the appropriate elastic buckling stress when only that type of loading is applied. Some specific interaction diagrams are given in [15].

4.6.2 Ultimate strength

The ultimate strength of a thick web in bearing depends chiefly on its yield stress f_y. Although yielding first occurs under the centre of the bearing plate, general yielding does not take place until the applied load is large enough to yield a web area defined by a dispersion of the applied stress through the flange. Even at this load the web does not collapse catastrophically, and some further yielding and redistribution is possible. When the web is also subjected to shear and bending, general yielding in bearing can be approximated by using the Hencky–Von Mises yield criterion (see Section 1.3.1)

$$\sigma_p^2 + \sigma_b^2 - \sigma_p\sigma_b + 3\tau^2 = f_y^2 \tag{4.47}$$

in which σ_p is the bearing stress, σ_b the bending stress and τ the shear stress.

Thin stiffened web panels in bearing have a reserve of strength after elastic buckling which is due to a redistribution of stress from the more flexible central region to the stiffeners. Studies of this effect suggest that the reserve of strength decreases as the bearing loads become more concentrated [15]. The collapse behaviour of stiffened panels is described in [16, 17].

4.7 Design against local buckling

4.7.1 Compression members

The cross-sections of compression members must be designed so that the design compression force N_{Ed} does not exceed the design compressive resistance $N_{c,Rd}$, whence

$$N_{Ed} \leq N_{c,Rd}. \tag{4.48}$$

This ignores overall member buckling, as discussed in Chapter 3, for which the non-dimensional slenderness $\bar{\lambda} \leq 0.2$. The EC3 design expression for cross-section resistance under uniform compression is

$$N_{c,Rd} = \frac{A_{eff}f_y}{\gamma_{M0}} \tag{4.49}$$

in which A_{eff} is the effective area of the cross-section, f_y is the nominal yield strength for the section and $\gamma_{M0}(=1)$ is the partial section resistance factor.

The cross-section of a compression member may buckle locally before it reaches its yield stress f_y, in which case it is defined in EC3 as a Class 4 cross-section. Cross-sections of compression members which yield prior to local buckling are 'effective', and are defined in EC3 as being either of Class 1, Class 2, or Class 3. For these cross-sections which are unaffected by local bucking, the effective area A_{eff} in equation 4.49 is taken as the gross area A. A cross-section composed of flat-plate elements has no local buckling effects when the width–thickness ratio

b/t of every element of the cross-section satisfies

$$\frac{b}{t} \leq \lambda_3 \varepsilon \tag{4.50}$$

where $\lambda_3 \varepsilon$ is the appropriate Class 3 slenderness limit of Table 5.2 of EC3 and ε is a constant given by

$$\varepsilon = \sqrt{235/f_y} \tag{4.51}$$

in which f_y is in N/mm^2 units. Some values of $\lambda_3 \varepsilon$ are given in Figures 4.27 and 4.28.

Class 4 cross-sections are those containing at least one slender element for which equation 4.50 is not satisfied. For these, the effective area is reduced to

$$A_{eff} = \sum A_{c,eff} \tag{4.52}$$

in which $A_{c,eff}$ is the effective area of a flat compression element comprising the cross-section, which is obtained from its gross area A_c by

$$A_{c,eff} = \rho A_c \tag{4.53}$$

where ρ is a reduction factor given by

$$\rho = (\bar{\lambda}_p - 0.22)/\bar{\lambda}_p^2 \leq 1.0 \tag{4.54}$$

Section description	UB, UC	CHS	Box	RHS
Section and element widths	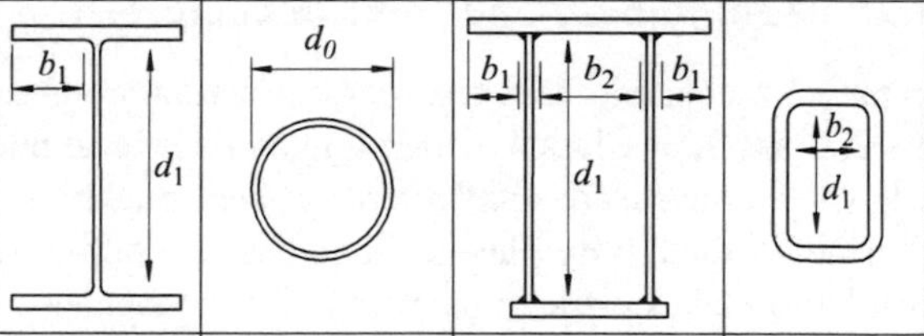			
Flange outstand b_1	14ε	–	14ε	–
Flange b_2 supported along both edges	–	–	42ε	42ε
Web d_1 supported along both edges	42ε	–	42ε	42ε
Diameter d_0	–	$90\varepsilon^2$	–	–

Figure 4.27 Yield limits for some compression elements.

for element supported on both edges, and

$$\rho = (\overline{\lambda}_p - 0.188)/\overline{\lambda}_p^2 \leq 1.0 \tag{4.55}$$

for outstands, and where

$$\overline{\lambda}_p = \sqrt{\frac{f_y}{\sigma_{cr}}} = \frac{b}{t} \frac{1}{28.4\varepsilon\sqrt{k_\sigma}} \tag{4.56}$$

is the modified plate slenderness (see equation 4.5). Values of the local buckling coefficient k_σ are given in Table 4.2 of EC3-1-5 and are similar to those in Figures 4.10 and 4.11.

A circular hollow section member has no local buckling effects when its diameter–thickness ratio d/t satisfies

$$\frac{d}{t} \leq 90\varepsilon^2. \tag{4.57}$$

If the diameter–thickness ratio does not satisfy equation 4.57, then it has a Class 4 cross-section whose effective area is reduced to

$$A_{eff} = \sqrt{\frac{90}{d/(t\varepsilon^2)}} A. \tag{4.58}$$

Worked examples for checking the section capacities of compression members are given in Sections 4.9.1 and 3.12.1–3.12.3.

4.7.2   Beam flanges and webs in compression

Compression elements in a beam cross-section are classified in EC3 as Class 1, Class 2, Class 3, or Class 4, depending on their local buckling resistance.

Class 1 elements are unaffected by local buckling, and are able to develop and maintain their fully plastic capacities (Section 5.5) while inelastic moment redistribution takes place in the beam. Class 1 elements satisfy

$$b/t \leq \lambda_1\varepsilon \tag{4.59}$$

in which $\lambda_1\varepsilon$ is the appropriate plasticity slenderness limit given in Table 5.2 of EC3 (some values of $\lambda_1\varepsilon$ are shown in Figures 4.28 and 5.33). These limits are closely related to equations 4.12 and 4.13 for the strain-hardening buckling stresses of inelastic plates.

Class 2 elements are unaffected by local buckling in the development of their fully plastic capacities (Section 5.5), but may be unable to maintain these capacities

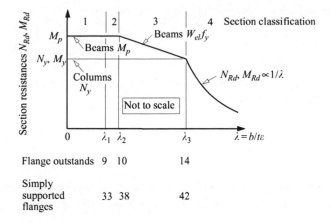

Figure 4.28 EC3 section resistances.

while inelastic moment redistribution takes place in the beam. Class 2 sections satisfy

$$\lambda_1 \varepsilon \leq b/t \leq \lambda_2 \varepsilon \qquad (4.60)$$

in which $\lambda_2 \varepsilon$ is the appropriate Class 2 slenderness limit given in Table 5.2 of EC3 (some values of $\lambda_2 \varepsilon$ are shown in Figures 4.28 and 5.33). These limits are slightly greater than those implied in equations 4.12 and 4.13 for the strain-hardening buckling of inelastic plates.

Class 3 elements are able to reach first yield, but buckle locally before they become fully plastic. Class 3 elements satisfy

$$\lambda_2 \varepsilon \leq b/t \leq \lambda_3 \varepsilon \qquad (4.61)$$

in which $\lambda_3 \varepsilon$ is the appropriate yield slenderness limit given in Table 5.2 of EC3 (some values of $\lambda_3 \varepsilon$ are shown in Figures 4.28 and 5.33). These limits are closely related to equations 4.18 used to define the effective widths of flange plates in uniform compression or modified from equation 4.37 for web plates in bending. Class 4 elements buckle locally before they reach first yield, and satisfy

$$\lambda_3 \varepsilon < b/t. \qquad (4.62)$$

Beam cross-sections are classified as being Class 1, Class 2, Class 3, or Class 4, depending on the classification of their elements. A Class 1 cross-section has all of its elements being Class 1. A Class 2 cross-section has no Class 3 or Class 4 elements and has at least one Class 2 element, while a Class 3 cross-section has no Class 4 elements and at least one Class 3 element. A Class 4 cross-section has at least one Class 4 element.

A beam cross-section must be designed so that its factored design moment M_{Ed} does not exceed the design section moment resistance, so that

$$M_{Ed} \leq M_{c,Rd}. \tag{4.63}$$

Beams must also be designed for shear (Section 4.7.3) and against lateral buckling (Section 6.9). The section resistance $M_{c,Rd}$ is given by

$$M_{c,Rd} = \frac{Wf_y}{\gamma_{M0}} \tag{4.64}$$

in which W is the appropriate section modulus, f_y the yield strength and $\gamma_{M0}(=1)$ the partial section resistance factor. For Class 1 and 2 beam cross-sections

$$W = W_{pl}, \tag{4.65}$$

where W_{pl} is the plastic section modulus and which allows many beams to be designed for the full plastic moment

$$M_p = W_{pl}f_y \tag{4.66}$$

as indicated in Figure 4.28.

For Class 3 beam cross-sections, the effective section modulus is taken as the minimum elastic section modulus $W_{el,min}$ which is that based on the extreme fibre that reaches yield first. Some I-sections have Class 1 or 2 flanges, but contain a Class 3 web and so based on the EC3 classification they would be Class 3 cross-sections and unsuitable for plastic design. However, the EC3 permits these sections to be classified as effective Class 2 cross-sections by neglecting part of the compression portion of the web. This simple procedure for hot-rolled or welded sections conveniently replaces the compressed portion of the web by a part of width $20\varepsilon t_w$ adjacent to the compression flange, and with another part of width $20\varepsilon t_w$ adjacent to the plastic neutral axis of the effective cross-section.

For a beam with a slender compression flange supported along both edges, the effective section modulus W_{el} may be determined by calculating the elastic section modulus of an effective cross-section obtained by using a compression flange effective width b_{eff} obtained using equation 4.53. The calculation of W_{el} must incorporate the possibility that the effective section is not symmetric, and that the centroids of the gross and effective sections do not coincide.

Worked examples of classifying the section and checking the section moment capacity are given in Sections 4.9.2–4.9.4, 5.12.15 and 5.12.17.

4.7.3 Members in compression and bending

For cross-sections subjected to a design compressive force N_{Ed} and a design major axis bending moment $M_{y,Ed}$, EC3 requires the interaction equation

$$\frac{N_{Ed}}{N_{Rd}} + \frac{M_{y,Ed}}{M_{y,Rd}} \leq 1 \qquad (4.67)$$

to be satisfied, where N_{Rd} is the compression resistance of the cross-section determined from equation 4.49 and $M_{y,Rd}$ is the bending resistance of the cross-section determined according to equation 4.63. Generally, the provision of equation 4.67 is overly conservative and is of use only for preliminary member sizing, and so EC3 provides less conservative and more detailed member checks depending on the section classification.

If the cross-section is Class 1 or 2, EC3 reduces the design bending resistance $M_{y,Rd}$ to a value $M_{y,N,Rd}$, dependent on the coincident axial force N_{Ed}. Provided that

$$N_{Ed} \leq 0.25N_{pl,Rd} \quad \text{and} \qquad (4.68)$$

$$N_{Ed} \leq \frac{0.5h_w t_w f_y}{\gamma_{M0}} \qquad (4.69)$$

where $N_{pl,Rd}$ is the resistance based on yielding given by equation 4.49 and $\gamma_{M0}(=1)$ is the partial section resistance factor, no reduction in the plastic moment capacity $M_{y,Rd} = M_{pl,y,Rd}$ is needed since for small axial loads the theoretical reduction in the plastic moment is offset by strain hardening. If either of equations 4.68 or 4.69 is not satisfied, EC3 requires that

$$M_{N,y,Rd} = M_{pl,y,Rd} \left(\frac{1-n}{1-0.5a} \right) \qquad (4.70)$$

where

$$n = N_{Ed}/N_{pl,Rd} \qquad (4.71)$$

is the ratio of the applied compression load to the plastic compression resistance of the cross-section, and

$$a = (A - 2b_f t_f)/A \leq 0.5 \qquad (4.72)$$

is the ratio of the area of the web to the total area of the cross-section.

For Class 3 cross-sections, EC3 allows only a linear interaction of the stresses arising from the combined bending moment and axial force, so that the maximum longitudinal stress is limited to the yield stress, that is

$$\sigma_{x,Ed} \leq f_y/\gamma_{M0}. \qquad (4.73)$$

As for Class 3 cross-sections, the longitudinal stress in Class 4 sections subjected to combined compression and bending is calculated based on the effective properties of the cross-section so that equation 4.73 is satisfied. The resulting expression that satisfies equation 4.73 is

$$\frac{N_{Ed}}{A_{eff}\,f_y/\gamma_{M0}} + \frac{M_{y,Ed} + N_{Ed}e_{Ny}}{W_{eff,y,min}\,f_y/\gamma_{M0}} \leq 1. \tag{4.74}$$

Equation 4.74 accounts for the additional bending moment caused by a shift e_{Ny} in the compression force N_{Ed} from the geometric centroid of the net cross-section to the centroid of the effective cross-section.

4.7.4 Longitudinal stiffeners

A logical basis for the design of a web in pure bending is to limit its proportions so that its maximum elastic bending strength can be used, so that the web is a Class 3 element. When this is done, the section resistance $M_{y,Rd}$ of the beam is governed by the slenderness of the flanges. Thus EC3 requires a fully effective unstiffened web to satisfy

$$h_w/t_w \leq 124\varepsilon. \tag{4.75}$$

This limit is close to the value of 126.9 at which the elastic buckling stress is equal to the yield stress (see equation 4.37).

Unstiffened webs whose slenderness exceeds this limit are Class 4 elements, but the use of one or more longitudinal stiffeners can delay local buckling so that they become Class 3 elements, and the cross-section can be designed as a Class 2 section as discussed in Section 4.7.2. EC3 does not specify the minimum stiffness of longitudinal stiffeners to prevent the web plate from deflecting at the stiffener location during local buckling, such as in equation 4.38. Rather, it allows the plate slenderness $\overline{\lambda}_p$ in equation 4.56 to be determined from the local buckling coefficient $k_{\sigma,p}$ which incorporates both the area and second moment of the area of the stiffener. If the stiffened web plate is proportioned such that $\overline{\lambda}_p \leq 0.874$, the reduction factor in equation 4.53 is unity and the longitudinal stress in the stiffened web can reach its yield strength prior to local buckling.

The Australian standard AS4100 [18] gives a simpler method of design in which a first longitudinal stiffener whose second moment of area is at least that of equation 4.28 is placed one-fifth of the web depth from the compression when the overall depth–thickness ratio of the web exceeds the limit of

$$h_w/t_w = 194\varepsilon \tag{4.76}$$

which is greater than that of equation 4.75 because it considers the effect of the flanges on the local buckling stress of the web. An additional stiffener whose second moment of area exceeds

$$I_{st} = h_w t_w^3 \tag{4.77}$$

is required at the neutral axis when h_w/t_w exceeds 242ε.

4.7.5   Beam webs in shear

4.7.5.1   Stocky webs

The factored design shear force V_{Ed} on a cross-section must satisfy

$$V_{Ed} \leq V_{c,Rd} \tag{4.78}$$

in which $V_{c,Rd}$ is the design uniform shear resistance which may be calculated based on a plastic ($V_{pl,Rd}$) or elastic distribution of shear stress.

I-sections have distributions of shear stress through their webs that are approximately uniform (Section 5.4.2) and for stocky webs with $h_w/t_w < 72\varepsilon$, the yield stress in shear $\tau_y = f_y/\sqrt{3}$ is reached before local buckling. Hence EC3 requires that

$$V_{c,Rd} = V_{pl,Rd} = \frac{A_v(f_y/\sqrt{3})}{\gamma_{M0}} \tag{4.79}$$

where $\gamma_{M0} = 1$ and A_v is the shear area of the web which is defined in Clause 6.2.6 of EC3. This resistance is close to, but more conservative than, the resistance given in equation 4.28, since the shear area A_v is reduced for hot-rolled sections by including the root radius in its calculation, and because equation 4.79 ignores the effects of strain hardening. Some sections, such as a monosymmetric I-section, have non-uniform distributions of the shear stress τ_{Ed} in their webs. Based on an elastic stress distribution, the maximum value of τ_{Ed} may be determined as in Section 5.4.2, and EC3 then requires that

$$\tau_{Ed} \leq \frac{f_y/\sqrt{3}}{\gamma_{M0}}. \tag{4.80}$$

Cross-sections for which $h_w/t_w \leq 72\varepsilon$ are stocky, and the webs of all UB's and UC's in Grade 275 steel satisfy $h_w/t_w \leq 72\varepsilon$.

4.7.5.2   Slender webs

The shear resistance of slender unstiffened webs for which $h_w/t_w > 72\varepsilon$ decreases rapidly from the value in equation 4.79 as the slenderness h_w/t_w increases. For these

$$V_{c,Rd} = V_{bw,Rd} \leq \eta \frac{f_{yw}/\sqrt{3}}{\gamma_{M1}} h_w t_w \tag{4.81}$$

where $V_{bw,Rd}$ is the design resistance governed by buckling of the web in shear, f_{yw} is the yield strength of the web, η is a factor for the shear area that can be taken as 1.2 for steels up to S460 and 1.0 otherwise, and $\gamma_{M1}(= 1)$ is a partial resistance factor based on buckling. The buckling resistance of the web is given in EC3 as

$$V_{bw,Rd} = \chi_w \frac{f_{yw}/\sqrt{3}}{\gamma_{M1}} h_w t_w \tag{4.82}$$

in which the web reduction factor on the yield strength $h_w t_w f_{yw}/\sqrt{3}$ due to buckling is

$$\chi_w = \begin{cases} 1 & \overline{\lambda}_w < 0.83 \\ 0.83/\overline{\lambda}_w & \overline{\lambda}_w \geq 0.83 \end{cases} \tag{4.83}$$

where the modified web plate slenderness is

$$\overline{\lambda}_w = \sqrt{\frac{f_{yw}/\sqrt{3}}{\tau_{cr}}} \approx 0.76 \sqrt{\frac{f_{yw}}{\tau_{cr}}} \tag{4.84}$$

and τ_{cr} is the elastic local buckling stress in shear given in equation 4.21. equation 4.84 is of the same form as equation 4.5. For an unstiffened web, $\overline{\lambda}_w = (h_w/t_w)/86.4\varepsilon$, and so it can be seen from equation 4.82 that the resistance of an unstiffened web decreases with an increase in its depth–thickness ratio h_w/t_w.

The shear resistance of a slender web may be increased by providing transverse stiffeners which increase the resistance to elastic buckling (Section 4.3.1) and also permit the development of tension field action (Section 4.3.2.2). Thus

$$V_{c,Rd} = V_{bw,Rd} + V_{bf,Rd} \leq \eta \frac{f_{yw}/\sqrt{3}}{\gamma_{M1}} h_w t_w \tag{4.85}$$

in which $V_{bw,Rd}$ is the web resistance given in equation 4.81 and $V_{bf,Rd}$ is the flange shear contribution provided by the understressed flanges during the development of the tension field in the web, which further enhances the shear resistance. In determining the web resistance reduction factor due to buckling χ_w in equation 4.82, the modified plate slenderness is given by

$$\overline{\lambda}_w = \frac{h_w}{t_w} \frac{1}{37.4\varepsilon\sqrt{k_\tau}} \tag{4.86}$$

where k_τ is the local buckling coefficient given in equations 4.22 or 4.23, with L equal to the stiffener spacing a and d equal to the web depth h_w. The additional post-buckling reserve of capacity due to tension field action for web plates with $\overline{\lambda}_w > 1.08$ is achieved by the use of an enhanced reduction factor of $\chi_w = 1.37/(0.7+\overline{\lambda}_w)$ instead of the more conservative $\chi_w = 0.83/\overline{\lambda}_w$. The presence of tension field action is also accounted for implicitly in the additional flange resistance

$$V_{bf,Rd} = \frac{b_f t_f^2}{c} \frac{f_{yf}}{\gamma_{M1}} \left[1 - \left(\frac{M_{Ed}}{M_{f,Rd}} \right)^2 \right] \tag{4.87}$$

where M_{Ed} is the design bending moment, $M_{f,Rd}$ is the moment of resistance of the cross-section consisting of the area of the effective flanges only, f_{yf} is the yield strength of the flanges, and where

$$c = a \left(0.25 + 1.6 \frac{b_f t_f^2}{t_w h_w^2} \frac{f_{yf}}{f_{yw}} \right). \tag{4.88}$$

Worked examples of checking the shear capacity of a slender web are given in Sections 4.9.5 and 4.9.6.

4.7.5.3 Transverse stiffeners

Provisions are made in EC3 for a minimum value for the second moment of area I_{st} of intermediate transverse stiffeners. These must be stiff enough to ensure that the elastic buckling stress τ_{cr} of a panel can be reached, and strong enough to transmit the tension field stiffener force. The stiffness requirement of EC3 is that the second moment of area of a transverse stiffener must satisfy

$$I_{st} \geq 0.75 h_w t_w^3, \tag{4.89}$$

when $a/h_w > \sqrt{2}$, in which h_w is the depth of the web, and

$$I_{st} \geq 1.5 h_w^3 t_w^3 / a^2 \tag{4.90}$$

when $a/h_w \leq \sqrt{2}$. These limits are shown in Figure 4.29, and are close to those of equations 4.27 and 4.26.

The strength requirement of EC3 for transverse stiffeners is that they should be designed as compression members of length h_w with an initial imperfection

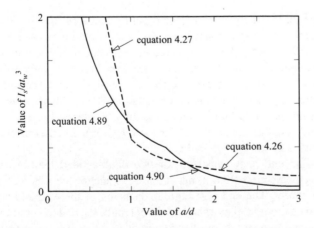

Figure 4.29 Stiffness requirements for web stiffeners.

$w_0 = h_w/300$, using second-order elastic analysis, for which the maximum stress is limited by

$$\sigma_{z,Rd} = f_y/\gamma_{M1}. \tag{4.91}$$

The effective stiffener cross-section for this design check consists of the cross-section of the stiffener itself plus a width of the web $15\varepsilon t_w$ each side of the stiffener, and the force to be resisted by the effective stiffener cross-section is

$$N_{z,Ed} = V_{Ed} - \frac{h_w t_w f_{yw}/\sqrt{3}}{\bar{\lambda}_w^2} \frac{1}{\gamma_{M1}} \tag{4.92}$$

where $\bar{\lambda}_w$ is given by equation 4.85. Using equation 3.8 for the bending of an initially crooked compression member, this procedure implies that the maximum stress in the effective stiffener cross-section is

$$\sigma_{z,Rd} = \frac{N_{z,Ed}}{A_{st}} \left[1 + \frac{h_w A_{st}}{300 W_{x,st,min}} \left(\frac{N_{z,Ed}/N_{cr}}{1 - N_{z,Ed}/N_{cr}} \right) \right] \tag{4.93}$$

where

$$N_{cr} = \pi^2 EI_{st}/h_w^2 \tag{4.94}$$

is the elastic flexural buckling load of the effective stiffener area and A_{st} and I_{st} are the area and second moment of area of the effective stiffener area, and $W_{x,st,min}$ is its minimum elastic section modulus.

A worked example of checking a pair of transverse stiffeners is given in Section 4.9.8.

4.7.5.4 End panels

At the end of a plate girder, there is no adjacent panel to absorb the horizontal component of the tension field in the last panel, and so this load may have to be resisted by the end stiffener. A rigid end post is required if tension field action is to be utilised in design. Special treatment of this end stiffener may be avoided if the length of the last (anchor) panel is reduced so that the tension field contribution τ_{tf} to the ultimate stress τ_{ult} is not required (see equation 4.29). This is achieved by designing the anchor panel so that its shear buckling resistance $V_{bw,Rd}$ given by equation 4.82 is not less than the design shear force V_{Ed}. A worked example of checking an end panel is given in Section 4.9.7.

Alternatively, a rigid end post consisting of two double-sided load-bearing transverse stiffeners (see Section 4.7.6.2) may be used to anchor the tension field at the end of a plate girder. The end post region consisting of the web and twin pairs of stiffeners can be thought of as a short beam of length h_w, and EC3 requires this beam to be designed to resist the in-plane stresses in the web resulting from the shear V_{Ed}.

4.7.6 Beam webs in shear and bending

Where the design moment M_{Ed} and shear V_{Ed} are both high, the beam must be designed against combined shear and bending. For stocky webs, when the shear force is less than half the plastic resistance $V_{pl,Rd}$, EC3 allows the effect of the shear on the moment resistance to be neglected. When $V_{Ed} > 0.5V_{pl,Rd}$, EC3 requires the bending resistance to be determined using a reduced yield strength

$$f_{yr} = (1 - \rho)f_{yw} \tag{4.95}$$

for the shear area, where

$$\rho = \left(\frac{2V_{Ed}}{V_{pl,Rd}} - 1 \right)^2. \tag{4.96}$$

The interaction curve between the design shear force V_{Ed} and design moment M_{Ed} is shown in Figure 4.30. These equations are less conservative than the von-Mises yield condition (see Section 1.3.1)

$$f_{yr}^2 + 3\tau_{Ed}^2 = f_y^2. \tag{4.97}$$

Equations 4.95 and 4.96 may be used conservatively for the well-known proportioning method of design, for which only the flanges are used to resist the moment and the web to resist the shear force, and which is equivalent to using a reduced yield strength $f_{yr} = 0$ in equation 4.95 for the web.

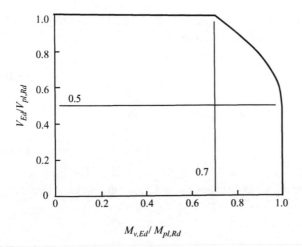

Figure 4.30 Section resistances of beams under bending and shear.

EC3 also uses a simpler reduced design plastic resistance $M_{y,V,Rd}$ for I-sections with a stocky web in lieu of the more tedious use of equations 4.95 and 4.96. Hence

$$M_{y,V,Rd} = \frac{(W_{pl,y} - \rho A_w^2/4t_w)f_y}{\gamma_{M0}},$$ (4.98)

where ρ is given in equation 4.96, $A_w = h_w t_w$ is the area of the web and $W_{pl,y}$ is the major axis plastic section modulus. If the web is slender so that its resistance is governed by shear buckling, EC3 allows the effect of shear on the bending resistance to be neglected when $V_{Ed} < 0.5V_{bw,Rd}$. When this is not the case, the reduced bending resistance is

$$M_{V,Rd} = M_{pl,Rd}(1 - \rho^2) + M_{f,Rd}\rho^2,$$ (4.99)

where ρ is given in equation 4.96 using the shear buckling resistance $V_{bw,Rd}$ instead of the plastic resistance $V_{pl,Rd}$, $M_{f,Rd}$ is the plastic moment of resistance consisting of the effective area of the flanges and $M_{pl,Rd}$ is the plastic resistance of the cross-section consisting of the effective area of the flanges and the fully effective web, irrespective of the section class. The reduced bending strength in equation 4.99 is similar to that implied in equation 4.44. If the contribution of the web is conservatively neglected in determining $M_{pl,Rd}$ for the cross-section, equation 4.99 leads to $M_{V,Rd} = M_{f,Rd}$ which is the basis of the proportioning method of design.

4.7.7   Beam webs in bearing

4.7.7.1   Unstiffened webs

For the design of webs in bearing according to EC3, the bearing resistance F_{Rd} based on yielding and local buckling is taken as

$$F_{Rd} = \chi_F \frac{f_{yw}t_w\ell_y}{\gamma_{M1}}$$ (4.100)

in which ℓ_y is the effective loaded length of the web and

$$\chi_F = \frac{0.5}{\bar{\lambda}_F}$$ (4.101)

is a reduction factor for the yield strength $F_y = f_{yw}t_w\ell_y$ due to local buckling in bearing, in which

$$\bar{\lambda}_F = \sqrt{\frac{F_y}{F_{cr}}}$$ (4.102)

is a modified plate slenderness for bearing buckling (which is of the same form as equation 4.5) and

$$F_{cr} = \frac{k_F \pi^2 E}{12(1 - v^2)} \left(\frac{t_w}{h_w}\right)^2 h_w t_w \qquad (4.103)$$

is the elastic buckling load for a web of area $h_w t_w$ in bearing. The buckling coefficients for plates in bearing are similar to those in Figure 4.26, with $k_F = 6 + 2/(a/h_w)^2$ being used for a web of the type shown in Figure 4.25a. The effective loaded length used in determining the yield resistance F_y in bearing is

$$\ell_y = s_s + 2n t_f \qquad (4.104)$$

in which s_s is the stiff bearing length and $n t_f$ is the additional length assuming a dispersion of the bearing at $1 : n$ through the flange thickness. More conveniently, the dispersion can be thought of as being at a slope 1:1 through a depth of $(n-1)t_f$ into the web. For a thick web (for which $\overline{\lambda}_F < 0.5$), $n = 1 + \sqrt{(b_f/t_w)}$, while for a slender web (for which $\overline{\lambda}_F > 0.5$), $n = 1 + \sqrt{[b_f/t_w + 0.02(h_w/t_f)^2]}$.

4.7.7.2 Stiffened webs

When a web alone has insufficient bearing capacity, it may be strengthened by adding one or more pairs of load-bearing stiffeners. These stiffeners increase the yield and buckling resistances by increasing the effective section to that of the stiffeners together with the web lengths $15t_w\varepsilon$ on either side of the stiffeners, if available. The effective length of the compression member is taken as the stiffener length h_w, or as $0.75h_w$ if flange restraints act to reduce the stiffener end rotations during buckling, and curve c of Figure 3.13 for compression members should be used.

A worked example of checking load-bearing stiffeners is given in Section 4.9.9.

4.8 Appendix – elastic buckling of plate elements in compression

4.8.1 Simply supported plates

A simply supported rectangular plate element of length L, width b, and thickness t is shown in Figure 4.5b. Applied compressive loads N are uniformly distributed over both edges b of the plate. The elastic buckling load N_{cr} can be determined by finding a deflected position such as that shown in Figure 4.5b which is one of equilibrium. The differential equation for this equilibrium position is [1–4]

$$bD \left(\frac{\partial^4 v}{\partial x^4} + 2\frac{\partial^4 v}{\partial x^2 \partial z^2} + \frac{\partial^4 v}{\partial z^4}\right) = -N_{cr} \frac{\partial^2 v}{\partial x \partial z} \qquad (4.105)$$

where

$$bD = \frac{Ebt^3}{12(1-v^2)} \tag{4.106}$$

is the flexural rigidity of the plate.

Equation 4.105 is compared in Figure 4.5 with the corresponding differential equilibrium equation for a simply supported rectangular section column (obtained by differentiating equation 3.56 twice). It can be seen that bD for the plate corresponds to the flexural rigidity EI of the column, except for the $(1-v^2)$ term which is due to the Poisson's ratio effect in wide plates. Thus the term $bD\partial^4 v/\partial x^4$ represents the resistance generated by longitudinal flexure of the plate to the disturbing effect $-N\partial^2 v/\partial x^2$ of the applied load. The additional terms $2bD\partial^4 v/\partial x^2 \partial z^2$ and $bD\partial^4 v/\partial z^4$ in the plate equation represent the additional resistances generated by twisting and lateral bending of the plate.

A solution of equation 4.105 which satisfies the boundary conditions along the simply supported edges [1–4] is

$$v = \delta \sin \frac{m\pi x}{L} \sin \frac{n\pi z}{b}, \tag{4.1}$$

where δ is the undetermined magnitude of the deflected shape. When this is substituted into equation 4.105, an expression for the elastic buckling load N_{cr} is obtained as

$$N_{cr} = \frac{n^2 \pi^2 bD}{L^2} \left[1 + 2 \left(\frac{nL}{mb} \right)^2 + \left(\frac{nL}{mb} \right)^4 \right],$$

which has its lowest values when $n = 1$ and the buckled shape has one half wave across the width of the plate b. Thus the elastic buckling stress

$$\sigma_{cr} = \frac{N_{cr}}{bt} \tag{4.2}$$

can be expressed as

$$\sigma_{cr} = \frac{\pi^2 E}{12(1-v^2)} \frac{k_\sigma}{(b/t)^2} \tag{4.3}$$

in which the buckling coefficient k_σ is given by

$$k_\sigma = \left[\left(\frac{mb}{L} \right)^2 + 2 + \left(\frac{L}{mb} \right)^2 \right]. \tag{4.107}$$

The variation of the buckling coefficient k_σ with the aspect ratio L/b of the plate and the number of half waves m along the plate is shown in Figure 4.6. It can be seen that the minimum value of k_σ is 4, and that this occurs whenever the buckles are square ($mb/L = 1$), as shown in Figure 4.7. In most structural steel members, the aspect ratio L/b of the plate elements is large so that the value of the buckling coefficient k_σ is always close to the minimum value of 4. The elastic buckling stress can therefore be closely approximated by

$$\sigma_{cr} = \frac{\pi^2 E/3(1 - v^2)}{(b/t)^2}. \tag{4.108}$$

4.9 Worked examples

4.9.1 Example 1 – compression resistance of a Class 4 compression member

Problem. Determine the compression resistance of the cross-section of the member shown in Figure 4.31a. The weld size is 8 mm.

Classifying the section plate elements.

$$t_f = 10 \text{ mm}, \ t_w = 10 \text{ mm}, \ f_y = 355 \text{ N/mm}^2 \qquad \text{EN10025-2}$$

$$\varepsilon = \sqrt{(235/355)} = 0.814. \qquad \text{T5.2}$$

$$c_f/(t_f\varepsilon) = (400/2 - 10/2 - 8)/(10 \times 0.814) \qquad \text{T5.2}$$

$$= 23.0 > 14 \text{ and so the flange is Class 4.} \qquad \text{T5.2}$$

$$c_w/(t_w\varepsilon) = (420 - 2 \times 10 - 2 \times 8)/(10 \times 0.814) \qquad \text{T5.2}$$

$$= 47.2 > 42 \text{ and so the web is Class 4.} \qquad \text{T5.2}$$

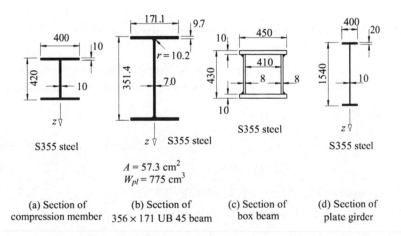

(a) Section of compression member

(b) Section of 356 × 171 UB 45 beam

$A = 57.3 \text{ cm}^2$
$W_{pl} = 775 \text{ cm}^3$

(c) Section of box beam

(d) Section of plate girder

Figure 4.31 Worked examples.

Effective flange area.

$$k_{\sigma,f} = 0.43 \qquad\qquad \text{EC3-1-5 T4.2}$$

$$\overline{\lambda}_{p,f} = \frac{(400/2 - 10/2 - 8)/10}{28.4 \times 0.814 \times \sqrt{0.43}} = 1.23 > 0.748 \qquad \text{EC3-1-5 4.4(2)}$$

$$\rho_f = (1.23 - 0.188)/1.23^2 = 0.687 \qquad\qquad \text{EC3-1-5 4.4(2)}$$

$$A_{eff,f} = 0.687 \times 4 \times (400/2 - 10/2 - 8) \times 10$$
$$+ (10 + 2 \times 8) \times 10 \times 2 \qquad\qquad \text{EC3-1-5 4.4(1)}$$
$$= 5658 \text{ mm}^2$$

Effective web area.

$$k_{\sigma,w} = 4.0 \qquad\qquad \text{EC3-1-5 T4.1}$$

$$\overline{\lambda}_{p,w} = \frac{(420 - 2 \times 10 - 2 \times 8)/10}{28.4 \times 0.814 \times \sqrt{4.0}} = 0.831 > 0.673 \qquad \text{EC3-1-5 4.4(2)}$$

$$\rho_w = \{0.831 - 0.055 \times (3 + 1)\}/0.831^2 = 0.885 \qquad \text{EC3-1-5 4.4(2)}$$

$$A_{eff,w} = 0.885 \times (420 - 2 \times 10 - 2 \times 8) \times (10 + 8 \times 10 \times 2)$$
$$= 3558 \text{ mm}^2 \qquad\qquad \text{EC3-1-5 4.4(1)}$$

Compression resistance.

$$A_{eff} = 5658 + 3558 = 9216 \text{ mm}^2$$
$$N_{c,Rd} = 9216 \times 355/1.0 \text{ N} = 3272 \text{ kN}.$$

4.9.2 Example 2 – section moment resistance of a Class 3 I-beam

Problem. Determine the section moment resistance and examine the suitability for plastic design of the 356 × 171 UB 45 of S355 steel shown in Figure 4.31b.

Classifying the section-plate elements.

$$t_f = 9.7 \text{ mm}, \ t_w = 7.0 \text{ mm}, \ f_y = 355 \text{ N/mm}^2 \qquad \text{EN10025-2}$$

$$\varepsilon = \sqrt{(235/355)} = 0.814 \qquad\qquad \text{T5.2}$$

$$c_f/(t_f\varepsilon) = (171.1/2 - 7.0/2 - 10.2)/(9.7 \times 0.814) \qquad \text{T5.2}$$

$$= 9.1 > 9 \text{ but } < 10, \text{ and so the flange is Class 2.} \qquad \text{T5.2}$$

$$c_w/(t_w\varepsilon) = (351.4 - 2 \times 9.7 - 2 \times 10.2)/(7.0 \times 0.814) \qquad \text{T5.2}$$

$$= 54.7 < 72 \text{ and so the web is Class 1.} \qquad \text{T5.2}$$

Section moment resistance.
The cross-section is Class 2 and therefore unsuitable for plastic design.

$$M_{c,Rd} = 775 \times 10^3 \times 355/1.0 \text{ N mm} = 275.1 \text{ kNm}. \qquad 6.2.5(2)$$

4.9.3 Example 3 – section moment resistance of a Class 4 box beam

Problem. Determine the section moment resistance of the welded-box section beam of S355 steel shown in Figure 4.31c. The weld size is 6 mm.

Classifying the section-plate elements.

$$t_f = 10 \text{ mm}, \ t_w = 8 \text{ mm}, \ f_y = 355 \text{ N/mm}^2 \qquad \text{EN10025-2}$$

$$\varepsilon = \sqrt{(235/355)} = 0.814 \qquad \text{T5.2}$$

$$c_f/(t_f\varepsilon) = 410/(10 \times 0.814) \qquad \text{T5.2}$$

$$= 50.4 > 42 \text{ and so the flange is Class 4.} \qquad \text{T5.2}$$

$$c_w/(t_w\varepsilon) = (430 - 2 \times 10 - 2 \times 6)/(8 \times 0.814) \qquad \text{T5.2}$$

$$= 61.1 < 72 \text{ and so the web is Class 1.} \qquad \text{T5.2}$$

The cross-section is therefore Class 4 since the flange is Class 4.

Effective cross-section.

$$k_\sigma = 4.0 \qquad \text{EC3-1-5 T4.2}$$

$$\bar{\lambda}_p = \frac{410}{10 \times 28.4 \times 0.814 \times \sqrt{4.0}} = 0.887 > 0.673$$
$$\text{EC3-1-5 4.4(2)}$$

$$\psi = 1 \qquad \text{EC3-1-5 T4.1}$$

$$\rho = (0.887 - 0.055 \times (3 + 1))/0.887^2 = 0.848 \quad \text{EC3-1-5 4.4(2)}$$

$$b_{c,eff} = 0.848 \times 410 = 347.5 \text{ mm}$$

$$A_{eff} = (450 - 410 + 347.5) \times 10 + (450 \times 10)$$
$$+ 2 \times (430 - 2 \times 10) \times 8$$
$$= 14\,935 \text{ mm}^2$$

$$14\,935 \times z_c = (450 - 410 + 347.5) \times 10 \times (430 - 10/2)$$
$$+ 450 \times 10 \times 10/2 + 2 \times (430 - 2 \times 10) \times 8 \times 430/2$$

$$z_c = 206.2 \text{ mm}$$

$$I_{eff} = (450 - 410 + 347.5) \times 10 \times (430 - 10/2 - 206.2)^2$$
$$+ 450 \times 10 \times (206.2 - 10/2)^2 + 2 \times (430 - 2 \times 10)^3 \times 8/12$$
$$+ 2 \times (430 - 2 \times 10) \times 8 \times (430/2 - 206.2)^2 \text{ mm}^4$$
$$= 460.1 \times 10^6 \text{ mm}^4$$

Section moment resistance.

$$W_{eff,min} = 460.1 \times 10^6/(430 - 206.2) = 2.056 \times 10^6 \text{ mm}^3$$

$$M_{c,Rd} = 2.056 \times 10^6 \times 355/1.0 \text{ Nmm} = 729.9 \text{ kNm}. \qquad 6.2.5(2)$$

4.9.4 *Example 4 – section moment resistance of a slender plate girder*

Problem. Determine the section moment resistance of the welded plate girder of S355 steel shown in Figure 4.31d. The weld size is 6 mm.

Solution.

$$t_f = 20 \text{ mm}, \ t_w = 10 \text{ mm}, \ f_{yf} = 345 \text{ N/mm}^2 \qquad \text{EN10025-2}$$

$$\varepsilon = \sqrt{(235/345)} = 0.825 \qquad \text{T5.2}$$

$$c_f/(t_f \varepsilon) = (400/2 - 10/2 - 6)/(20 \times 0.825) \qquad \text{T5.2}$$

$$= 11.5 > 10 \text{ but } < 14, \text{ and so the flange is Class 3.} \qquad \text{T5.2}$$

$$c_w/(t_w \varepsilon) = (1540 - 2 \times 20 - 2 \times 6)/(10 \times 0.825) \qquad \text{T5.2}$$

$$= 180.3 > 124 \text{ and so the web is Class 4.} \qquad \text{T5.2}$$

A conservative approximation for the cross-section moment resistance may be obtained by ignoring the web completely, so that

$$M_{c,Rd} = M_f = (400 \times 20) \times (1540 - 20) \times 345/1.0 \text{ Nmm} = 4195 \text{ kNm}.$$

A higher resistance may be calculated by determining the effective width of the web.

$$\psi = -1 \qquad \text{EC3-1-5 T4.1}$$

$$k_\sigma = 23.9 \qquad \text{EC3-1-5 T4.1}$$

$$\overline{\lambda}_p = \frac{(1540 - 2 \times 20 - 2 \times 6)/10}{28.4 \times 0.825 \times \sqrt{23.9}} = 1.299 \qquad \text{EC3-1-5 4.4(2)}$$

$$\rho = (1.299 - 0.055 \times (3 - 1))/1.299^2 = 0.705 \qquad \text{EC3-1-5 4.4(2)}$$

$$b_c = (1540 - 2 \times 20 - 2 \times 6)/\{1 - (-1)\} = 744.0 \text{ mm}. \qquad \text{EC3-1-5 T4.1}$$

$$b_{eff} = 0.705 \times 744.0 = 524.4 \text{ mm}. \qquad \text{EC3-1-5 T4.1}$$

$$b_{e1} = 0.4 \times 524.4 = 209.8 \text{ mm}. \qquad \text{EC3-1-5 T4.1}$$

$$b_{e2} = 0.6 \times 524.4 = 314.6 \text{ mm}. \qquad \text{EC3-1-5 T4.1}$$

and the ineffective width of the web is

$$b_c - b_{e1} - b_{e2} = 744.0 - 209.8 - 314.6 = 219.6 \text{ mm.} \qquad \text{EC3-1-5 T4.1}$$

$$A_{eff} = (1540 - 2 \times 20 - 219.6) \times 10 + 2 \times 400 \times 20$$

$$= 28\,804 \text{ mm}^2$$

$$28\,804 \times z_c = (2 \times 400 \times 20 + (1540 - 2 \times 20) \times 10) \times (1540/2)$$

$$- 219.6 \times 10 \times (1540 - 20 - 6 - 209.8 - 219.6/2)$$

$$z_c = 737.6 \text{ mm}$$

$$I_{eff} = (400 \times 20) \times (1540 - 10 - 737.6)^2$$

$$+ (400 \times 20) \times (737.6 - 10)^2$$

$$+ (1540 - 2 \times 20)^3 \times 10/12 + (1540 - 2 \times 20)$$

$$\times 10 \times (1540/2 - 737.6)^2$$

$$- 219.6^3 \times 10/12 - 219.6$$

$$\times 10 \times (1540 - 20 - 6 - 209.8 - 219.6/2 - 737.6)^2 \text{ mm}^4$$

$$= 11.62 \times 10^9 \text{ mm}^4$$

$$W_{eff} = 11.62 \times 10^9/(1540 - 737.6) = 14.48 \times 10^6 \text{ mm}^3$$

$$M_{c,Rd} = 14.48 \times 10^6 \times 345/1.0 \text{ Nmm} = 4996 \text{ kNm.}$$

4.9.5 Example 5 – shear buckling resistance of an unstiffened plate girder web

Problem. Determine the shear buckling resistance of the unstiffened plate girder web of S355 steel shown in Figure 4.31d.

Solution.

$$t_w = 10 \text{ mm, } f_{yw} = 355 \text{ N/mm}^2 \qquad \qquad \text{EN10025-2}$$

$$\varepsilon = \sqrt{(235/355)} = 0.814 \qquad \qquad \text{T5.2}$$

$$\eta = 1.2 \qquad \qquad \text{EC3-1-5 5.1(2)}$$

$$h_w = 1540 - 2 \times 20 = 1500 \text{ mm.} \qquad \qquad \text{EC3-1-5 F5.1}$$

$$\eta h_w/(t_w \varepsilon) = 1.2 \times 1500/(10 \times 0.814)$$

$$= 221.2 > 72 \text{ and so the web is slender.} \qquad \text{EC3-1-5 5.1(2)}$$

$$a/h_w = \infty/h_w = \infty, \; k_{\tau st} = 0 \qquad \qquad \text{EC3-1-5 A.3(1)}$$

$$k_\tau = 5.34 \qquad \qquad \text{EC3-1-5 A.3(1)}$$

$$\tau_{cr} = 5.34 \times 190\,000 \times (10/1500)^2 = 45.1 \text{ N/mm}^2$$
EC3-1-5 A.1(2)

$$\overline{\lambda}_w = 0.76 \times \sqrt{(355/45.1)} = 2.132 > 1.08 \qquad \text{EC3-1-5 5.3(3)}$$

Assuming that there is a non-rigid end post, then

$$\chi_w = 0.83/2.132 = 0.389 \qquad\qquad \text{EC3-1-5 T5.1}$$

Neglecting any contribution from the flanges,

$$V_{b,Rd} = V_{bw,Rd} = \frac{0.389 \times 355 \times 1500 \times 10}{\sqrt{3} \times 1.0} \text{ N} = 1196 \text{ kN.} \quad \text{EC3-1-5 2(1)}$$

4.9.6   Example 6 – shear buckling resistance of a stiffened plate girder web

Problem. Determine the shear buckling resistance of the plate girder web of S355 steel shown in Figure 4.31d if intermediate transverse stiffeners are placed at 1800 mm spacing.

Solution.

$$t_w = 10 \text{ mm, } f_{yw} = 355 \text{ N/mm}^2 \qquad\qquad \text{EN10025-2}$$

$$\varepsilon = \sqrt{(235/355)} = 0.814 \qquad\qquad \text{T5.2}$$

$$h_w = 1540 - 2 \times 20 = 1500 \text{ mm.}$$

$$a/h_w = 1800/1500 = 1.20, \, k_{\tau st} = 0 \qquad\qquad \text{EC3-1-5 A.3(1)}$$

$$k_\tau = 5.34 + 4.00/1.20^2 = 8.12 \qquad\qquad \text{EC3-1-5 A.3(1)}$$

$$\tau_{cr} = 8.12 \times 190\,000 \times (10/1500)^2 = 68.6 \text{ N/mm}^2 \qquad \text{EC3-1-5 A.1(2)}$$

Assuming that there is a rigid end post, then

$$\overline{\lambda}_w = 0.76 \times \sqrt{(355/68.6)} = 1.73 > 1.08 \qquad \text{EC3-1-5 5.3(3)}$$

$$\chi_w = 1.37/(0.7 + 1.73) = 0.564 \qquad\qquad \text{EC3-1-5 T5.1}$$

Neglecting any contribution from the flanges,

$$V_{b,Rd} = V_{bw,Rd} = \frac{0.564 \times 355 \times 1500 \times 10}{\sqrt{3} \times 1.0} \text{ N} = 1734 \text{ kN.}$$
EC3-1-5 5.2(1)

If the flange contribution is considered and $M_{Ed} = 0$,

$$2 \times (15\varepsilon t_f) + t_w = 2 \times (15 \times 0.814 \times 20) + 10 \qquad \text{EC3-1-5 5.4(1)}$$

$$= 498 \text{ mm} > 400 \text{ mm and so } b_f = 400 \text{ mm.} \qquad \text{EC3-1-5 5.4(1)}$$

$$c = 1800 \times \left(0.25 + \frac{1.6 \times 400 \times 20^2 \times 345}{10 \times 1500^2 \times 355} \right) = 469.9 \text{ mm}$$
$$\text{EC3-1-5 5.4(1)}$$

$$V_{bf,Rd} = (400 \times 20^2 \times 345 \times 1.0)/(469.9 \times 1.0) \text{ N} = 117 \text{ kN}$$
$$\text{EC3-1-5 5.4(1)}$$

$$V_{b,Rd} = 1734 + 117 = 1851 \text{ kN.} \qquad \text{EC3-1-5 5.2(1)}$$

4.9.7 Anchor panel in a stiffened plate girder web

Problem. Determine the shear elastic buckling resistance of the end anchor panel of the welded stiffened plate girder of Section 4.9.6 if the width of the panel is 1800 mm.

Solution.

$$t_w = 10 \text{ mm}, \ f_{yw} = 355 \text{ N/mm}^2 \qquad \text{EN10025-2}$$

Using $\bar{\lambda}_w = 1.73$ (Section 4.9.6), the shear elastic buckling resistance of the end panel is

$$V_{p,Rd} = (1/1.73^2) \times 355 \times 1500 \times 10/(10^3 \times \sqrt{3} \times 1.0) \text{ N} = 1027 \text{ kN.}$$
$$\text{EC3-1-5 9.3.3(3)}$$

4.9.8 Example 8 – intermediate transverse stiffener

Problem. Check the adequacy of a pair of intermediate transverse web stiffeners 100×16 of S460 steel for the plate girder of Section 4.9.6.

Solution.
   Using $V_{Ed} = 1734$ kN (Section 4.9.6) and $V_{p,Rd} = 1027$ kN (Section 4.9.7), the stiffener section must resist an axial force of

$$N_{Ed,s} = 1734 - 1027 = 706.4 \text{ kN.} \qquad \text{EC3-1-5 9.3.3(3)}$$

The stiffener section consists of the two stiffener plates and a length of the web defined in Figure 9.1 of EC3-1-5. For the stiffener plates,

$$t_p = 16 \text{ mm}, \ f_{yp} = 460 \text{ N/mm}^2 \qquad \text{EN10025-2}$$

$$\varepsilon = \sqrt{(235/460)} = 0.715 \qquad \text{T5.2}$$

$$c_p/(t_p\varepsilon) = 100/(16 \times 0.715) = 8.74 < 9 \qquad \text{T5.1}$$

and so the plates are fully effective and the stiffener section is Class 1.

Using the lower of the web and plate yield stresses of $f_{yw} = 355$ N/mm^2 and $\varepsilon = 0.814$,

$$A_{eff,s} = 2 \times (16 \times 100) + (2 \times 15 \times 0.814 \times 10 + 16) \times 10$$
$$= 5801 \text{ mm}^2 \qquad\qquad \text{EC3-1-5 F9.1}$$

$$I_{eff,s} = (2 \times 100 + 10)^3 \times 16/12 = 12.35 \times 10^6 \text{ mm}^4 \qquad \text{EC3-1-5 F9.1}$$

$$i_{eff,s} = \sqrt{(12.35 \times 10^6/5801)} = 46.1 \text{ mm}.$$

$$\lambda_1 = 93.9 \times 0.814 = 76.4 \qquad\qquad 6.3.1.3(1)$$

$$L_{cr} = 1.0 \times 1500 = 1500 \text{ mm} \qquad\qquad \text{EC3-1-5 9.4(2)}$$

$$\overline{\lambda} = 1500/(46.1 \times 76.4) = 0.426 \qquad\qquad 6.3.1.3(1)$$

For buckling curve c, $\alpha = 0.49$ $\qquad\qquad$ EC3-1-5 9.4(2), T6.1

$$\Phi = 0.5 \times [1 + 0.49 \times (0.426 - 0.2) + 0.426^2] = 0.646$$
$$6.3.1.2(1)$$

$$\chi = 1/[0.646 + \sqrt{(0.646^2 - 0.426^2)}] = 0.884 \qquad 6.3.1.2(1)$$

$$N_{b,Rd} = 0.884 \times 5801 \times 355/1.0 \text{ N} \qquad\qquad 6.3.1.1(3)$$

$$= 1820 \text{ kN} > 706.4 \text{ kN} = N_{Ed,s} \qquad\qquad \text{OK.}$$

$$a/h_w = 1800/1500 = 1.20 < \sqrt{2} \qquad\qquad \text{EC3-1-5 9.3.3(3)}$$

$$1.5 h_w^3 t_w^3/a^2 = 1.5 \times 1500^3 \times 10^3/1800^2 \qquad \text{EC3-1-5 9.3.3(3)}$$

$$= 1.563 \times 10^6 \text{ mm}^4 < 12.35 \times 10^6 \text{ mm}^4 = I_s \qquad \text{OK.}$$

4.9.9   Example 9 – load-bearing stiffener

Problem. Check the adequacy of a pair of load-bearing stiffeners 100×16 of S460 steel which are above the support of the plate girder of Section 4.9.6. The flanges of the girder are not restrained by other structural elements against rotation. The girder is supported on a stiff bearing $s_s = 300$ mm long, the end panel width is 1000 mm, and the design reaction is 1400 kN.

Stiffener yield resistance.

$$t_s = 16 \text{ mm}, \ f_y = 460 \text{ N/mm}^2 \qquad\qquad \text{EN10025-2}$$

$$\varepsilon = \sqrt{(235/460)} = 0.715 \qquad\qquad \text{T5.2}$$

$$c_s/(t_s\varepsilon) = 100/(16 \times 0.715) \qquad\qquad \text{T5.1}$$

$$= 8.74 < 9 \text{ and so the stiffeners are fully effective.} \qquad \text{T5.1}$$

$$F_{s,Rd} = 2 \times (100 \times 16) \times 460/1.0 \text{ N} = 1472 \text{ kN}.$$

Unstiffened web resistance.

$k_F = 2 + 6 \times (300 + 0)/1500 = 3.2$ EC3-1-5 F6.1

$F_{cr} = 0.9 \times 3.2 \times 210\,000 \times 10^3/1500 \text{ N} = 403.2 \text{ kN}$ EC3-1-5 6.4(1)

$$\ell_e = \frac{3.2 \times 210\,000 \times 10^2}{2 \times 355 \times 1500} = 63.1 \text{ mm} < 300 \text{ mm}$$

$\qquad = s_s + c.$ EC3-1-5 6.5(3)

$m_1 = (355 \times 400)/(355 \times 10) = 40$ EC3-1-5 6.5(1)

$m_2 = 0.02 \times (1500/20)^2 = 112.5$ EC3-1-5 6.5(1)

$l_{y6.11} = 63.1 + 20 \times \sqrt{40/2 + (63.1/20)^2 + 112.5}$

$\qquad = 301.8 \text{ mm}$ EC3-1-5 6.5(3)

$l_{y6.12} = 63.1 + 20 \times \sqrt{40 + 112.5}$

$\qquad = 310.1 \text{ mm} > 301.8 \text{ mm} = l_{y6.11}$ EC3-1-5 6.5(3)

$l_y = 301\ 8 \text{ mm}$ EC3-1-5 6.5(3)

$$\overline{\lambda}_F = \sqrt{\frac{301.8 \times 10 \times 355}{403.2 \times 10^3}} = 1.630(>0.5)$$ EC3-1-5 6.4(1)

$\chi_F = 0.5/1.630 = 0.307$ EC3-1-5 6.4(1)

$L_{eff} = 0.307 \times 301.8 = 92.6 \text{ mm}$ EC3-1-5 6.4(1)

$F_{Rd} = 355 \times 92.6 \times 10/1.0 \text{ N}$ EC3-1-5 6.2(1)

$\qquad = 328.6 \text{ kN} < 1400 \text{ kN}$, and so load bearing stiffeners are required.

Stiffener buckling resistance.

$A_{eff,s} = 2 \times (16 \times 100) + (2 \times 15 \times 0.814 \times 10 + 16) \times 10$

$\qquad = 5801 \text{ mm}^2$ EC3-1-5 F9.1

$I_{eff,s} = (2 \times 100 + 10)^3 \times 16/12 = 12.35 \times 10^6 \text{ mm}^4$ EC3-1-5 F9.1

$i_{eff,s} = \sqrt{(12.35 \times 10^6/5801)} = 46.1 \text{ mm}$

$\lambda_1 = 93.9 \times 0.814 = 76.4$ 6.3.1.3(1)

$L_{cr} = 1.0 \times 1500 \text{ mm} = 1500 \text{ mm}$ EC3-1-5 9.4(2)

$\overline{\lambda} = 1500/(46.1 \times 76.4) = 0.426$ 6.3.1.3(1)

For buckling curve c, $\alpha = 0.49$ EC3-1-5 9.4(2), T6.1

$\Phi = 0.5 \times [1 + 0.49 \times (0.426 - 0.2) + 0.426^2] = 0.646$ 6.3.1.2(1)

$\chi = 1/[0.646 + \sqrt{(0.646^2 - 0.426^2)}] = 0.884$ 6.3.1.2(1)

$N_{b,Rd} = 0.884 \times 5801 \times 355/1.0 \text{ N} = 1820 \text{ kN} > 1400 \text{ kN}$ OK.
6.3.1.1(3)

4.9.10   Example 10 – shear and bending of a Class 2 beam

Problem. Determine the design resistance of the 356×171 UB 45 of S355 steel shown in Figure 4.31b at a point where the design moment is $M_{Ed} = 230 \, \text{kNm}$.

Solution.

As in Section 4.9.2, $f_y = 355 \, \text{N/mm}^2$, $\varepsilon = 0.814$, $\eta = 1.2$, the flange is Class 2 and the web is Class 1, the section is Class 2, and $W_{pl} = 775 \, \text{cm}^3$.

$$A_v = 57.3 \times 10^2 - 2 \times 171.1 \times 9.7 + (7.0 + 2 \times 10.2) \times 9.7$$

$$= 2676 \, \text{mm}^2 \hspace{3cm} 6.2.6(3)$$

$$\eta = 1.2 \hspace{3cm} \text{EC3-1-5 5.1(2)}$$

$$\eta h_w t_w = 1.2 \times (351.4 - 2 \times 9.7) \times 7.0 = 2789 \, \text{mm}^2 > 2676 \, \text{mm}^2$$
$$6.2.6(3)$$

$$V_{pl,Rd} = 2789 \times (355/\sqrt{3})/1.0 \, \text{N} = 571.6 \, \text{kN} \hspace{2cm} 6.2.6(2)$$

$$\eta h_w/(t_w \varepsilon) = 1.2 \times (351.4 - 2 \times 9.7)/(7.0 \times 0.814) = 70.0 < 72$$
$$\text{EC3-1-5 5.1(2)}$$

so that shear buckling need not be considered.

Using a reduced bending resistance corresponding to $M_{V,Rd} = M_{Ed} = 230$ kNm, then

$$(1 - \rho) \times 355 \times 775 \times 10^3 = 230 \times 10^6, \text{ so that} \hspace{1.5cm} 6.2.8(3)$$

$$\rho = 0.164$$

Now $\rho = (2V_{Ed}/V_{pl,Rd} - 1)^2$, which leads to \hspace{2cm} 6.2.8(3)

$$V_{c,M,Rd}/V_{pl,Rd} = [\sqrt{(0.164)} + 1]/2 = 0.702, \text{ so that}$$

$$V_{c,M,Rd} = 0.702 \times 571.6 = 401.5 \, \text{kN}.$$

4.9.11   Example 11 – shear and bending of a stiffened plate girder

Problem. Determine the shear resistance of the stiffened plate web girder of Section 4.9.6 shown and shown in Figure 4.31d at the point where the design moment is $M_{Ed} = 4000 \, \text{kNm}$.

Solution.

As previously, $f_{yf} = 345 \, \text{N/mm}^2$, $\varepsilon_f = 0.825$, and the flanges are Class 3 (Section 4.9.4), $f_{yw} = 345 \, \text{N/mm}^2$ and $\varepsilon_w = 0.814$ (Section 4.9.5), and $V_{bw,Rd} = 1733 \, \text{kNm}$ (Section 4.9.6).

The resistance of the flanges to bending is

$$M_{f,Rd} = (400 \times 20) \times (1540 - 20) \times 345/1.0 \text{ Nmm}$$
$$= 4195 \text{ kNm} > 4000 \text{ kNm} = M_{Ed} \qquad \text{EC3-1-5 5.4(1)}$$

and so the flange resistance is not completely utilised in resisting the bending moment.

$$2 \times (15\varepsilon t_f) + t_w = 2 \times (15 \times 0.814 \times 20) + 10 = 498 \text{ mm} > 400 \text{ mm}$$
$$\text{EC3-1-5 5.4(1)}$$

$$b_f = 400 \text{ mm} < 498 \text{ mm} \qquad \text{EC3-1-5 5.4(1)}$$

$$c = 1800 \times \left(0.25 + \frac{1.6 \times 400 \times 20^2 \times 345}{10 \times 1500^2 \times 355} \right) = 469.9 \text{ mm}$$
$$\text{EC3-1-5 5.4(1)}$$

$$V_{bf,Rd} = \frac{400 \times 20^2 \times 355}{469.9 \times 1.0} \times \left[1 - \left(\frac{4000}{4195} \right)^2 \right] \text{ N} = 11 \text{ kN}$$
$$\text{EC3-1-5 5.4(1)}$$

$$V_{bw,Rd} + V_{bf,Rd} = 1733 + 11 = 1744 \text{ kN.} \qquad \text{EC3-1-5 5.2(1)}$$

4.10 Unworked examples

4.10.1 Example 12 – elastic local buckling of a beam

Determine the elastic local buckling moment for the beam shown in Figure 4.32a.

4.10.2 Example 13 – elastic local buckling of a beam-column

Determine the elastic local buckling load for the beam-column shown in Figure 4.32b when $M/N = 20$ mm.

4.10.3 Example 14 – section capacity of a welded box beam

Determine the section moment resistance for the welded box beam of S275 steel shown in Figure 4.32c.

4.10.4 Example 15 – designing a plate web girder

The overall depth of the laterally supported plate girder of S275 steel shown in Figure 4.32d must not exceed 1800 mm. Design a constant section girder for the

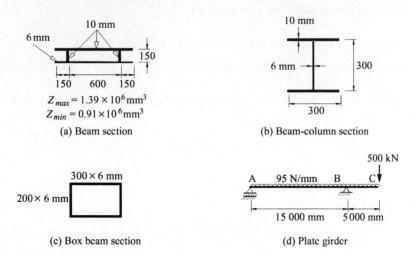

Figure 4.32 Unworked examples.

factored loads shown, and determine:

(a) The flange proportions;
(b) The web thickness;
(c) The distribution of any intermediate stiffeners;
(d) The stiffener proportions; and
(e) The proportions and arrangements of any load-bearing stiffeners.

References

1. Timoshenko, S.P. and Woinowsky-Krieger, S. (1959) *Theory of Plates and Shells*, 2nd edition, McGraw-Hill, New York.
2. Timoshenko, S.P. and Gere, J.M. (1961) *Theory of Elastic Stability*, 2nd edition, McGraw-Hill, New York.
3. Bleich, F. (1952) *Buckling Strength of Metal Structures*, McGraw-Hill, New York.
4. Bulson, P.S. (1970) *The Stability of Flat Plates*, Chatto and Windus, London.
5. Column Research Committee of Japan (1971) *Handbook of Structural Stability*, Corona, Tokyo.
6. Allen, H.G. and Bulson, P.S. (1980) *Background to Buckling*, McGraw-Hill (UK).
7. Basler, K. (1961) Strength of plate girders in shear, *Journal of the Structural Division, ASCE*, **87**, No. ST7, pp. 151–80.
8. Bradford, M.A., Bridge, R.Q., Hancock, G.J., Rotter, J.M., and Trahair, N.S. (1987) Australian limit state design rules for the stability of steel structures, *Proceedings*, First Structural Engineering Conference, Institution of Engineers, Australia, Melbourne, pp. 209–16.

9. Evans, H.R. (1983) Longitudinally and transversely reinforced plate girders, Chapter 1 in *Plated Structures: Stability and Strength* (ed. R. Narayanan), Applied Science Publishers, London, pp. 1–37.

10. Rockey, K.C. and Skaloud, M. (1972) The ultimate load behaviour of plate girders loaded in shear, *The Structural Engineer*, **50**, No. 1, pp. 29–47.

11. Usami, T. (1982) Postbuckling of plates in compression and bending, *Journal of the Structural Division, ASCE*, **108**, No. ST3, pp. 591–609.

12. Kalyanaraman, V. and Ramakrishna, P. (1984) Non-uniformly compressed stiffened elements, *Proceedings*, Seventh International Specialty Conference Cold-Formed Structures, St Louis, Department of Civil Engineering, University of Missouri-Rolla, pp. 75–92.

13. Merrison Committee of the Department of Environment (1973) *Inquiry into the Basis of Design and Method of Erection of Steel Box Girder Bridges*, Her Majesty's Stationery Office, London.

14. Rockey, K.C. (1971) An ultimate load method of design for plate girders, *Developments in Bridge Design and Construction*, (eds K.C. Rockey, J.L. Bannister, and H.R. Evans), Crosby Lockwood and Son, London, pp. 487–504.

15. Rockey, K.C., El-Gaaly, M.A. and Bagchi, D.K. (1972) Failure of thin-walled members under patch loading, *Journal of the Structural Division, ASCE*, **98**, No. ST12, pp. 2739–52.

16. Roberts, T.M. (1981) Slender plate girders subjected to edge loading, *Proceedings*, Institution of Civil Engineers, **71**, Part 2, September, pp. 805–19.

17. Roberts, T.M. (1983) Patch loading on plate girders, Chapter 3 in *Plated Structures: Stability and Strength* (ed. R. Narayanan), Applied Science Publishers, London, pp. 77–102.

18. SA (1998) *AS4100–1998 Steel Structures*, Standards Australia, Sydney, Australia.

Chapter 5

In-plane bending of beams

5.1 Introduction

Beams are structural members which transfer the transverse loads they carry to the supports by bending and shear actions. Beams generally develop higher stresses than axially loaded members with similar loads, while the bending deflections are much higher. These bending deflections of a beam are often, therefore, a primary design consideration. On the other hand, most beams have small shear deflections, and these are usually neglected.

Beam cross-sections may take many different forms, as shown in Figure 5.1, and these represent various methods of obtaining an efficient and economical member. Thus most steel beams are not of solid cross-section, but have their material distributed more efficiently in thin walls. Thin-walled sections may be open, and while these tend to be weak in torsion, they are often cheaper to manufacture than the stiffer closed sections. Perhaps the most economic method of manufacturing steel beams is by hot-rolling, but only a limited number of open cross-sections is available. When a suitable hot-rolled beam cannot be found, a substitute may be fabricated by connecting together a series of rolled plates, and this has become increasingly common. Fabricating techniques also allow the production of beams compounded from hot-rolled members and plates, and of hybrid members in which the flange material is of a higher yield stress than the web. Tapered and castellated beams can also be fabricated from hot-rolled beams. In many cases, a steel beam is required to support a reinforced concrete slab, and in this case its strength may be increased by connecting the steel and concrete together so that they act compositely. The fire resistance of a steel beam may also be increased by encasing it in concrete. The final member cross-section chosen will depend on its suitability for the use intended, and on the overall economy.

The strength of a steel beam in the plane of loading depends on its section properties and on its yield stress f_y. When bending predominates in a determinate beam, the effective ultimate strength is reached when the most highly stressed cross-section becomes fully yielded so that it forms a plastic hinge. The moment M_p at which this occurs is somewhat higher than the first yield moment M_y at which elastic behaviour nominally ceases, as shown in Figure 5.2, and for hot-rolled

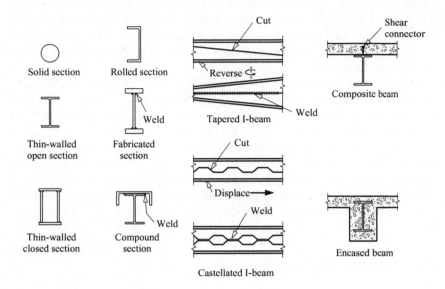

Figure 5.1 Beam types.

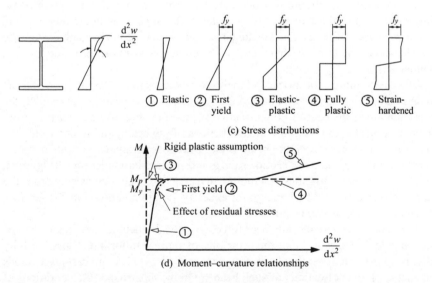

(c) Stress distributions

(d) Moment–curvature relationships

Figure 5.2 Moment–curvature relationships for steel beams.

I-beams this margin varies between 10 and 20% approximately. The attainment of the first plastic hinge forms the basis for the traditional method of design of beams in which the bending moment distribution is calculated from an elastic analysis.

However, in an indeterminate beam, a substantial redistribution of bending moment may occur after the first hinge forms, and failure does not occur until sufficient plastic hinges have formed to cause the beam to become a mechanism. The load which causes this mechanism to form provides the basis for the more rational method of plastic design.

When shear predominates, as in some heavily loaded deep beams of short span, the ultimate strength is controlled by the shear force which causes complete plastification of the web. In the more common sections, this is close to the shear force which causes the nominal first yield in shear, and so the shear design is carried out for the shear forces determined from an elastic analysis. In this chapter, the in-plane behaviour and design of beams are discussed. It is assumed that neither local buckling (which is treated in Chapter 4) nor lateral buckling (which is treated in Chapter 6) occurs. Beams with axial loads are discussed in Chapter 7, while the torsion of beams is treated in Chapter 10.

5.2   Elastic analysis of beams

The design of a steel beam is often preceded by an elastic analysis of the bending of the beam. One purpose of such an analysis is to determine the bending moment and shear force distributions throughout the beam, so that the maximum bending moments and shear forces can be found and compared with the moment and shear capacities of the beam. An elastic analysis is also required to determine the deflections of the beam so that these can be compared with the desirable limiting values.

The data required for an elastic analysis include both the distribution and the magnitudes of the applied loads and the geometry of the beam. In particular, the variation along the beam of the effective second moment of area I of the cross-section is needed to determine the deflections of the beam, and to determine the moments and shears when the beam is statically indeterminate. For this purpose, local variations in the cross-section such as those due to bolt holes may be ignored, but more general variations, including any general reductions arising from the use of the effective width concept for excessively thin compression flanges (see Section 4.2.2.2), should be allowed for.

The bending moments and shear forces in statically determinate beams can be determined by making use of the principles of static equilibrium. These are fully discussed in standard textbooks on structural analysis [1, 2], as are various methods of analysing the deflections of such beams. On the other hand, the conditions of statics are not sufficient to determine the bending moments and shear forces in statically indeterminate beams, and the conditions of compatibility between the various elements of the beam or between the beam and its supports must also be used. This is done by analysing the deflections of the statically indeterminate beam. Many methods are available for this analysis, both manual and computer, and these are fully described in standard textbooks [3–8]. These methods can also

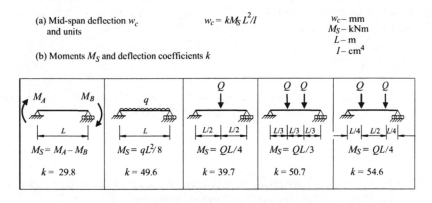

(a) Mid-span deflection w_c
   and units

$$w_c = kM_S L^2/I$$

w_c– mm
M_S– kNm
L– m
I– cm^4

(b) Moments M_S and deflection coefficients k

M_A M_B	q	Q	Q Q	Q Q
L	L	$L/2$ $L/2$	$L/3$ $L/3$ $L/3$	$L/4$ $L/2$ $L/4$
$M_S = M_A - M_B$	$M_S = qL^2/8$	$M_S = QL/4$	$M_S = QL/3$	$M_S = QL/4$
$k = 29.8$	$k = 49.6$	$k = 39.7$	$k = 50.7$	$k = 54.6$

Figure 5.3 Mid-span deflections of steel beams.

be used to analyse the behaviour of structural frames in which the member axial forces are small.

While the deflections of beams can be determined accurately by using the methods referred to above, it is often sufficient to use approximate estimates for comparison with the desirable maximum values. Thus in simply supported or continuous beams, it is usually accurate enough to use the mid-span deflection (see Figure 5.3). The mid-span deflection w_c (measured relative to the level of the left-hand support) depends on the distribution of the applied load, the end moments M_A and M_B (taken as clockwise positive), and the sinking w_{AB} of the right-hand support below the left-hand support, and can be expressed as

$$w_c = \{kM_S + 29.8(M_A - M_B)\}\frac{L^2}{I} + \frac{w_{AB}}{2}. \qquad (5.1)$$

In this equation, the deflections w_c and w_{AB} are expressed in mm, the moments M_S, M_A, M_B are in kNm, the span L is in m and the second moment of area I is in cm^4. Expressions for the simple beam moments M_S and values of the coefficients k are given in Figure 5.3 for a number of loading distributions. These can be combined together to find the central deflections caused by many other loading distributions.

5.3   Bending stresses in elastic beams

5.3.1   Bending in a principal plane

The distribution of the longitudinal bending stresses σ in an elastic beam bent in a principal plane xz can be deduced from the experimentally confirmed assumption that plane sections remain substantially plane during bending, provided any shear lag effects which may occur in beams with very wide flanges (see Section 5.4.5) are negligible. Thus the longitudinal strains ε vary linearly through the depth of the beam, as shown in Figure 5.4, as do the longitudinal stresses σ. It is shown in Section 5.8.1 that the moment resultant M_y of these stresses is

$$M_y = -EI_y \frac{d^2w}{dx^2} \tag{5.2}$$

where the sign conventions for the moment M_y and the deflection w are as shown in Figure 5.5, and that the stress at any point in the section is

$$\sigma = \frac{M_y z}{I_y} \tag{5.3}$$

in which tensile stresses are positive. In particular, the maximum stresses occur at the extreme fibres of the cross-section, and are given by

$$\left. \begin{aligned} \sigma_{max} &= \frac{M_y z_T}{I_y} = -\frac{M_y}{W_{el,yT}}, \\ \sigma_{max} &= \frac{M_y z_B}{I_y} = \frac{M_y}{W_{el,yB}}, \end{aligned} \right\} \tag{5.4}$$

in compression, and

in tension, in which $W_{el,yT} = -I_y/z_T$ and $W_{el,yB} = I_y/z_B$ are the elastic section moduli for the top and bottom fibres, respectively, for bending about the y axis.

The corresponding equations for bending in the principal plane xy are

$$M_z = EI_z \frac{d^2v}{dx^2} \tag{5.5}$$

where the sign conventions for the moment M_z and the deflection v are also shown in Figure 5.5,

$$\sigma = -M_z y / I_z \tag{5.6}$$

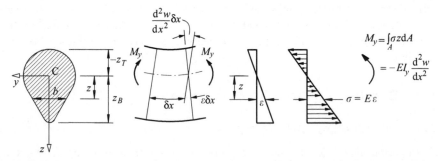

(a) Cross-section   (b) Bending   (c) Strains ε (d) Stresses σ (e) Moment resultant

Figure 5.4 Elastic bending of beams.

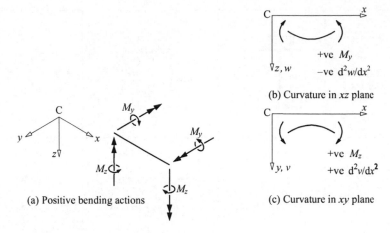

Figure 5.5 Bending sign conventions.

$$\left.\begin{array}{l} \sigma_{max} = M_z/W_{el,zL} \\[2mm] \sigma_{max} = -M_z/W_{el,zR} \end{array}\right\} \qquad (5.7)$$

and

in which $W_{el,zL} = -I_z/y_L$ and $W_{el,zR} = I_z/y_R$ are the elastic section moduli for bending about the z axis.

Values of $I_y, I_z, W_{el,z}, W_{el,z}$ for hot-rolled steel sections are given in [9], while values for other sections can be calculated as indicated in Section 5.9 or in standard textbooks [1, 2]. Expressions for the properties of some thin-walled sections are given in Figure 5.6 [10]. When a section has local holes, or excessive widths (see Section 4.2.2.2), these properties may need to be reduced accordingly.

Worked examples of the calculation of cross-section properties are given in Sections 5.12.1–5.12.4.

5.3.2   Biaxial bending

When a beam deflects only in a plane xz_1 which is not a principal plane (see Figure 5.7a), so that its curvature d^2v_1/dx^2 in the perpendicular xy_1 plane is zero, then the bending stresses

$$\sigma = -Ez_1 d^2w_1/dx^2 \qquad (5.8)$$

have moment resultants

$$\left.\begin{array}{l} M_{y1} = -EI_{y1}d^2w_1/dx^2 \\[2mm] M_{z1} = -EI_{y1z1}d^2w_1/dx^2 \end{array}\right\} \qquad (5.9)$$

and

Section Geometry	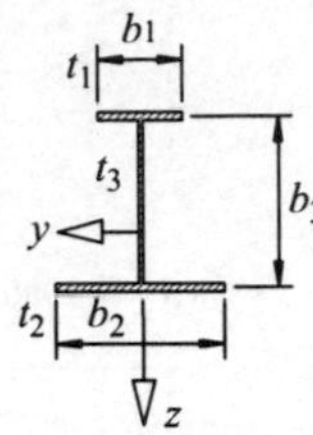	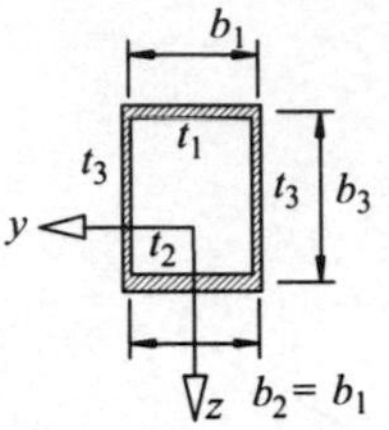	
A	$A_1 + A_2 + A_3$	$A_1 + A_2 + 2A_3$	$\pi d t$
I_y	$\sum_3 A_n z_n^2 + I_3$	$A_1 z_1^2 + A_2 z_2^2 + 2A_3 z_3^2 + 2I_3$	$\pi d^3 t/8$
I_z	$I_1 + I_2$	$I_1 + I_2 + A_3 b_1^2/2$	$\pi d^3 t/8$
$W_{el.y,1,2}$	$I_y/z_{1,2}$	I_y/z_n	$\pi d^2 t/4$
$W_{el.z,1,2}$	$2 I_z/b_{1,2}$	$2 I_z/b_1$	$\pi d^2 t/4$
$W_{pl.y,1,2}$	$\sum_2 (A_n z_{pn} + z_{pn}^2 t_3/2)$	$\sum_2 (A_n z_{pn} + z_{pn}^2 t_3)$	$d^2 t$
$W_{pl.z,1,2}$	$\sum_2 A_n b_n/4$	$\sum_2 A_n b_n/4 + A_3 b_1$	$d^2 t$
y_0	0	0	0
z_0	$b_3\left\{\dfrac{(I_2-I_1)}{2I_z} - \dfrac{(A_2-A_1)}{2A}\right\}$	$b_3\left\{\dfrac{(I_2-I_1)}{2I_z} - \dfrac{(A_2-A_1)}{2A}\right\}$	0

$$A_n = b_n t_n , \quad I_n = b_n^3 t_n/12$$

$$z_{1,2} = b_3 (A_{2,1} + A_3/2)/A$$

$$z_3 = (z_2 - z_1)/2$$

$$z_{pn} = [0 \le (A - 2A_n)/2t_3 \le b_3]$$

Figure 5.6a Thin-walled section properties.

where I_{y1z1} is the product second moment of area (see Section 5.9). Thus the resultant bending moment

$$M = \surd(M_{y1}^2 + M_{z1}^2), \tag{5.10}$$

is inclined to the xz_1 plane of the bending deflections, as shown in Figure 5.7a.

The simplest method of analysing this or any other biaxial bending situation is to replace the moments M_{y1}, M_{z1} by their principal plane static equivalents M_y,

Section Geometry	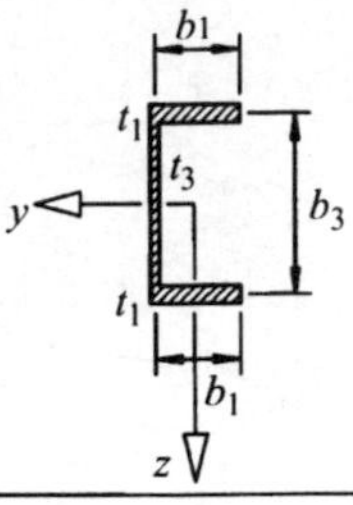	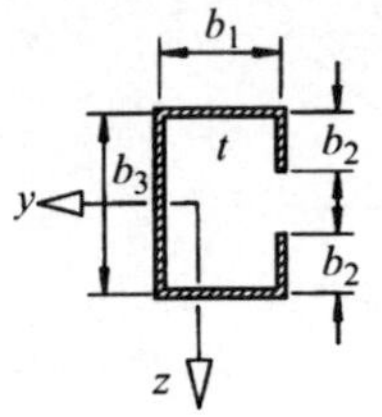	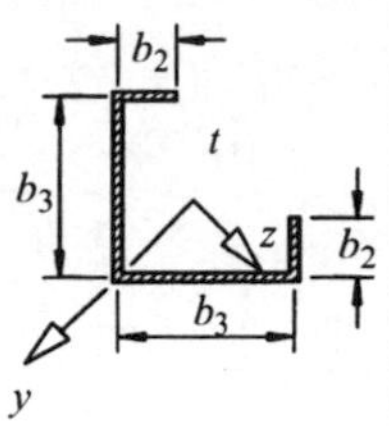
A	$2A_1 + A_3$	$2A_1 + 2A_2 + 2A_3$	$\dfrac{2\Sigma A_n}{2}$
I_y	$A_1 b_3^2/2 + I_3$	$\dfrac{A_1 b_3^2 + I_3 + I_2 + A_2(b_3-b_2)^2}{2}$	$A_2(b_3-b_2/2)^2 + A_3 b_3^2/4 + \dfrac{\Sigma A_n}{2}$
I_z	$2A_1 y_1^2 + A_3 y_3^2 + 2I_1$	$2A_1 y_1^2 + 2A_2 y_2^2 + A_3 y_3^2 + 2I_1$	$A_2 b_3^2/4 + A_3 b_2^2/4 + \dfrac{\Sigma I_n}{2}$
$W_{el,y}$	$2I_y/b_3$	$2I_y/b_3$	$\sqrt{2}I_y/b_3$
$W_{el,z}$	$I_z/(b_1-y_3)$ and I_z/y_3	I_z/y_2 and I_z/y_3	$2\sqrt{2}I_z/(b_2+b_3)$
$W_{pl,y}$	$A_1 b_3 + A_3 b_3/4$	$A_1 b_3 + A_2(b_3-b_2) + A_3 b_3/4$	$(b_3^2 + 2b_2 b_3 - b_2^2)t/2\sqrt{2}$
$W_{pl,z}$	$(b_1-y_p)^2 t_1 + y_p^2 t_1 + A_3 y_p$	$\{(b_1-y_p)^2 + y_p^2 + 2b_2(b_1-y_p) + (b_3 y_p)\}t$	$(b_2+b_3)^2 t/2\sqrt{2}$
y_0	$y_3 + \dfrac{A_1 b_3^2 b_1}{4I_y}$	$y_3 + \{b_3^2(b_1+2b_2)\}\dfrac{A_1}{4I_y} - 8b_2^3/3$	$\dfrac{(b_2+b_3)/2\sqrt{2} + (3b_3-2b_2)b_2^2 b_3 t}{3\sqrt{2}I_y}$
z_0	0	0	0
y_1	$b_1 A_3/2A$	$b_1/2 - y_3$	—
y_2	—	$b_1 - y_3$	—
y_3	$b_1 A_1/A$	$(b_1+2b_2)A_1/A$	—
y_p	$(A-2A_3)/4t_1 \geq 0$	$(A-2A_3)/4t \geq 0$	—
	$A_n = b_n t_n$,	$I_n = b_n^3 t_n/12$	

Figure 5.6b Thin-walled section properties.

M_z calculated from

$$\left.\begin{aligned} M_y &= M_{y1}\cos\alpha + M_{z1}\sin\alpha \\ M_z &= -M_{y1}\sin\alpha + M_{z1}\cos\alpha, \end{aligned}\right\} \tag{5.11}$$

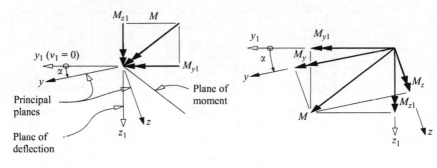

(a) Deflection in a non-principal plane   (b) Principal plane moments M_y, M_z

Figure 5.7 **Bending in a non-principal plane.**

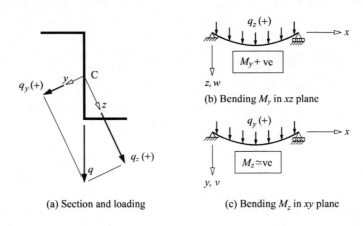

(a) Section and loading   (c) Bending M_z in xy plane

Figure 5.8 **Biaxial bending of a zed beam.**

in which α is the angle between the y_1, z_1 axes and the principal y, z axes, as shown in Figure 5.7b. The bending stresses can then be determined from

$$\sigma = \frac{M_y z}{I_y} - \frac{M_z y}{I_z},\qquad(5.12)$$

in which I_y, I_z are the principal second moments of area. If the values of α, I_y, I_z are unknown, they can be determined from I_{y1}, I_{z1}, I_{y1z1} as shown in Section 5.9.

Care needs to be taken to ensure that the correct signs are used for M_y and M_z in equation 5.12, as well as for y and z. For example, if the zed section shown in Figure 5.8a is simply supported at both ends with a uniformly distributed load q acting in the plane of the web, then its principal plane components q_y, q_z cause positive bending M_y and negative bending M_z, as indicated in Figure 5.8b and c.

The deflections of the beam can also be determined from the principal plane moments by solving equations 5.2 and 5.5 in the usual way [1–8] for the principal plane deflections w and v, and by adding these vectorially.

Worked examples of the calculation of the elastic stresses and deflections in an angle section cantilever are given in Section 5.12.5.

5.4 Shear stresses in elastic beams

5.4.1 Solid cross-sections

A vertical shear force V_z acting parallel to the minor principal axis z of a section of a beam (see Figure 5.9a) induces shear stresses τ_{xy}, τ_{xz} in the plane of the section. In solid section beams, these are usually assumed to act parallel to the shear force (i.e. $\tau_{xy} = 0$), and to be uniformly distributed across the width of the section, as shown in Figure 5.9a. The distribution of the vertical shear stresses τ_{xz} can be determined by considering the horizontal equilibrium of an element of the beam as shown in Figure 5.9b. Because the bending normal stresses σ vary with x, they create an imbalance of force in the x direction, which can only be compensated for by the horizontal shear stresses $\tau_{zx} = \tau_v$ which are equal to the vertical shear stresses τ_{xz}. It is shown in Section 5.10.1 that the stress τ_v at a distance z_2 from the centroid where the section width is b_2 is given by

$$\tau_v = -\frac{V_z}{I_y b_2} \int_{z_T}^{z_2} bz\,\mathrm{d}z. \tag{5.13}$$

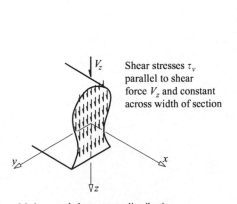

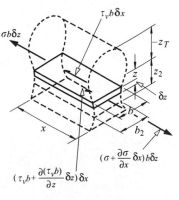

(a) Assumed shear stress distribution   (b) Horizontal equilibrium

Figure 5.9 Shear stresses in a solid section.

This can also be expressed as

$$\tau_v = -\frac{V_z A_2 \bar{z}_2}{I_y b_2} \tag{5.14}$$

in which A_2 is the area above b_2 and $\bar{z}_2$ is the height of its centroid.

For the particular example of the rectangular section beam of width b and depth d shown in Figure 5.10a, the width is constant, and so equation 5.13 becomes

$$\tau_v = \frac{V_z}{bd}\left(1.5 - \frac{6z^2}{d^2}\right). \tag{5.15}$$

This shear stress distribution is parabolic, as shown in Figure 5.10b, and has a maximum value at the y axis of 1.5 times of average shear stress V_z/bd.

The shear stresses calculated from equations 5.13 or 5.14 are reasonably accurate for solid cross-sections, except near the unstressed edges where the shear stress is parallel to the edge rather than to the applied shear force.

5.4.2   Thin-walled open cross-sections

The shear stress distributions in thin-walled open-section beams differ from those given by equations 5.13 or 5.14 for solid section beams in that the shear stresses are parallel to the wall of the section as shown in Figure 5.11a instead of parallel to the applied shear force. Because of the thinness of the walls, it is quite accurate to assume that the shear stresses are uniformly distributed across the thickness t of the thin-walled section. Their distribution can be determined by considering the horizontal equilibrium of an element of the beam, as shown in Figure 5.11b. It is shown in Section 5.10.2 that the stress τ_v at a distance s from the end of the section

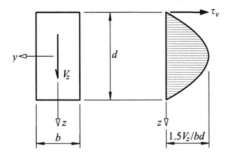

(a) Cross-section   (b) Shear stress distribution

Figure 5.10 Shear stress distribution in a rectangular section.

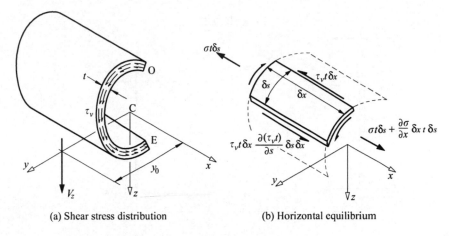

(a) Shear stress distribution   (b) Horizontal equilibrium

Figure 5.11 **Shear stresses in a thin-walled open section.**

caused by a vertical shear force V_z can be obtained from the shear flow

$$\tau_v t = -\frac{V_z}{I_y} \int_0^s zt \, \mathrm{d}s. \tag{5.16}$$

This can also be expressed as

$$\tau_v = -\frac{V_z A_s \bar{z}_s}{I_y t} \tag{5.17}$$

in which A_s is the area from the free end to the point s and $\bar{z}_s$ is the height of the centroid of this area above the point s.

At a junction in the cross-section wall, such as that of the top flange and the web of the I-section shown in Figures 5.12 and 5.13, horizontal equilibrium of the zero area junction element requires

$$(\tau_v t_f)_{21} \delta x + (\tau_v t_f)_{23} \delta x = (\tau_v t_w)_{24} \delta x \tag{5.18}$$

This can be thought of as an analogous flow continuity condition for the corresponding shear flows in each cross-section element at the junction, as shown in Figure 5.13c, so that

$$(\tau_v t_f)_{21} + (\tau_v t_f)_{23} = (\tau_v t_w)_{24} \tag{5.19}$$

which is a particular example of the general junction condition

$$\sum (\tau_v t) = 0 \tag{5.20}$$

in which each shear flow is now taken as positive if it acts towards the junction.

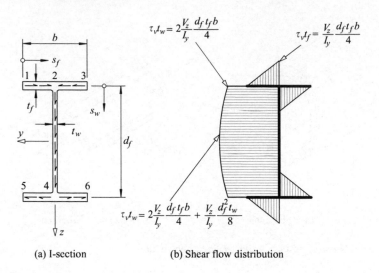

(a) I-section   (b) Shear flow distribution

Figure 5.12 Shear flow distribution in an I-section.

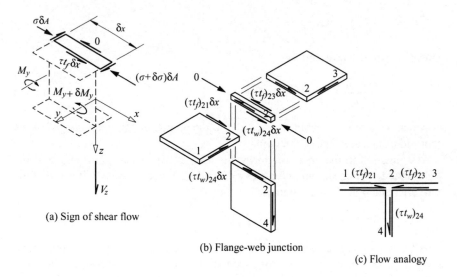

(a) Sign of shear flow

(b) Flange-web junction

(c) Flow analogy

Figure 5.13 Horizontal equilibrium considerations for shear flow.

A free end of a cross-section element is the special case of a one-element junction, for which equation 5.20 reduces to

$$\tau_v t = 0 \tag{5.21}$$

This condition has already been used (at $s = 0$) in deriving equation 5.16.

An example of the analysis of the shear flow distribution in the I-section beam shown in Figure 5.12a is given in Section 5.12.6. The shear flow distribution is shown in Figure 5.12b. It can be seen that the shear stress in the web is nearly constant and equal to its average value

$$\tau_{v(av)} = \frac{V_z}{d_f t_w},$$

(5.22)

which provides the basis for the commonly used assumption that the applied shear is resisted only by the web. A similar result is obtained for the shear stress in the web of the channel section shown in Figure 5.14 and analysed in Section 5.12.7.

When a horizontal shear force V_y acts parallel to the y axis, additional shear stresses τ_h parallel to the walls of the section are induced. These can be obtained from the shear flow

$$\tau_h t = -\frac{V_y}{I_z} \int_0^s yt ds.$$

(5.23)

which is similar to equation 5.17 for the shear flow due to a vertical shear force. A worked example of the calculation of the shear flow distribution caused by a horizontal shear force is given in Section 5.12.8.

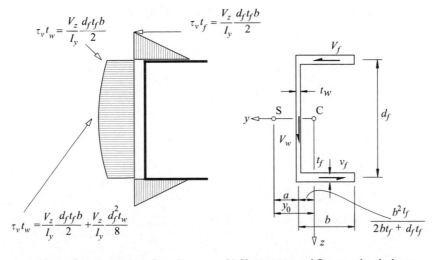

(a) Shear flow due to shear force V_z (b) Shear centre and flange and web shears

Figure 5.14 Shear flow distribution in a channel section.

5.4.3　Shear centre

The shear stresses τ_v induced by the vertical shear force V_z exert a torque equal to $\int_0^E \tau_v t \rho \, ds$ about the centroid C of the thin-walled cross-section, as shown in Figure 5.15, and are therefore statically equivalent to a vertical shear force V_z which acts at a distance y_0 from the centroid equal to

$$y_0 = \frac{1}{V_z} \int_0^E \tau_v t \rho \, ds. \tag{5.24}$$

Similarly, the shear stresses τ_h induced by the horizontal shear force V_y are statically equivalent to a horizontal shear force V_y which acts at a distance z_0 from the centroid equal to

$$z_0 = -\frac{1}{V_y} \int_0^E \tau_h t \rho \, ds. \tag{5.25}$$

The coordinates y_0, z_0 define the position of the shear centre S of the cross-section, through which the resultant of the bending shear stresses must act. When any applied load does not act through the shear centre, as shown in Figure 5.16, then it induces another set of shear stresses in the section which are additional to those caused by the changes in the bending normal stresses described above (see equations 5.16 and 5.23). These additional shear stresses are statically equivalent to the torque exerted by the eccentric applied load about the shear centre. They can be calculated as shown in Sections 10.2.1.4 and 10.3.1.2.

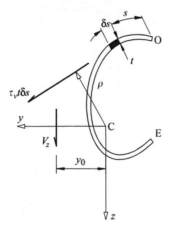

Figure 5.15　Moment of shear stress τ_v about centroid.

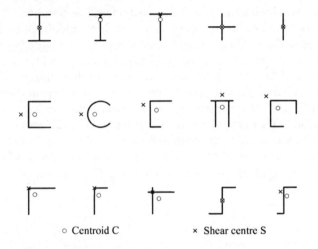

Figure 5.16 Off shear centre loading.

Off shear centre loading ≡ Pure bending Bending shear stresses + Pure torsion Torsion shear stresses

○ Centroid C   × Shear centre S

Figure 5.17 Centroids and shear centres.

The position of the shear centre of the channel section shown in Figure 5.14 is determined in Section 5.12.8. Its y_0 coordinate is given by

$$y_0 = \frac{d_f^2 t_f b^2}{4I_y} + \frac{b^2 t_f}{2bt_f + d_f t_w} \tag{5.26}$$

while its z_0 coordinate is zero because the section is symmetrical about the y axis.

In Figure 5.17, the shear centres of a number of thin-walled open sections are compared with their centroids. It can be seen that

(a) if the section has an axis of symmetry, then the shear centre and centroid lie on it,

(b) if the section is of a channel type, then the shear centre lies outside the web
     and the centroid inside it, and
(c) if the section consists of a set of concurrent rectangular elements (tees, angles,
     and cruciforms), then the shear centre lies at the common point.

A general matrix method for analysing the shear stress distributions and for
determining the shear centres of thin-walled open-section beams has been prepared
[11, 12].

   Worked examples of the determination of the shear centre position are given in
Sections 5.12.8 and 5.12.10.

5.4.4 Thin-walled closed cross-sections

The shear stress distribution in a thin-walled, closed-section beam is similar to
that in an open-section beam, except that there is an additional constant shear
flow $\tau_{vc}t$ around the section. This additional shear flow is required to prevent any
discontinuity in the longitudinal warping displacements u which arise from the
shear straining of the walls of the closed sections. To show this, consider the slit
rectangular box whose shear flow distribution $\tau_{vo}t$ due to a vertical shear force V_z is
as shown in Figure 5.18a. Because the beam is not twisted, the longitudinal fibres
remain parallel to the centroidal axis, so that the transverse fibres rotate through
angles τ_{vo}/G equal to the shear strain in the wall, as shown in Figure 5.19. These
rotations lead to the longitudinal warping displacements u shown in Figure 5.18b,
the relative warping displacement at the slit being $\int_0^E (\tau_{vo}/G)ds$.

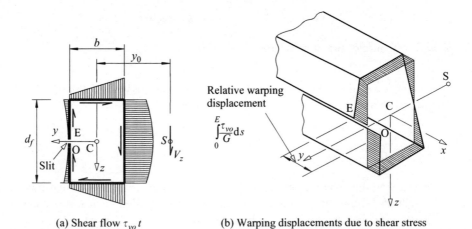

(a) Shear flow $\tau_{vo}t$ (b) Warping displacements due to shear stress

Figure 5.18 Warping of a slit box.

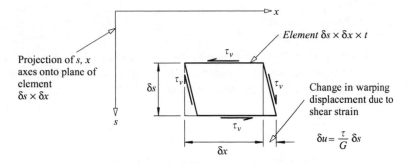

Figure 5.19 Warping displacements due to shear.

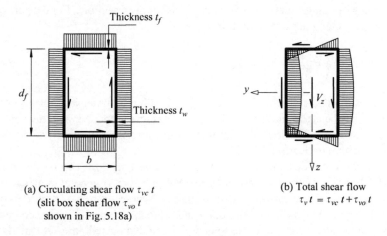

(a) Circulating shear flow $\tau_{vc}\,t$
(slit box shear flow $\tau_{vo}\,t$
shown in Fig. 5.18a)

(b) Total shear flow
$\tau_v t = \tau_{vc}\,t + \tau_{vo}\,t$

Figure 5.20 Shear stress distribution in a rectangular box.

However, when the box is not slit, as shown in Figure 5.20, this relative warping displacement must be zero. Thus

$$\oint \frac{\tau_v}{G}\,\mathrm{d}s = 0,$$ (5.27)

in which total shear stress τ_v can be considered as the sum

$$\tau_v = \tau_{vc} + \tau_{vo}$$ (5.28)

of the slit box shear stress τ_{vo} (see Figure 5.18a) and the shear stress τ_{vc} due to a constant shear flow $\tau_{vc}t$ circulating around the closed section (see Figure 5.20a). Alternatively, equation 5.16 for the shear flow in an open section can be modified for the closed section to

$$\tau_v t = \tau_{vc} t - \frac{V_z}{I_y} \int_0^s zt\,ds, \tag{5.29}$$

since it can no longer be said that τ_v is zero at $s = 0$, this not being a free end (the closed section has no free ends). Mathematically, $\tau_{vc}t$ is a constant of integration. Substituting equation 5.28 or 5.29 into equation 5.27 leads to

$$\tau_{vc}t = -\frac{\oint \tau_{vo}ds}{\oint(1/t)ds}, \tag{5.30}$$

which allows the circulating shear flow $\tau_{vc}t$ to be determined.

The shear stress distribution in any single-cell closed section can be obtained by using equations 5.29 and 5.30. The shear centre of the section can then be determined by using equations 5.24 and 5.25 as for open sections. For the particular case of the rectangular box shown in Figure 5.20,

$$\tau_{vc}t = \frac{V_z}{I_y}\left(\frac{d_f^2 t_w}{8} + \frac{d_f t_f b}{4}\right), \tag{5.31}$$

and the resultant shear flow shown in Figure 5.20b is symmetrical because of the symmetry of the cross-section, while the shear centre coincides with the centroid.

A worked example of the calculation of the shear stresses in a thin-walled closed section is given in Section 5.12.11.

The shear stress distributions in multi-cell closed sections can be determined by extending this method, as indicated in Section 5.10.3. A general matrix method of analysing the shear flows in thin-walled closed sections has been described [11, 12]. This can be used for both open and closed sections, including composite and asymmetric sections, and sections with open and closed parts. It can also be used to determine the centroid, principal axes, section constants, and the bending normal stress distribution.

5.4.5   Shear lag

In the conventional theory of bending, shear strains are neglected so that it can be assumed that plane sections remain plane after loading. From this assumption follow the simple linear distributions of the bending strains and stresses discussed in Section 5.3, and from these the shear stress distributions discussed in Sections 5.4.1–5.4.4. The term shear lag [13] is related to some of the discrepancies between this approximate theory of the bending of beams and their real behaviour, and in particular, refers to the increases of the bending stresses near the flange-to-web junctions, and the corresponding decreases in the flange stresses away from these junctions.

The shear lag effects near the mid-point of the simply supported centrally loaded I-section beam shown in Figure 5.21a are illustrated in Figure 5.22. The shear

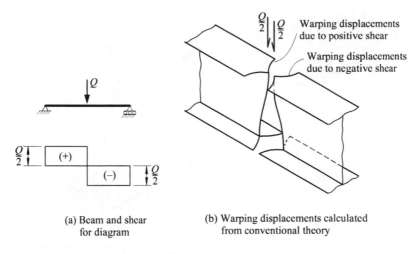

(a) Beam and shear
for diagram

(b) Warping displacements calculated
from conventional theory

Figure 5.21 Incompatible warping displacements at a shear discontinuity.

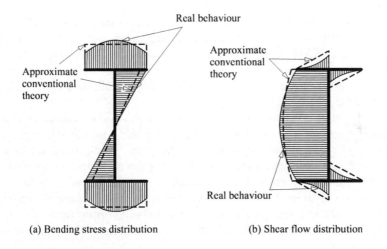

(a) Bending stress distribution

(b) Shear flow distribution

Figure 5.22 Shear lag effects in an I-section beam.

stresses calculated by the conventional theory are as shown in Figure 5.23a, and these induce the warping (longitudinal) displacements shown in Figure 5.23b. The warping displacements of the web are almost linear, and these are responsible for the shear deflections [13] which are usually neglected when calculating the beam's deflections. The warping displacements of the flanges vary parabolically, and it is these which are responsible for most of the shear lag effect. The application of the conventional theory of bending at either side of a point in a beam where

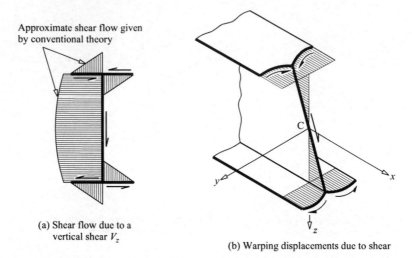

(a) Shear flow due to a
vertical shear V_z

(b) Warping displacements due to shear

Figure 5.23 Warping of an I-section beam.

there is a sudden change in the shear force, such as at the mid-point of the beam of Figure 5.21a, leads to two different distributions of warping displacement, as shown in Figure 5.21b. These are clearly incompatible, and so changes are required in the bending stress distribution (and consequently in the shear stress distribution) to remove the incompatibility. These changes, which are illustrated in Figure 5.22, constitute the shear lag effect.

Shear lag effects are usually very small except near points of high concentrated load or at reaction points in short-span beams with thin wide flanges. In particular, shear lag effects may be significant in light-gauge, cold-formed sections [14] and in stiffened box girders [15–17]. Shear lag has no serious consequences in a ductile structure in which any premature local yielding leads to a favourable redistribution of stress. However, the increased stresses due to shear lag may be of consequence in a tension flange which is liable to brittle fracture or fatigue damage, or in a compression flange whose strength is controlled by its resistance to local buckling.

An approximate method of dealing with shear lag is to use an effective width concept, in which the actual width b of a flange is replaced by a reduced width b_{eff} given by

$$\frac{b_{eff}}{b} = \frac{\text{nominal bending stress}}{\text{maximum bending stress}}. \tag{5.32}$$

This is equivalent to replacing the actual flange bending stresses by constant stresses which are equal to the actual maximum stress and distributed over the effective flange area $b_{eff} \times t$. Some values of effective widths are given in [14–16].

This approach is similar to that used to allow for the redistribution of stress which takes place in a thin compression flange after local buckling (see Chapter 4). However, the two effects of shear lag and local buckling are quite distinct, and should not be confused.

5.5 Plastic analysis of beams

5.5.1 General

As the load on a ductile steel beam is increased, the stresses in the beam also increase, until the yield stress is reached. With further increases in the load, yielding spreads through the most highly strained cross-section of the beam until it becomes fully plastic at a moment M_p. At this stage the section forms a plastic hinge which allows the beam segments on either side to rotate freely under the moment M_p. If the beam was originally statically determinate, this plastic hinge reduces it to a mechanism, and prevents it from supporting any additional load.

However, if the beam was statically indeterminate, the plastic hinge does not reduce it to a mechanism, and it can support additional load. This additional load causes a redistribution of the bending moment, during which the moment at the plastic hinge remains fixed at M_p, while the moment at another highly strained cross-section increases until it forms a plastic hinge. This process is repeated until enough plastic hinges have formed to reduce the beam to a mechanism. The beam is then unable to support any further increase in load, and its ultimate strength is reached.

In the plastic analysis of beams, this mechanism condition is investigated to determine the ultimate strength. The principles and methods of plastic analysis are fully described in many textbooks [18–24], and so only a brief summary is given in the following sub-sections.

5.5.2 The plastic hinge

The bending stresses in an elastic beam are distributed linearly across any section of the beam, as shown in Figure 5.4, and the bending moment M is proportional to the curvature $-d^2w/dx^2$ (see equation 5.2). However, once the yield strain $\varepsilon_y = f_y/E$ (see Figure 5.24) of a steel beam is exceeded, the stress distribution is no longer linear, as indicated in Figure 5.2c. Nevertheless, the strain distribution remains linear, and so the inelastic bending stress distributions are similar to the basic stress–strain relationship shown in Figure 5.24, provided the influence of shear on yielding can be ignored (which is a reasonable assumption for many I-section beams). The moment resultant M of the bending stresses is no longer proportional to the curvature $-d^2w/dx^2$, but varies as shown in Figure 5.2d. Thus the section becomes elastic-plastic when the yield moment $M_y = f_y W_{el}$ is exceeded, and the curvature increases rapidly as yielding progresses through the section. At high curvatures, the limiting situation is approached for which the section is completely

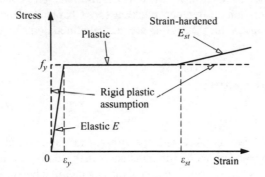

Figure 5.24 Idealised stress–strain relationships for structural steel.

yielded at the fully plastic moment

$$M_p = f_y W_{pl},$$
(5.33)

where W_{pl} is the plastic section modulus. Methods of calculating the fully plastic moment M_p are discussed in Section 5.11.1, and worked examples are given in Sections 5.12.12 and 5.12.13. For solid rectangular sections, the shape factor W_{pl}/W_{el} is 1.5, but for rolled I-sections, W_{pl}/W_{el} varies between 1.1 and 1.2 approximately. Values of W_{pl} for hot-rolled I-section members are given in [9].

In real beams, strain-hardening commences just before M_p is reached, and the real moment–curvature relationship rises above the fully plastic limit of M_p, as shown in Figure 5.2d. On the other hand, high shear forces cause small reductions in M_p below $f_y W_{pl}$, due principally to reductions in the plastic bending capacity of the web. This effect is discussed in Section 4.5.2.

The approximate approach of the moment–curvature relationship shown in Figure 5.2d to the fully plastic limit forms the basis of the simple rigid-plastic assumption for which the basic stress–strain relationship is replaced by the rect-angular block shown by the dashed line in Figure 5.24. Thus elastic strains are completely ignored, as are the increased stresses due to strain-hardening. This assumption ignores the curvature of any elastic and elastic–plastic regions $(M < M_p)$, and assumes that the curvature becomes infinite at any point where $M = M_p$.

The consequences of the rigid-plastic assumption on the theoretical behaviour of a simply supported beam with a central concentrated load are shown in Figures 5.25 and 5.26. In the real beam shown in Figure 5.25a, there is a finite length of the beam which is elastic-plastic and in which the curvatures are large, while the remaining portions are elastic, and have small curvatures. However, according to the rigid-plastic assumption, the curvature becomes infinite at mid-span when this section becomes fully plastic $(M = M_p)$, while the two halves of the beam have zero curvature and remain straight, as shown in Figure 5.25b. The

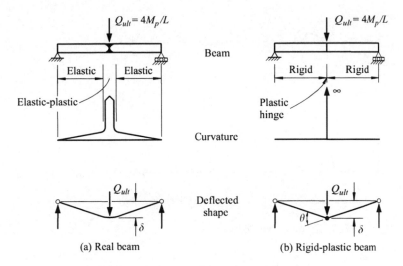

(a) Real beam

(b) Rigid-plastic beam

Figure 5.25 Beam with full plastic moment M_p at mid-span.

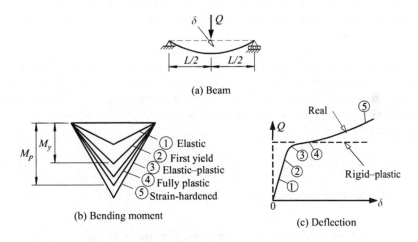

(a) Beam

(b) Bending moment

(c) Deflection

Figure 5.26 Behaviour of a simply supported beam.

infinite curvature at mid-span causes a finite change θ in the beam slope, and so the deflected shape of the rigid-plastic beam closely approximates that of the real beam, despite the very different curvature distributions. The successive stages in the development of the bending moment diagram and the central deflection of the beam as the load is increased are summarised in Figure 5.26b and c.

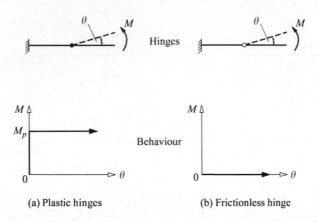

(a) Plastic hinges   (b) Frictionless hinge

Figure 5.27 Plastic hinge behaviour.

The infinite curvature at the point of full plasticity and the finite slope change θ predicted by the rigid-plastic assumption lead to the plastic hinge concept illustrated in Figure 5.27a. The plastic hinge can assume any slope change θ once the full plastic moment M_p has been reached. This behaviour is contrasted with that shown in Figure 5.27b for a frictionless hinge, which can assume any slope change θ at zero moment.

It should be noted that when the full plastic moment M_p of the simply supported beam shown in Figure 5.26a is reached, the rigid-plastic assumption predicts that a two-bar mechanism will be formed by the plastic hinge and the two frictionless support hinges, as shown in Figure 5.25b, and that the beam will deform freely without any further increase in load, as shown in Figure 5.26c. Thus the ultimate load of the beam is

$$Q_{ult} = 4M_p/L, \tag{5.34}$$

which reduces the beam to a plastic collapse mechanism.

5.5.3   Moment redistribution in indeterminate beams

The increase in the full plastic moment M_p of a beam over its nominal first yield moment M_y is accounted for in elastic design (i.e. design based on an elastic analysis of the bending of the beam) by the use of M_p for the moment capacity of the cross-section. Thus elastic design might be considered to be 'first hinge' design. However, the feature of plastic design which distinguishes it from elastic design is that it takes into account the favourable redistribution of the bending moment which takes place in an indeterminate structure after the first hinge forms. This redistribution may be considerable, and the final load at which the collapse mechanism forms may be significantly higher than that at which the first hinge is developed. Thus, a first hinge design based on an elastic analysis of the bending moment may significantly underestimate the ultimate strength.

The redistribution of bending moment is illustrated in Figure 5.28 for a built-in beam with a concentrated load at a third point. This beam has two redundancies, and so three plastic hinges must form before it can be reduced to a collapse mechanism. According to the rigid-plastic assumption, all the bending moments remain proportional to the load until the first hinge forms at the left-hand support A at $Q = 6.75 M_p/L$. As the load increases further, the moment at this hinge remains constant at M_p, while the moments at the load point and the right-hand support increase until the second hinge forms at the load point B at $Q = 8.68 M_p/L$. The moment at this hinge then remains constant at M_p while the moment at the right-hand support C increases until the third and final hinge forms at this point at $Q = 9.00 M_p/L$. At this load the beam becomes a mechanism, and so the ultimate load is $Q_{ult} = 9.00 M_p/L$ which is 33% higher than the first hinge load of $6.75 M_p/L$. The redistribution of bending moment is shown in Figure 5.28b and d, while the deflection of the load point (derived from an elastic–plastic assumption) is shown in Figure 5.28c.

5.5.4 Plastic collapse mechanisms

A number of examples of plastic collapse mechanisms in cantilevers and single- and multi-span beams is shown in Figure 5.29. Cantilevers and overhanging beams

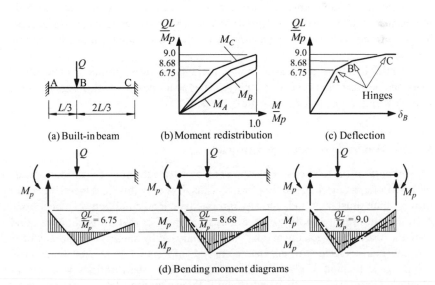

(a) Built-in beam   (b) Moment redistribution   (c) Deflection

(d) Bending moment diagrams

Figure 5.28 Moment redistribution in a built-in beam.

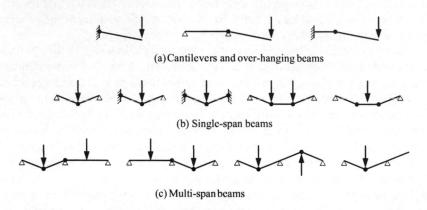

(a) Cantilevers and over-hanging beams

(b) Single-span beams

(c) Multi-span beams

Figure 5.29 Beam plastic collapse mechanisms.

generally collapse as single-bar mechanisms with a plastic hinge at the support. When there is a reduction in section capacity, then the plastic hinge may form in the weaker section.

Single-span beams generally collapse as two-bar mechanisms, with a hinge (plastic or frictionless) at each support and a plastic hinge within the span. Sometimes general plasticity may occur along a uniform moment region.

Multi-span beams generally collapse in one span only, as a local two-bar mechanism, with a hinge (plastic or frictionless) at each support, and a plastic hinge within the span. Sometimes two adjacent spans may combine to form a three-bar mechanism, with the common support acting as a frictionless pivot, and one plastic hinge forming within each span. Similar mechanisms may form in over-hanging beams.

Potential locations for plastic hinges include supports, points of concentrated load, and points of cross-section change. The location of a plastic hinge in a beam with distributed load is often not well defined.

5.5.5 Methods of plastic analysis

The purpose of the methods of plastic analysis is to determine the ultimate load at which a collapse mechanism first forms. Thus, it is only this final mechanism condition which must be found, and any intermediate load conditions can be ignored. A further important simplification arises from the fact that in its collapse condition, the beam is a mechanism, and can be analysed by statics, without any of the difficulties associated with the elastic analysis of a statically indeterminate beam.

The basic method of plastic analysis is to assume the locations of a series of plastic hinges and to investigate whether the three conditions of equilibrium, mechanism and plasticity are satisfied. The equilibrium condition is that the bending moment distribution defined by the assumed plastic hinges must be in static equilibrium with the applied loads and reactions. This condition applies to all beams, elastic or plastic. The mechanism condition is that there must be a sufficient number of plastic and frictionless hinges for the beam to form a mechanism. This condition is usually satisfied directly by the choice of hinges. The plasticity condition is that the full plastic moment of every cross-section must not be exceeded, so that

$$-M_p \leq M \leq M_p. \tag{5.35}$$

If the assumed plastic hinges satisfy these three conditions, they are the correct ones, and they define the collapse mechanism of the beam, so that

$$\text{(Collapse mechanism)} \quad \text{satisfies} \quad \begin{pmatrix} \text{Equilibrium} \\ \text{Mechanism} \\ \text{Plasticity} \end{pmatrix} \tag{5.36}$$

The ultimate load can then be determined directly from the equilibrium conditions.

However, it is usually possible to assume more than one series of plastic hinges, and so while the assumed plastic hinges may satisfy the equilibrium and mechanism conditions, the plasticity condition (equation 5.35) may be violated. In this case, the load calculated from the equilibrium condition is greater than the true ultimate load ($Q_m > Q_{ult}$). This forms the basis of the mechanism method of plastic analysis

$$\begin{pmatrix} \text{Mechanism method} \\ Q_m \geq Q_{ult} \end{pmatrix} \quad \text{satisfies} \quad \begin{pmatrix} \text{Equilibrium} \\ \text{Mechanism} \end{pmatrix}. \tag{5.37}$$

which provides an upper-bound solution for the true ultimate load.

A lower bound solution ($Q_s \leq Q_{ult}$) for the true ultimate load can be obtained by reducing the loads and bending moments obtained by the mechanism method proportionally (which ensures that the equilibrium condition remains satisfied) until the plasticity condition is satisfied everywhere. These reductions decrease the number of plastic hinges, and so the mechanism condition is not satisfied. This is the statical method of plastic analysis

$$\begin{pmatrix} \text{Statical Method} \\ Q_s \leq Q_{ult} \end{pmatrix} \quad \text{satisfies} \quad \begin{pmatrix} \text{Equilibrium} \\ \text{Plasticity} \end{pmatrix} \tag{5.38}$$

If the upper and lower bound solutions obtained by the mechanism and statical methods coincide or are sufficiently close, the beam may be designed immediately. If, however, these bounds are not precise enough, the original series of hinges must be modified (and the bending moments determined in the statical analysis will provide some indication of how to do this), and the analysis repeated.

The use of the methods of plastic analysis is demonstrated in Sections 5.11.2 and 5.11.3 for the examples of the built-in beam and the propped cantilever shown in Figures 5.30a and 5.31a. A worked example of the plastic collapse analysis of a non-uniform beam is given in Section 5.12.14.

The limitations on the use of the method of plastic analysis in the EC3 strength design of beams are discussed in Section 5.6.2.

The application of the methods of plastic analysis to rigid-jointed frames is discussed in Section 8.3.5.5. This application can only be made provided the effects of any axial forces in the members are small enough to preclude any instability effects and provided any significant decreases in the full plastic moments due to axial forces are accounted for (see Section 7.2.2).

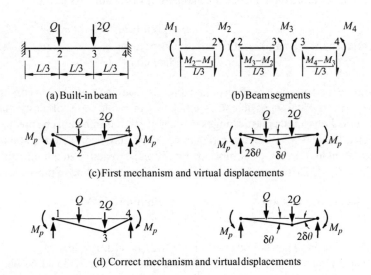

(a) Built-in beam   (b) Beam segments

(c) First mechanism and virtual displacements

(d) Correct mechanism and virtual displacements

Figure 5.30 Plastic analysis of a built-in beam.

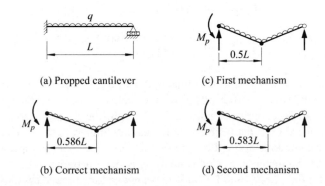

(a) Propped cantilever   (c) First mechanism

(b) Correct mechanism   (d) Second mechanism

Figure 5.31 Plastic analysis of a propped cantilever.

5.6 Strength design of beams

5.6.1 Elastic design of beams

5.6.1.1 General

For the elastic method of designing a beam for the strength limit state, the strength design loads are first obtained by multiplying the nominal loads (dead, imposed, or wind) by the appropriate load combination factors (see Section 1.5.6). The distributions of the design bending moments and shear forces in the beam under the strength design load combinations are determined by an elastic bending analysis (when the structure is statically indeterminate), or by statics (when it is statically determinate). EC3 also permits a moment redistribution to be made in each adequately braced Class 1 or Class 2 span of a continuous beam of up to 15% of the span's peak elastic moment, provided equilibrium is maintained. The beam must be designed to be able to resist the design moments and shears, as well as any concentrated forces arising from the factored loads and their reactions.

The processes of checking a specified member or of designing an unknown member for the design actions are summarised in Figure 5.32. When a specified

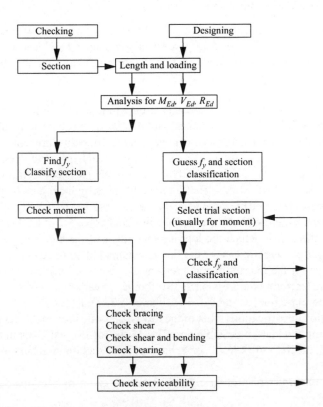

Figure 5.32 Flow chart for the elastic design of beams.

member is to be checked, then the cross-section should first be classified as Class 1, 2, 3, or 4. This allows the effective section modulus to be determined and the design section moment resistance to be compared with the maximum design moment. Following this, the lateral bracing should be checked, then the web shear resistance, and then combined bending and shear should be checked at any cross-sections where both the design moment and shear are high. Finally, the web bearing resistance should be checked at reactions and concentrated load points.

When the member size is not known, then a target value for the effective section modulus may be determined from the maximum design moment, and a trial section chosen whose effective section modulus exceeds this target. The remainder of the design process is then the same as that for checking a specified member. Usually the moment resistance governs the design, but when it doesn't, then an iterative process may need to be followed, as indicated in Figure 5.32.

The following sub-sections describe each of the EC3 checking processes summarised in Figure 5.32. A worked example of their application is given in Section 5.12.15.

5.6.1.2   Section classification

The moment, shear, and concentrated load bearing resistances of beams whose plate elements are slender may be significantly influenced by local buckling considerations (Chapter 4). Because of this, beam cross-sections are classified as Class 1, 2, 3, or 4, depending on the ability of the elements to resist local buckling (Section 4.7.2).

Class 1 sections are unaffected by local buckling and are able to develop and maintain their fully plastic resistances until a collapse mechanism forms. Class 2 sections are able to form a first plastic hinge, but local buckling prevents subsequent moment redistribution. Class 3 sections are able to reach the yield stress, but local buckling prevents full plastification of the cross-section. Class 4 sections have their resistances reduced below their first yield resistances by local buckling effects.

Sections are classified by comparing the slenderness $\lambda = (c/t)\sqrt{(f_y/235)}$ of each compression element with the appropriate limits of Table 5.2 of EC3. These limits depend on the way in which the longitudinal edges of the element are supported (either one edge supported as for the flange outstand of an I-section, or two edges supported as for the internal flange element of a box section), the bending stress distribution (uniform compression as in a flange, or varying stresses as in a web), and the type of section, as indicated in Figures 5.33 and 4.29.

The section classification is that of the lowest classification of its elements, with Class 1 being the highest possible and Class 4 the lowest. All UB's in S275 steel are Class 1, all UC's are Class 1 or 2, but welded I-section members may also be Class 3 or 4.

Worked examples of section classification are given in Sections 5.12.15 and 4.9.2–4.9.4.

Section description	Hot-rolled UB,UC Welded PWG		Welded box	RHS
Section and element widths	c_{f1} c_{f1} c_{f1} c_w	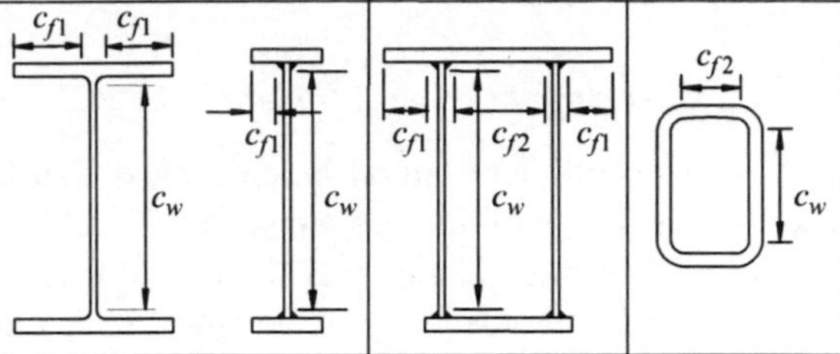 c_w	c_{f1} c_{f2} c_{f1} c_w	c_{f2} c_w

		Hot-rolled UB,UC Welded PWG	Welded box	RHS
Flange outstand c_{f1}	Class 1	9	9	–
	Class 2	10	10	–
	Class 3	14	14	–
Flange supported along both edges c_{f2}	Class 1	–	33	33
	Class 2	–	38	38
	Class 3	–	42	42
Web c_w	Class 1	72	72	72
	Class 2	83	83	83
	Class 3	124	124	124

Figure 5.33 Local buckling limits $\lambda = (c/t)/\varepsilon$ for some beam sections.

5.6.1.3 Checking the moment resistance

For the moment resistance check of a beam that has adequate bracing against lateral buckling, the inequality

$$M_{Ed} \leq M_{c,Rd} \tag{5.39}$$

must be satisfied, in which M_{Ed} is the maximum design moment, and

$$M_{c,Rd} = Wf_y/\gamma_{M0} \tag{5.40}$$

is the design section moment resistance, in which W is the appropriate section modulus, f_y the nominal yield strength for the section, and $\gamma_{M0}(=1)$ the partial factor for section resistance.

For Class 1 and Class 2 sections, the appropriate section modulus is the plastic section modulus W_{pl}, so that

$$W = W_{pl}, \tag{5.41}$$

For Class 3 sections, the appropriate section modulus may simply be taken as the minimum elastic section modulus $W_{el,min}$. For Class 4 sections, the appropriate section modulus is reduced below the elastic section modulus $W_{el,min}$, as discussed in Section 4.7.2.

Worked examples of checking the moment resistance are given in Sections 5.12.15 and 4.9.2–4.9.4, and a worked example of using the moment resistance check to design a suitable section is given in Section 5.12.16.

5.6.1.4   Checking the lateral bracing

Beams with insufficient lateral bracing must also be designed to resist lateral buckling. The design of beams against lateral buckling is discussed in Section 6.9. It is assumed in this chapter that there is sufficient bracing to prevent lateral buckling. While the adequacy of bracing for this purpose may be checked by determining the lateral buckling moment resistance as in Section 6.9, EC3 also gives simpler (and often conservative) rules for this purpose. For example, lateral buckling may be assumed to be prevented if the geometrical slenderness of each segment between restraints satisfies

$$\frac{L_c}{i_{f,z}} \leq \frac{37.56}{k_c} \frac{M_{c,Rd}}{M_{y,Ed}} \sqrt{\frac{235}{f_y}} \tag{5.42}$$

in which $i_{f,z}$ is the radius of gyration about the z axis of an equivalent compression flange, k_c is a correction factor which allows for the moment distribution, $M_{c,Rd}$ is the design resistance of a fully braced segment, and $M_{y,Ed}$ is the maximum design moment in the segment.

A worked example of checking the lateral bracing is given in Section 5.12.15.

5.6.1.5   Checking the shear resistance

For the shear resistance check, the inequality

$$V_{Ed} \leq V_{c,Rd} \tag{5.43}$$

must be satisfied, in which V_{Ed} is the maximum design shear force, and $V_{c,Rd}$ the design shear resistance.

For a stocky web with $d_w/t_w \leq 72\sqrt{(235/f_y)}$ and for which the elastic shear stress distribution is approximately uniform (as in the case of an equal flanged I-section), the uniform shear resistance $V_{c,Rd}$ is usually given by

$$V_{c,Rd} = A_v(f_y/\sqrt{3}) \tag{5.44}$$

in which $f_y/\sqrt{3}$ is the shear yield stress τ_y (Section 1.3.1) and A_v is the shear area of the web defined in Clause 6.2.6(2) of EC3. All the webs of UB's and UC's in S275 steel satisfy $d_w/t_w \leq 72\sqrt{(235/f_y)}$. A worked example of checking the uniform shear capacity of a compact web is given in Section 5.12.15.

For a stocky web with $d_w/t_w \leq 72\sqrt{(235/f_y)}$ and with a non-uniform elastic shear stress distribution (such as an unequal flanged I-section), the inequality

$$\tau_{Ed} \leq f_y/\sqrt{3} \tag{5.45}$$

must be satisfied, in which τ_{Ed} is the maximum design shear stress induced in the web by V_{Ed}, as determined by an elastic shear stress analysis (Section 5.4).

Webs for which $d_w/t_w \geq 72\sqrt{(235/f_y)}$ have their shear resistances reduced by local buckling effects, as discussed in Section 4.7.4. The shear design of thin webs is governed by EC3-1-5 [25]. Worked examples of checking the shear capacity of such webs are given in Sections 4.9.5 and 4.9.6.

5.6.1.6 Checking for bending and shear

High shear forces may reduce the ability of a web to resist bending moment, as discussed in Section 4.5. To allow for this, Class 1 and Class 2 beams under combined bending and shear, in addition to satisfying equations 5.39 and 5.43, must also satisfy

$$M_{Ed} \leq M_{c,Rd} \left\{ 1 - \left(\frac{2V_{Ed}}{V_{pl,Rd}} - 1 \right)^2 \right\} \tag{5.46}$$

in which V_{Ed} is the design shear at the cross-section being checked and $V_{pl,Rd}$ is the design plastic shear resistance. This condition need only be considered when $V_{Ed} > 0.5V_{pl,Rd}$, as indicated in Figure 4.32. It therefore need only be checked at cross-sections under high bending and shear, such as at the internal supports of continuous beams. Even so, many such beams have $V_{Ed} \leq 0.5V_{pl,Rd}$ and so this check for combined bending and shear rarely governs.

5.6.1.7 Checking the bearing resistance

For the bearing check, the inequality

$$F_{Ed} \leq F_{Rd} \tag{5.47}$$

must be satisfied at all supports and points of concentrated load. In this equation, F_{Ed} is the design-concentrated load or reaction, and F_{Rd} the bearing resistance of the web and its load-bearing stiffeners, if any. The bearing design of webs and load-bearing stiffeners is governed by EC3-1-5 [25] and discussed in Sections 4.6 and 4.7.6. Stocky webs often have sufficient bearing resistance and do not require the addition of load-bearing stiffeners.

A worked example of checking the bearing resistance of a stocky unstiffened web is given in Section 5.12.15, while a worked example of checking the bearing resistance of a slender web with load-bearing stiffeners is given in Section 4.9.9.

5.6.2 Plastic design of beams

In the plastic method of designing for the strength limit state, the distributions of bending moment and shear force in a beam under factored loads are determined by a plastic bending analysis (Section 5.5.5). Generally, the material properties for plastic design should be limited to ensure that the stress–strain curve has an adequate plastic plateau and exhibits strain-hardening, so that plastic hinges can be formed and maintained until the plastic collapse mechanism is fully developed. The use of the plastic method is restricted to beams which satisfy certain limitations on cross-section form and lateral bracing (additional limitations are imposed on members with axial forces, as discussed in Section 7.2.4.2).

EC3 permits only Class 1 sections which are symmetrical about the axis perpendicular to the axis of bending to be used at plastic hinge locations, in order to ensure that local buckling effects (Sections 4.2 and 4.3) do not reduce the ability of the section to maintain the full plastic moment until the plastic collapse mechanism is fully developed, while Class 1 or Class 2 sections should be used elsewhere. The slenderness limits for Class 1 and Class 2 beam flanges and webs are shown in Figure 5.33.

The spacing of lateral braces is limited in plastic design to ensure that inelastic lateral buckling (see Section 6.3) does not prevent the complete development of the plastic collapse mechanism. EC3 requires the spacing of braces at both ends of a plastic hinge segment to be limited. A conservative approximation for the spacing limit for sections with $h/t_f \leq 40\sqrt{(235/f_y)}$ is given by

$$L_{stable} \leq 35 i_z \sqrt{(235/f_y)} \tag{5.48}$$

in which i_z is the radius of gyration about the minor z axis, although higher values may be calculated by allowing for the effects of moment gradient.

The resistance of a beam designed plastically is based principally on its plastic collapse load Q_{ult} calculated by plastic analysis using the full plastic moment M_p of the beam. The plastic collapse load must satisfy $Q_{Ed} \leq Q_{ult}$ in which Q_{Ed} is the corresponding design load.

The shear capacity of a beam analysed plastically is based on the fully plastic shear capacity of the web (see Section 4.3.2.1). Thus the maximum shear V_{Ed} determined by plastic analysis under the design loads is limited to

$$V_{Ed} \leq V_{pl,Rd} = A_v f_y / \sqrt{3} \tag{5.49}$$

EC3 also requires web stiffeners to be provided near all plastic hinge locations where an applied load or reaction exceeds 10% of the above limit.

The ability of a beam web to transmit concentrated loads and reactions is discussed in Section 4.6.2. The web should be designed as in Section 4.7.6 using the design loads and reactions calculated by plastic analysis.

A worked example of the plastic design of a beam is given in Section 5.12.17.

5.7 Serviceability design of beams

The design of a beam is often governed by the serviceability limit state, for which the behaviour of the beam should be so limited as to give a high probability that the beam will provide the serviceability necessary for it to carry out its intended function. The most common serviceability criteria are associated with the stiffness of the beam, which governs its deflections under load. These may need to be limited so as to avoid a number of undesirable situations: an unsightly appearance; cracking or distortion of elements fixed to the beam such as cladding, linings, and partitions; interference with other elements such as crane girders; or vibration under dynamic loads such as traffic, wind, or machinery loads.

A serviceability design should be carried out by making appropriate assumptions for the load types, combinations, and levels; for the structural response to load; and for the serviceability criteria. Because of the wide range of serviceability limit states, design rules are usually of limited extent. The rules given are often advisory rather than mandatory because of the uncertainties as to the appropriate values to be used. These uncertainties result from the variable behaviour of real structures, and what is often seen to be the non-catastrophic (in terms of human life) nature of serviceability failures.

Serviceability design against unsightly appearance should be based on the total sustained load, and so should include the effects of dead and long-term imposed loads. The design against cracking or distortion of elements fixed to a beam should be based on the loads imposed after their installation, and will often include the unfactored imposed load or wind load, but exclude the dead load. The design against interference may need to include the effects of dead load as well as of imposed and wind loads. Where imposed and wind loads act simultaneously, it may be appropriate to multiply the unfactored loads by combination factors such as 0.8. The design against vibration will require an assessment to be made of the dynamic nature of the loads.

Most stiffness serviceability limit states are assessed by using an elastic analysis to predict the static deflections of the beam. For vibration design, it may be sufficient in some cases to represent the dynamic loads by static equivalents and carry out an analysis of the static deflections. More generally, however, a dynamic elastic analysis will need to be made.

The serviceability deflection limits depend on the criterion being used. To avoid unsightly appearance, a value of $L/200$ ($L/180$ for cantilevers) might be used after making allowances for any pre-camber. Values as low as $L/360$ have been used for design against the cracking of plaster finishes, while a limit of $L/600$ has been used for crane gantry girders. A value of $H/300$ has been used for horizontal storey drift in buildings under wind load.

A worked example of the serviceability design is given in Section 5.12.18.

5.8 Appendix – bending stresses in elastic beams

5.8.1 Bending in a principal plane

When a beam bends in the xz principal plane, plane cross-sections rotate as shown in Figure 5.4, so that sections δx apart become inclined to each other at $-(\mathrm{d}^2 w/\mathrm{d}x^2)\delta x$, where w is the deflection in the z principal direction as shown in Figure 5.5, and δx the length along the centroidal axis between the two cross-sections. The length between the two cross-sections at a distance z from the axis is greater than δx by $z(-\mathrm{d}^2 w/\mathrm{d}x^2)\delta x$, so that the longitudinal strain is

$$\varepsilon = -z\frac{\mathrm{d}^2 w}{\mathrm{d}x^2}$$

The corresponding tensile stress $\sigma = E\varepsilon$ is

$$\sigma = -Ez\frac{\mathrm{d}^2 w}{\mathrm{d}x^2} \tag{5.50}$$

which has the stress resultants

$$N = \int_A \sigma \mathrm{d}A = 0$$

since $\int_A z\mathrm{d}A = 0$ for centroidal axes (see Section 5.9), and

$$M_y = \int_A \sigma z \mathrm{d}A$$

whence

$$M_y = -EI_y\frac{\mathrm{d}^2 w}{\mathrm{d}x^2} \tag{5.2}$$

since $\int_A z^2 \mathrm{d}A = I_y$ (see Section 5.9). If this is substituted into equation 5.50, then the bending tensile stress σ is obtained as

$$\sigma = \frac{M_y z}{I_y}. \tag{5.3}$$

5.8.2 Alternative formulation for biaxial bending

When bending moments M_{y1}, M_{z1} acting about non-principal axes y_1, z_1 cause curvatures $\mathrm{d}^2 w_1 / \mathrm{d}x^2$, $\mathrm{d}^2 v_1 / \mathrm{d}x^2$ in the non-principal planes xz_1, xy_1, the stress σ at a point y_1, sz_1 is given by

$$\sigma = -Ez_1 \mathrm{d}^2 w_1 / \mathrm{d}x^2 - Ey_1 \mathrm{d}^2 v_1 / \mathrm{d}x^2 \tag{5.51}$$

and the moment resultants of these stresses are

and

$$\left. \begin{aligned} M_{y1} &= -EI_{y1}\mathrm{d}^2 w_1 / \mathrm{d}x^2 - EI_{y1z1}\mathrm{d}^2 v_1 / \mathrm{d}x^2 \\ M_{z1} &= EI_{y1z1}\mathrm{d}^2 w_1 / \mathrm{d}x^2 + EI_{z1}\mathrm{d}^2 v_1 / \mathrm{d}x^2 \end{aligned} \right\} \tag{5.52}$$

These can be rearranged as

and

$$\left. \begin{aligned} -E\frac{\mathrm{d}^2 w_1}{\mathrm{d}x^2} &= \frac{M_{y1}I_{z1} + M_{z1}I_{y1z1}}{I_{y1}I_{z1} - I_{y1z1}^2} \\ E\frac{\mathrm{d}^2 v_1}{\mathrm{d}x^2} &= \frac{M_{y1}I_{y1z1} + M_{z1}I_{y1}}{I_{y1}I_{z1} - I_{y1z1}^2} \end{aligned} \right\} \tag{5.53}$$

which are much more complicated than the principal plane formulations of equations 5.9.

If equations 5.53 are substituted, then equation 5.51 can be expressed as

$$\sigma = \left(\frac{M_{y1}I_{z1} + M_{z1}I_{y1z1}}{I_{y1}I_{z1} - I_{y1z1}^2} \right) z_1 - \left(\frac{M_{y1}I_{y1z1} + M_{z1}I_{y1}}{I_{y1}I_{z1} - I_{y1z1}^2} \right) y_1 \tag{5.54}$$

which is much more complicated than the principal plane formulation of equation 5.12.

5.9 Appendix – thin-walled section properties

The cross-section properties of many steel beams can be analysed with sufficient accuracy by using the thin-walled assumption demonstrated in Figure 5.34, in which the cross-section is replaced by its mid-thickness line. While the thin-walled assumption may lead to small errors in some properties due to junction and thickness effects, its consistent use will avoid anomalies that arise when a more accurate theory is combined with thin-walled theory, as might be the case in the determination of shear flows. The small errors are typically of the same order as those introduced by the common practice of ignoring the fillets or corner radii of hot-rolled sections.

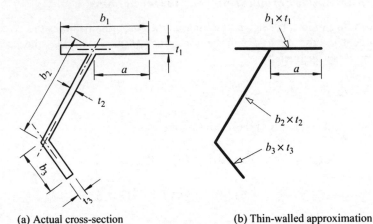

(a) Actual cross-section   (b) Thin-walled approximation

Figure 5.34 Thin-walled cross-section assumption.

The area A of a thin-walled section composed of a series of thin rectangles b_n wide and t_n thick ($b_n \gg t_n$) is given by

$$A = \int_A \mathrm{d}A = \sum_n A_n \tag{5.55}$$

in which $A_n = b_n t_n$. This approximation may make small errors of the order of $(t/b)^2$ at some junctions.

The centroid of a cross-section is defined by

$$\int_A y\mathrm{d}A = \int_A z\mathrm{d}A = 0$$

and for a thin-walled section these conditions become

$$\sum_n A_n y_n = \sum_n A_n z_n = 0 \tag{5.56}$$

in which y_n, z_n are the centroidal distances of the centre of the rectangular element, as demonstrated in Figure 5.35a. The position of the centroid can be found by adopting any convenient initial set of axes y_i, z_i as shown in Figure 5.35a, so that

$$y_n = y_{in} - y_{ic}$$
$$z_n = z_{in} - z_{ic}$$

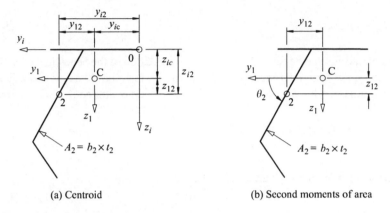

(a) Centroid   (b) Second moments of area

Figure 5.35 Calculation of cross-section properties.

When these are substituted into equation 5.56, the centroid position is found from

$$
\left.
\begin{aligned}
y_{ic} &= \sum A_n y_{in}/A \\
z_{ic} &= \sum A_n z_{in}/A
\end{aligned}
\right\}
\tag{5.57}
$$

The second moments of area about the centroidal axes are defined by

$$
\left.
\begin{aligned}
I_y &= \int_A z^2 \, dA \;=\; \sum (A_n z_n^2 + I_n \sin^2 \theta_n) \\
I_z &= \int_A y^2 \, dA \;=\; \sum (A_n y_n^2 + I_n \cos^2 \theta_n) \\
I_{yz} &= \int_A yz \, dA \;=\; \sum (A_n y_n z_n + I_n \sin \theta_n \cos \theta_n)
\end{aligned}
\right\}
\tag{5.58}
$$

in which θ_n is always the angle from the y axis to the rectangular element (positive from the y axis to the z axis), as shown in Figure 5.35b, and

$$
I_n = b_n^3 t_n/12.
\tag{5.59}
$$

These thin-walled approximations ignore small terms $b_n t_n^3/12$, while the use of the angle θ_n adjusts for the inclination of the element to the y, z axes.

The principal axis directions y, z are defined by the condition that

$$I_{yz} = 0. \tag{5.60}$$

These directions can be found by considering a rotation of the centroidal axes from y_1, z_1 to y_2, z_2 through an angle α, as shown in Figure 5.36a. The y_2, z_2 coordinates of any point are related to its y_1, z_1 coordinates as demonstrated in Figure 5.36a by

$$\left. \begin{array}{l} y_2 = y_1 \cos\alpha + z_1 \sin\alpha \\ z_2 = -y_1 \sin\alpha + z_1 \cos\alpha \end{array} \right\} \tag{5.61}$$

The second moments of area about the y_2, z_2 axes are

$$\left. \begin{array}{ll} I_{y2} &= \displaystyle\int_A z_2^2 \, dA = I_{y1} \cos^2\alpha - 2I_{y1z1} \sin\alpha \cos\alpha + I_{z1} \sin^2\alpha \\[2mm] I_{z2} &= \displaystyle\int_A y_2^2 \, dA = I_{y1} \sin^2\alpha + 2I_{y1z1} \sin\alpha \cos\alpha + I_{z1} \cos^2\alpha \\[2mm] I_{y2z2} &= \displaystyle\int_A y_2 z_2 \, dA = \dfrac{1}{2}(I_{y1} - I_{z1}) \sin 2\alpha + I_{y1z1} \cos 2\alpha \end{array} \right\} \tag{5.62}$$

The principal axis condition of equation 5.60 requires $I_{y2z2} = 0$, so that

$$\tan 2\alpha = -2I_{y1z1}/(I_{y1} - I_{z1}). \tag{5.63}$$

In this book, y and z are reserved exclusively for the principal centroidal axes.

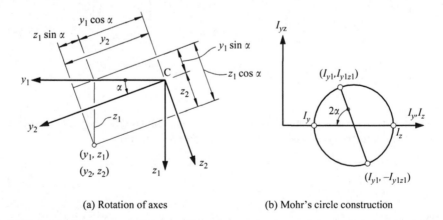

(a) Rotation of axes   (b) Mohr's circle construction

Figure 5.36 Calculation of principal axis properties.

The principal axis second moments of area can be obtained by using this value of α in equation 5.62, whence

$$\left. \begin{aligned} I_y &= \frac{1}{2}(I_{y1} + I_{z1}) + (I_{y1} - I_{z1})/2 \cos 2\alpha \\ I_z &= \frac{1}{2}(I_{y1} + I_{z1}) - (I_{y1} - I_{z1})/2 \cos 2\alpha \end{aligned} \right\} \tag{5.64}$$

These results are summarised by the Mohr's circle constructed in Figure 5.36b on the diameter whose ends are defined by (I_{y1}, I_{y1z1}) and $(I_{z1}, -I_{y1z1})$. The angles between this diameter and the horizontal axis are equal to 2α and $(2\alpha + 180°)$, while the intersections of the circle with the horizontal axis define the principal axis second moments of area I_y, I_z.

Worked examples of the calculation of cross-section properties are given in Sections 5.12.1–5.12.4.

5.10 Appendix – shear stresses in elastic beams

5.10.1 Solid cross-sections

A solid cross-section beam with a shear force V_z parallel to the z principal axis is shown in Figure 5.9a. It is assumed that the resulting shear stresses also act parallel to the z axis and are constant across the width b of the section. The corresponding horizontal shear stresses $\tau_{zx} = \tau_v$ are in equilibrium with the horizontal bending stresses σ. For the element $b \times \delta z \times \delta x$ shown in Figure 5.9b, this equilibrium condition reduces to

$$b\frac{\partial \sigma}{\partial x}\delta x \delta z + \frac{\partial (\tau_v b)}{\partial z}\delta z \delta x = 0$$

whence

$$\tau_v = -\frac{1}{b_2}\int_{z_T}^{z_2} b\frac{\partial \sigma}{\partial x}\mathrm{d}z, \tag{5.65}$$

which satisfies the condition that the shear stress is zero at the unstressed boundary z_T. The bending normal stresses σ are given by

$$\sigma = \frac{M_y z}{I_y}$$

and the shear force V_z is

$$V_z = \frac{\mathrm{d}M_y}{\mathrm{d}x}$$

and so

$$\frac{\partial \sigma}{\partial x} = \frac{V_z z}{I_y}. \tag{5.66}$$

Substituting this into equation 5.65 leads to

$$\tau_v = -\frac{V_z}{I_y b_2} \int_{z_T}^{z_2} bz \, dz. \tag{5.13}$$

5.10.2   Thin-walled open cross-sections

A thin-walled open cross-section beam with a shear force V_z parallel to the z principal axis is shown in Figure 5.11a. It is assumed that the resulting shear stresses act parallel to the walls of the section and are constant across the wall thickness. The corresponding horizontal shear stresses τ_v are in equilibrium with the horizontal bending stresses σ. For the element $t \times \delta s \times \delta x$ shown in Figure 5.11b, this equilibrium condition reduces to

$$\frac{\partial \sigma}{\partial x} \delta x t \delta s + \frac{\partial (\tau_v t)}{\partial x} \delta s \delta x = 0,$$

and substitution of equation 5.66 and integration leads to

$$\tau_v t = -\frac{V_z}{I_y} \int_0^s zt \, ds, \tag{5.16}$$

which satisfies the condition that the shear stress is zero at the unstressed boundary $s = 0$.

The direction of the shear stress can be determined from the sign of the right-hand side of equation 5.16. If this is positive, then the direction of the shear stress is the same as the positive direction of s, and vice versa. The direction of the shear stress may also be determined from the direction of the corresponding horizontal shear force required to keep the out-of-balance bending force in equilibrium, as shown for example in Figure 5.13a. Here a horizontal force $\tau_v t_f \delta x$ in the positive x direction is required to balance the resultant bending force $\delta \sigma \, \delta A$ (due to the shear force $V_z = dM_y/dx$), and so the direction of the corresponding flange shear stress is from the tip of the flange towards its centre.

5.10.3   Multi-cell thin-walled closed sections

The shear stress distributions in multi-cell closed sections can be determined by extending the method discussed in Section 5.4.4 for single-cell closed sections. If the section consists of n junctions connected by m walls, then there are $(m - n + 1)$ independent cells. In each wall there is an unknown circulating shear flow $\tau_{vc} t$.

There are $(m - n + 1)$ independent cell warping continuity conditions of the type (see equation 5.27)

$$\sum_{\text{cell}} \int_{\text{wall}} (\tau/G)\, ds = 0,$$

and $(n - 1)$ junction equilibrium equations of the type

$$\sum_{\text{junction}} (\tau t) = 0.$$

These equations can be solved simultaneously for the m circulating shear flows $\tau_{vc}t$.

5.11   Appendix – plastic analysis of beams

5.11.1   Full plastic moment

The fully plastic stress distribution in a section bent about a y axis of symmetry is antisymmetric about that axis, as shown in Figure 5.37a for a channel section example. The plastic neutral axis, which divides the cross-section into two equal areas so that there is no axial force resultant of the stress distribution, coincides with the axis of symmetry. The moment resultant of the fully plastic stress distribution is

$$M_{p,y} = f_y \sum_{T} A_n z_n - f_y \sum_{C} A_n z_n \tag{5.67}$$

in which the summations $\sum_T$, $\sum_C$ are carried out for the tension and compression areas of the cross-section, respectively. Note that z_n is negative for the compression areas of the channel section, so that the two summations in

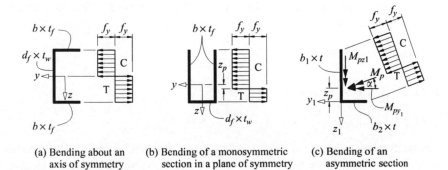

(a) Bending about an   (b) Bending of a monosymmetric   (c) Bending of an
    axis of symmetry   section in a plane of symmetry   asymmetric section

Figure 5.37 Full plastic moment.

equation 5.67 are additive. The plastic section modulus $W_{pl,y} = M_{p,y}/f_y$ is obtained from equation 5.67 as

$$W_{pl,y} = \sum_T A_n z_n - \sum_C A_n z_n. \tag{5.68}$$

A worked example of the determination of the full plastic moment of a doubly symmetric I-section is given in Section 5.12.12.

The fully plastic stress distribution in a monosymmetric section bent in a plane of symmetry is not antisymmetric, as can be seen in Figure 5.37b for a channel section example. The plastic neutral axis again divides the cross-section into two equal areas so that there is no axial force resultant, but the neutral axis no longer passes through the centroid of the cross-section. The fully plastic moment $M_{p,y}$ and the plastic section modulus $W_{pl,y}$ are given by equations 5.67 and 5.68. For convenience, any suitable origin may be used for computing z_n. A worked example of the determination of the full plastic moment of a monosymmetric tee-section is given in Section 5.12.13.

The fully plastic stress distribution in an asymmetric section is also not antisymmetric, as can be seen in Figure 5.37c for an angle section example. The plastic neutral axis again divides the cross-section into two equal areas. The direction of the plastic neutral axis is perpendicular to the plane in which the moment resultant acts. The full plastic moment is given by the vector addition of the moment resultants about any convenient y_1, z_1 axes, so that

$$M_p = \sqrt{(M_{p,y1}^2 + M_{p,z1}^2)} \tag{5.69}$$

where $M_{p,y1}$ is given by an equation similar to equation 5.67, and $M_{p,z1}$ is given by

$$M_{p,z1} = -f_y \sum_T A_n y_{1n} + f_y \sum_C A_n y_{1n}. \tag{5.70}$$

The inclination α of the plastic neutral axis to the y_1 axis (and of the plane of the full plastic moment to the zx plane) can be obtained from

$$\tan \alpha = M_{p,z1}/M_{p,y1}. \tag{5.71}$$

5.11.2   Built-in beam

5.11.2.1   General

The built-in beam shown in Figure 5.30a has two redundancies, and so three plastic hinges must form before it can be reduced to a collapse mechanism. In this case the only possible plastic hinge locations are at the support points 1 and 4 and at the load points 2 and 3, and so there are two possible mechanisms, one with a hinge at point 2 (Figure 5.30c), and the other with a hinge at point 3 (Figure 5.30d). Three different methods of analysing these mechanisms are presented in the following sub-sections.

5.11.2.2   Equilibrium equation solution

The equilibrium conditions for the beam express the relationships between the applied loads and the moments at these points. These relationships can be obtained by dividing the beam into the segments shown in Figure 5.30b and expressing the end shears of each segment in terms of its end moments, so that

$$V_{12} = (M_2 - M_1)/(L/3)$$

$$V_{23} = (M_3 - M_2)/(L/3)$$

$$V_{34} = (M_4 - M_3)/(L/3).$$

For equilibrium, each applied load must be equal to the algebraic sum of the shears at the load point in question, and so

$$Q = V_{12} - V_{23} = \frac{1}{L}(-3M_1 + 6M_2 - 3M_3)$$

$$2Q = V_{23} - V_{34} = \frac{1}{L}(-3M_2 + 6M_3 - 3M_4)$$

which can be rearranged as

$$\left. \begin{array}{l} 4QL = -6M_1 + 9M_2 - 3M_4 \\ 5QL = -3M_1 + 9M_3 - 6M_4 \end{array} \right\}. \tag{5.72}$$

If the first mechanism chosen (incorrectly, as will be shown later) is that with plastic hinges at points 1, 2, and 4 (see Figure 5.30c), so that

$$-M_1 = M_2 = -M_4 = M_p$$

then substitution of this into the first of equations 5.72 leads to $4QL = 18M_p$ and so

$$Q_{ult} \le 4.5M_p/L.$$

If this result is substituted into the second of equations 5.72, then

$$M_3 = 1.5\, M_p > M_p,$$

and the plasticity condition is violated. The statical method can now be used to obtain a lower bound by reducing Q until $M_3 = M_p$, whence

$$Q = \frac{4.5M_p}{L} \bigg/ \frac{M_3}{M_p} = \frac{3M_p}{L}$$

and so

$$3M_p/L < Q_{ult} < 4.5M_p/L.$$

The true collapse load can be obtained by assuming a collapse mechanism with plastic hinges at point 3 (the point of maximum moment for the previous mechanism) and at points 1 and 4. For this mechanism

$$-M_1 = M_3 = -M_4 = M_p,$$

and so, from the second of equations 5.72,

$$Q_{ult} \leq 3.6M_p/L.$$

If this result is substituted into the first of equations 5.72, then

$$M_2 = 0.6M_p < M_p,$$

and so the plasticity condition is also satisfied. Thus

$$Q_{ult} = 3.6M_p/L.$$

5.11.2.3   Graphical solution

A collapse mechanism can be analysed graphically by first plotting the known free moment diagram (for the applied loading on a statically determinate beam), and then the reactant bending moment diagram (for the redundant reactions) which corresponds to the collapse mechanism.

For the built-in beam (Figure 5.38a), the free moment diagram for a simply supported beam is shown in Figure 5.38c. It is obtained by first calculating the left-hand reaction as

$$R_L = (Q \times 2L/3 + 2Q \times L/3)/L = 4Q/3$$

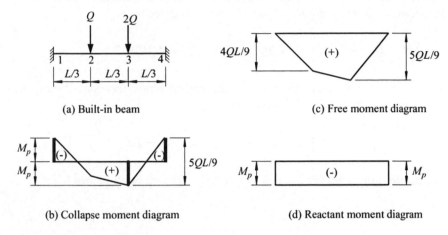

(a) Built-in beam

(c) Free moment diagram

(b) Collapse moment diagram

(d) Reactant moment diagram

Figure 5.38 Graphical solution for the plastic collapse of a built-in beam.

and then using this to calculate the moments

$$M_2 = (4Q/3) \times L/3 = 4QL/9$$

and

$$M_3 = (4Q/3) \times 2L/3 - QL/3 = 5QL/9$$

Both of the possible collapse mechanisms (Figure 5.29c and d) require plastic hinges at the supports, and so the reactant moment diagram is one of uniform negative bending, as shown in Figure 5.38d. When this is combined with the free moment diagram as shown in Figure 5.38b, it becomes obvious that there must be a plastic hinge at point 3 and not at point 2, and that at collapse

$$2M_p = 5Q_{ult}L/9$$

so that

$$Q_{ult} = 3.6M_p/L$$

as before.

5.11.2.4   *Virtual work solution*

The virtual work principle [4, 7, 8] can also be used to analyse each mechanism. For the first mechanism shown in Figure 5.30c, the virtual external work done by the forces Q, $2Q$ during the incremental virtual displacements defined by the incremental virtual rotations $\delta\theta$ and $2\delta\theta$ shown is given by

$$\delta W = Q \times (\delta\theta \times 2L/3) + 2Q \times (\delta\theta \times L/3) = 4QL\delta\theta/3$$

while the internal work absorbed at the plastic hinge locations during the incremental virtual rotations is given by

$$\delta U = M_p \times 2\delta\theta + M_p \times (2\delta\theta + \delta\theta) + M_p \times \delta\theta = 6M_p\delta\theta.$$

For equilibrium of the mechanism, the virtual work principle requires

$$\delta W = \delta U$$

so that the upper-bound estimate of the collapse load is

$$Q_{ult} \leq 4.5M_p/L$$

as in Section 5.11.2.2.

  Similarly, for the second mechanism shown in Figure 5.30d,

$$\delta W = 5QL\delta\theta/3$$
$$\delta U = 6M_p\delta\theta$$

so that

$$Q_{ult} \leq 3.6M_p/L$$

once more.

5.11.3   Propped cantilever

5.11.3.1   General

The exact collapse mechanisms for beams with distributed loads are not always as easily obtained as those for beams with concentrated loads only, but sufficiently accurate solutions can usually be found without great difficulty. Three different methods of determining these mechanisms are presented in the following sub-sections for the propped cantilever shown in Figure 5.31.

5.11.3.2 Equilibrium equation solution

If the plastic hinge is assumed to be at the mid-span of the propped cantilever as shown in Figure 5.31c (the other plastic hinge must obviously be located at the built-in support), then equilibrium considerations of the left and right segments of the beam allow the left and right reactions to be determined from

$$R_L L/2 = 2M_p + (qL/2)L/4$$

and

$$R_R L/2 = M_p + (qL/2)L/4$$

while for overall equilibrium

$$qL = R_L + R_R.$$

Thus

$$qL = (4M_p/L + qL/4) + (2M_p/L + qL/4)$$

so that

$$q_{ult} \leq 12M_p/L^2.$$

The left-hand reaction is therefore given by

$$R_L = 4M_p/L + (12M_p/L^2)L/4 = 7M_p/L.$$

The bending moment variation along the propped cantilever is therefore given by

$$M = -M_p + (7M_p/L)x - (12M_p/L^2)x^2/2$$

which has a maximum value of $25M_p/24 > M_p$ at $x = 7L/12$. Thus the lower bound is given by

$$q_{ult} \geq (12M_p/L^2)/(25/24) \approx 11.52M_p/L^2$$

and so

$$\frac{11.52M_p}{L^2} < q_{ult} < \frac{12M_p}{L^2}.$$

If desired, a more accurate solution can be obtained by assuming that the hinge is located at a distance $7L/12(= 0.583L)$ from the built-in support (i.e. at the point of maximum moment for the previous mechanism), whence

$$\frac{11.6567M_p}{L^2} < q_{ult} < \frac{11.6575M_p}{L^2}.$$

These are very close to the exact solution of

$$q_{ult} = (6+4\sqrt{2})M_p/L^2 \approx 11.6569M_p/L^2$$

for which the hinge is located at a distance $(2 - \sqrt{2})L \approx 0.5858L$ from the built-in support.

5.11.3.3   Graphical solution

For the propped cantilever of Figure 5.31, the parabolic free moment diagram for a simply supported beam is shown in Figure 5.39c. All the possible collapse mechanisms have a plastic hinge at the left-hand support and a frictionless hinge at the right-hand support, so that the reactant moment diagram is triangular, as shown in Figure 5.39d. A first attempt at combining this with the free moment diagram so as to produce a trial collapse mechanism is shown in Figure 5.39b. For this mechanism, the interior plastic hinge occurs at mid-span, so that

$$M_p + M_p/2 = qL^2/8$$

which leads to

$$q = 12M_p/L^2$$

as before.

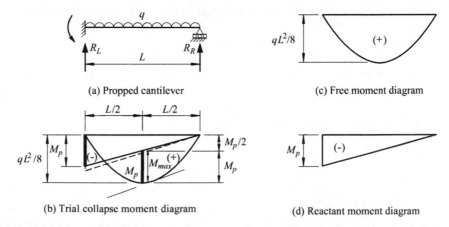

(a) Propped cantilever

(c) Free moment diagram

(b) Trial collapse moment diagram

(d) Reactant moment diagram

Figure 5.39 Graphical analysis of the plastic collapse of a propped cantilever.

Figure 5.39b indicates that in this case, the maximum moment occurs to the right of mid-span, and is greater than M_p, so that the solution above is an upper bound. A more accurate solution can be obtained by trial and error, by increasing the slope of the reactant moment diagram until the moment of the left-hand support is equal to the interior maximum moment, as indicated by the dashed line in Figure 5.39b.

5.11.3.4   Virtual work solution

For the first mechanism for the propped cantilever shown in Figure 5.31c, for virtual rotations $\delta\theta$,

$$\delta W = qL^2\delta\theta/4$$
$$\delta U = 3M_p\delta\theta$$

and

$$q_{ult} \leq 12M_p/L^2$$

as before.

5.12   Worked examples

5.12.1   Example 1 – properties of a plated UB section

Problem. The 610×229 UB 125 section shown in Figure 5.40a is strengthened by welding a 300 mm $\times$ 20 mm plate to each flange. Determine the section properties I_y and $W_{el,y}$.

Solution. Because of the symmetry of the section, the centroid of the plated UB is at the web centre. Adapting equation 5.58,

$$I_y = 98\,610 + 2 \times 300 \times 20 \times [(612.2 + 20)/2]^2/10^4 = 218\,500\,\text{cm}^4$$

$$W_{el,y} = 218\,500/[(612.2 + 2 \times 20)/(2 \times 10)] = 6701\,\text{cm}^3$$

5.12.2   Example 2 – properties of a tee-section

Problem. Determine the section properties I_y and $W_{el,y}$ of the welded tee-section shown in Figure 5.40b.

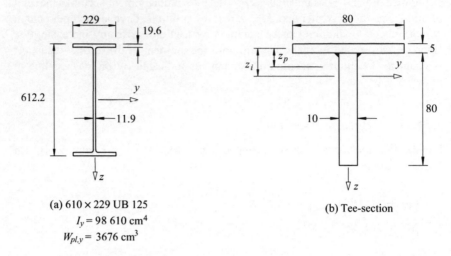

(a) 610 × 229 UB 125
$I_y = 98\ 610\ \text{cm}^4$
$W_{pl,y} = 3676\ \text{cm}^3$

(b) Tee-section

Figure 5.40 Worked examples 1, 2, and 18.

Solution. Using the thin-walled assumption and equations 5.57 and 5.58,

$$z_i = \frac{(80 + 5/2) \times 10 \times (80 + 5/2)/2}{(80 \times 5) + (80 + 5/2) \times 10} = 27.8\ \text{mm}$$

$$I_y = (80 \times 5) \times 27.8^2 + (82.5^3 \times 10/12) + (82.5 \times 10)$$
$$\times\ (82.5/2 - 27.8)^2\ \text{mm}^4$$
$$= 92.63\ \text{cm}^4$$

$$W_{el,y} = 92.63/\{(8.25 - 2.78)/10\} = 16.93\ \text{cm}^3$$

Alternatively, the equations of Figure 5.6 may be used.

5.12.3   Example 3 – properties of an angle section

Problem. Determine the properties $A, I_{y1}, I_{z1}, I_{y1z1}, I_y, I_z$, and α of the thin-walled angle section shown in Figure 5.41c.

Solution. Using the thin-walled assumption, the angle section is replaced by two rectangles 142.5(= 150 − 15/2) × 15 and 82.5(= 90 − 15/2) × 15, as shown in

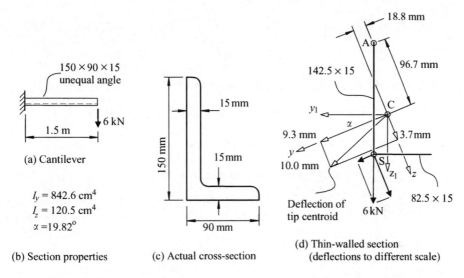

(a) Cantilever

150 × 90 × 15 unequal angle

6 kN

1.5 m

$I_y = 842.6$ cm^4
$I_z = 120.5$ cm^4
$\alpha = 19.82°$

(b) Section properties

(c) Actual cross-section

15 mm

15 mm

150 mm

90 mm

18.8 mm

142.5 × 15

96.7 mm

y_1

α

C

9.3 mm

3.7mm

y

10.0 mm

S

z_1 z

Deflection of tip centroid

6 kN

82.5 × 15

(d) Thin-walled section (deflections to different scale)

Figure 5.41 Worked examples 3 and 5.

Figure 5.41d. If the initial origin is taken at the intersection of the legs and the initial y_i, z_i axes parallel to the legs, then using equations 5.55 and 5.57,

$$A = (142.5 \times 15) + (82.5 \times 15) = 3375 \text{ mm}^2$$

$$y_{ic} = \{(142.5 \times 15) \times 0 + (82.5 \times 15) \times (-82.5/2)\}/3375$$
$$= -15.1 \text{ mm}$$

$$z_{ic} = \{(142.5 \times 15) \times (-142.5/2) + (82.5 \times 15) \times 0\}/3375$$
$$= -45.1 \text{ mm}$$

Transferring the origin to the centroid and using equations 5.58,

$$I_{y1} = (142.5 \times 15) \times (-142.5/2 + 45.1)^2 + (142.5^3 \times 15/12) \times \sin^2 90°$$

$$+ (82.5 \times 15) \times (45.1)^2 + (82.5^3 \times 15/12) \times \sin^2 180°$$

$$= 7.596 \times 10^6 \text{ mm}^4 = 759.6 \text{ cm}^4$$

$$I_{z1} = (142.5 \times 15) \times (15.1)^2 + (142.5^3 \times 15/12) \times \cos^2 90°$$

$$+ (82.5 \times 15) \times (-82.5/2 + 15.1)^2 + (82.5^3 \times 15/12) \times \cos^2 180°$$

$$= 2.035 \times 10^6 \text{ mm}^4 = 203.5 \text{ cm}^4$$

$$I_{y1z1} = (142.5 \times 15) \times (-142.5/2 + 45.1) \times 15.1 + (142.5^3 \times 15/12)$$

$$\times \sin 90° \cos 90° + (82.5 \times 15) \times (45.1) \times (-82.5/2 + 15.1)$$

$$+ (82.5^3 \times 15/12) \times \sin 180° \cos 180°$$

$$= -2.303 \times 10^6 \text{ mm}^4 = -230.3 \text{ cm}^4$$

Using equation 5.63,

$$\tan 2\alpha = -2 \times (-230.3)/(759.6 - 203.5) = 0.8283$$

whence $2\alpha = 39.63°$ and $\alpha = 19.82°$.
   Using Figure 5.30b, the Mohr's circle centre is at

$$(I_{y1} + I_{z1})/2 = (759.6 + 203.5)/2 = 481.6 \text{ cm}^4$$

and the Mohr's circle diameter is

$$\sqrt{\{(I_{y1} - I_{z1})^2 + (2I_{y1z1})^2\}} = \sqrt{\{(759.6 - 203.5)^2 + (2 \times 230.3)^2\}}$$

$$= 722.1 \text{ cm}^4$$

and so

$$I_y = 481.6 + 722.1/2 = 842.6 \text{ cm}^4$$

$$I_z = 481.6 - 722.1/2 = 120.5 \text{ cm}^4$$

5.12.4   Example 4 – properties of a zed section

Problem. Determine the properties I_{y1}, I_{z1}, I_{y1z1}, I_y, I_z, and α of the thin-walled Z-section shown in Figure 5.42a.
Solution.

$$I_{y1} = 2 \times (75 \times 10 \times 75^2) + (150^3 \times 5/12) = 9844 \times 10^3 \text{ mm}^4$$

$$= 984.4 \text{ cm}^4,$$

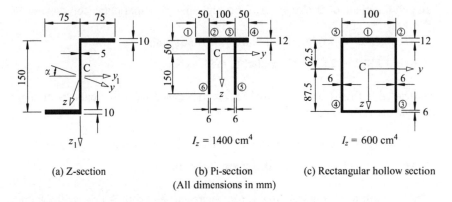

(a) Z-section

(b) Pi-section
(All dimensions in mm)

(c) Rectangular hollow section

$I_z = 1400 \text{ cm}^4$

$I_z = 600 \text{ cm}^4$

Figure 5.42 Worked examples 4, 9, 10, and 11.

$$I_{z1} = 2 \times (75 \times 10 \times 37.5^2) + 2 \times (75^3 \times 10/12)$$

$$= 2813 \times 10^3 \text{ mm}^4 = 281.3 \text{ cm}^4,$$

$$I_{y1z1} = 75 \times 10 \times 37.5 \times (-75) + 75 \times 10 \times (-37.5) \times 75$$

$$= -4219 \times 10^3 \text{ mm}^4 = -421.9 \text{ cm}^4.$$

The centre of the Mohr's circle (see Figure 5.37b) is at

$$\frac{I_{y1} + I_{z1}}{2} = \left[\frac{984.4 + 281.3}{2} \right] = 6328 \text{ cm}^4.$$

The diameter of the circle is

$$\sqrt{[(984.4 - 281.3)^2 + (2 \times 421.9)^2]} = 1098.3 \text{ cm}^4$$

and so

$$I_y = (632.8 + 1098.3/2) = 1182 \text{ cm}^4,$$

$$I_z = (632.8 - 1098.3/2) = 83.7 \text{ cm}^4,$$

$$\tan 2\alpha = \frac{2 \times 421.9}{(984.4 - 281.3)} = 1.200,$$

$$\alpha = 25.1°.$$

5.12.5 Example 5 – biaxial bending of an angle section cantilever

Problem. Determine the maximum elastic stress and deflection of the angle section cantilever shown in Figure 5.41 whose section properties were determined in Section 5.12.3.

Solution for maximum stress. The shear centre (see Section 5.4.3) load of 6 kN acting in the plane of the long leg of the angle has principal plane components of

$$6 \cos 19.82° = 5.645 \text{ kN parallel to the } z \text{ axis, and}$$

$$6 \sin 19.82° = 2.034 \text{ kN parallel to the } y \text{ axis,}$$

as indicated in Figure 5.41d. These load components cause principal axis moment components at the support of

$$M_y = -5.645 \times 1.5 = -8.467 \text{ kNm about the } y \text{ axis, and}$$

$$M_z = +2.034 \times 1.5 = +3.052 \text{ kNm about the } z \text{ axis.}$$

The maximum elastic bending stress occurs at the point A ($y_{1A} = 15.1$ mm, $z_{1A} = -97.4$ mm) shown in Figure 5.41d. The coordinates of this point can be obtained by using equation 5.61 as

$$y = 15.1 \cos 19.82° - 97.4 \sin 19.82° = -18.8 \text{ mm, and}$$

$$z = -15.1 \sin 19.82° - 97.4 \cos 19.82° = -96.7 \text{ mm.}$$

The maximum stress at A can be obtained by using equation 5.12 as

$$\sigma_{max} = \frac{(-8.467 \times 10^6) \times (-96.7)}{842.6 \times 10^4} - \frac{(3.052 \times 10^6) \times (-18.8)}{120.5 \times 10^4}$$

$$= 146.9 \text{ N/mm}^2$$

The stresses at the other leg ends should be checked to confirm that the maximum stress is at A.

Solution for maximum deflection. The maximum deflection occurs at the tip of the cantilever. Its components v and w can be calculated using $QL^3/3EI$. Thus

$$w = (5.645 \times 10^3) \times (1500)^3/(3 \times 210,000 \times 842.6 \times 10^4) = 3.6 \text{ mm,}$$

and

$$v = (2.034 \times 10^3) \times (1500)^3/(3 \times 210,000 \times 120.5 \times 10^4) = 9.0 \text{ mm.}$$

The resultant deflection can be obtained by vector addition as

$$\delta = \sqrt{(3.6^2 + 9.0^2)} = 9.7 \text{ mm}$$

Using non-principal plane properties. Problems of this type are sometimes incorrectly analysed on the basis that the plane of loading can be assumed to be a principal plane. When this approach is used, the maximum stress calculated for the cantilever is

$$\sigma_{max} = (6 \times 10^3) \times 1500 \times (97.4)/759.6 \times 10^4 = 115 \text{ N/mm}^2$$

which seriously underestimates the correct value by more than 20%. The correct non-principal plane formulations which are needed to solve the example in this way are given in Section 5.8.2. The use of these formulations should be avoided, since their excessive complication makes them very error-prone. The much simpler principal plane formulations of equations 5.9–5.12 should be used instead.

It should be noted that in this book y and z are reserved exclusively for the principal axis directions, instead of also being used for the directions parallel to the legs.

5.12.6 Example 6 – shear stresses in an I-section

Problem. Determine the shear flow distribution and the maximum shear stress in the I-section shown in Figure 5.12a.

Solution. Applying equation 5.16 to the left-hand half of the top flange,

$$\tau_v t_f = -(V_z/I_y) \int_0^{s_f} (-d_f/2)t_f \mathrm{d}s_f = (V_z/I_y)d_f\, t_f s_f/2,$$

which is linear, as shown in Figure 5.12b.

At the flange web junction, $(\tau_v t_f)_{12} = (V_z/I_y)d_f t_f b/4$, which is positive, and so the shear flow acts into the flange–web junction, as shown in Figure 5.12a. Similarly, for the right-hand half of the top flange, $(\tau_v t_f)_{23} = (V_z/I_y)d_f t_f b/4$, and again the shear flow acts in to the flange–web junction.

For horizontal equilibrium at the flange–web junction, the shear flow from the junction into the web is obtained from equation 5.20 as

$$(\tau_v t_w)_{24} = (V_z/I_y)d_f t_f b/4 + (V_z/I_y)d_f t_f b/4 = (V_z/I_y)d_f t_f b/2$$

Adapting equation 5.16 for the web,

$$\tau_v t_w = (V_z/I_y)(d_f t_f b/2) - (V_z/I_y)\int_0^{s_w}(-d_f/2 + s_w)t_w \mathrm{d}s_w$$

$$= (V_z/I_y)(d_f t_f b/2) + (V_z/I_y)(d_f s_w/2 - s_w^2/2)t_w$$

which is parabolic, as shown in Figure 5.12b. At the web centre, $(\tau_v t_w)_{d_f/2} = (V_z/I_y)(d_f t_f b/2 + d_f^2 t_w/8)$ which is the maximum shear flow.
At the bottom of the web,

$$(\tau_v t_w)_{d_f} = (V_z/I_y)(d_f t_f b/2 + d_f^2 t_w/2 - d_f^2 t_w/2)$$

$$= (V_z/I_y)(d_f t_f b/2) = (\tau_v t_w)_{24},$$

which provides a symmetry check.

The maximum shear stress is

$$(\tau_v)_{d_f/2} = (V_z/I_y)(d_f t_f b/2 + d_f^2 t_w/8)t_w$$

A vertical equilibrium check is provided by finding the resultant of the web shear flow as

$$V_w = \int_0^{d_f} (\tau_v t_w) ds_w$$

$$= (V_z/I_y)[d_f t_f b s_w/2 + d_f t_w s_w^2/4 - s_w^3 t_w/6]_0^{d_f}$$

$$= (V_z/I_y)(d_f^2 t_f b/2 + d_f^3 t_w/12)$$

$$= V_z$$

when $I_y = d_f^2 t_f b/2 + d_f^3 t_w/12$ is substituted.

5.12.7   Example 7 – shear stress in a channel section

Problem. Determine the shear flow distribution in the channel section shown in Figure 5.14.

Solution. Applying equation 5.15 to the top flange,

$$\tau_v t_f = -(V_z/I_y) \int_0^{s_f} (-d_f/2)t_f ds_f = (V_z/I_y)d_f t_f s_f/2$$

$$(\tau_v t_f)_b = (V_z/I_y)d_f t_f b/2$$

$$\tau_v t_w = (V_z/I_y)(d_f t_f b/2 - \int_0^{s_w} (-d_f/2 + s_w)t_w ds_w)$$

$$= (V_z/I_y)(d_f t_f b/2 + d_f s_w t_w/2 - s_w^2 t_w/2)$$

$$(\tau_v t_w)_{d_f} = (V_z/I_y)(d_f t_f b/2 + d_f^2 t_w/2 - d_f^2 t_w/2)$$

$$= (V_z/I_y)(d_f t_f b/2) = (\tau_v t_f)_b$$

which provides a symmetry check.

A vertical equilibrium check is provided by finding the resultant of the web flows as

$$V_w = \int_0^{d_f} (\tau_v t_w) ds_w$$

$$= (V_z/I_y)[d_f t_f b s_w/2 + d_f s_w^2 t_w/4 - s_w^3 t_w/6]_0^{d_f}$$

$$= (V_z/I_y)(d_f^2 t_f b/2 + d_f^3 t_w/12)$$

$$= V_z$$

when $I_y = d_f^2 t_f b/2 + d_f^3 t_w/12$ is substituted.

5.12.8   Example 8 – shear centre of a channel section

Problem. Determine the position of the shear centre of the channel section shown in Figure 5.14.

Solution. The shear flow distribution in the channel section was analysed in Section 5.12.7 and is shown in Figure 5.14a. The resultant flange shear forces shown in Figure 5.14b are obtained from

$$V_f = \int_0^b (\tau_v t_f) \, ds_f$$
$$= (V_z/I_y)[d_f t_f s_f^2/4]_0^b$$
$$= (V_z/I_y)(d_f t_f b^2/4),$$

while the resultant web shear force is $V_w = V_z$ (see Section 5.12.7).

These resultant shear forces are statically equivalent to a shear force V_z acting through the point S which is a distance

$$a = \frac{V_f d_f}{V_w} = \frac{d_f^2 t_f b^2}{4 I_y} \quad \text{from the web, and so}$$

$$y_0 = \frac{d_f^2 t_f b^2}{4 I_y} + \frac{b^2 t_f}{2 b t_f + d_f t_w}.$$

5.12.9   Example 9 – shear stresses in a pi-section

Problem. The pi shaped section shown in Figure 5.42b has a shear force of 100 kN acting parallel to the y axis. Determine the shear stress distribution.

Solution. Using equation 5.23, the shear flow in the segment 12 is

$$(\tau_h \times 12)_{12} = \frac{-100 \times 10^3}{1400 \times 10^4} \int_0^{s_1} (-100 + s_1) \times 12 \times ds_1,$$

whence $(\tau_h)_{12} = 0.007143 \times (100s_1 - s_1^2/2)$ N/mm².

Similarly, the shear flow in the segment 62 is

$$(\tau_h \times 6)_{62} = \frac{-100 \times 10^3}{1400 \times 10^4} \int_0^{s_6} (-50) \times 6 \times ds_1,$$

whence $(\tau_h)_{62} = 0.007143 \times (50s_6)$ N/mm².

For horizontal equilibrium of the longitudinal shear forces at the junction 2, the shear flows in the segments 12, 62, and 23 must balance (equation 5.20), and so

$$[(\tau_h \times 12)_{23}]^{s_3=0} = [(\tau_h \times 12)_{12}]^{s_1=50} + [(\tau_h \times 6)_{62}]^{s_6=200}$$

$$= 0.007143 \left[\left(100 \times 50 - \frac{50^2}{2} \right) \times 12 + 50 \times 200 \times 6 \right]$$

$$= 0.007143 \times 8750.$$

The shear flow in the segment 23 is

$$(\tau_h \times 12)_{23} = 0.007143 \times 12 \times 8750 - \frac{100 \times 10^3}{1400 \times 10^4}$$

$$\times \int_0^{s_2} (-50 + s_2) \times 12 \times \mathrm{d}s_2$$

whence $(\tau_h)_{23} = 0.007143(8750 + 50s_2 - s_2^2/2)$ N/mm^2.

As a check, the resultant of the shear stresses in the segments 12, 23, and 34 is

$$\int \tau_h t \mathrm{d}s = 2 \times 0.007143 \times 12 \int_0^{50} (100s_1 - s_1^2/2)\mathrm{d}s_1$$

$$+ 0.007143 \times 12 \times \int_0^{100} (8750 + 50s_2 - s_2^2/2)\mathrm{d}s_2$$

$$= 0.007143 \times 12\{2 \times (100 \times 50^2/2 - 50^3/6)$$

$$+ (8750 \times 100 + 50 \times 100^2/2 - 100^3/6)\}\mathrm{N}$$

$$= 100 \text{ kN}$$

which is equal to the shear force.

5.12.10   Example 10 – shear centre of a pi-section

Problem. Determine the position of the shear centre of the pi section shown in Figure 5.42b.

Solution. The resultant of the shear stresses in the segment 62 is

$$\int \tau_h t \mathrm{d}s = 0.007143 \times 6 \int_0^{200} (50s_6) \times \mathrm{d}s_6$$

$$= 0.007143 \times 6 \times 50 \times 200^2/2 \text{ N}$$

$$= 42.86 \text{ kN}.$$

The resultant of the shear stresses in the segment 53 is equal and opposite to this.

The torque about the centroid exerted by the shear stresses is equal to the sum of the torques of the force in the flange 1234 and the forces in the webs 62 and 53. Thus

$$\int_A \tau_h t \rho \, ds = (100 \times 50) + (2 \times 42.86 \times 50) = 9286 \, \text{kNmm}.$$

For the applied shear to be statically equivalent to the shear stresses, it must exert an equal torque about the centroid, and so

$$-100 \times z_0 = 9286$$

whence $z_0 = -92.9$ mm.

Because of symmetry, the shear centre lies on the z axis, and so $y_0 = 0$.

5.12.11   Example 11 – shear stresses in a box section

Problem. The rectangular box section shown in Figure 5.42c has a shear force of 100 kN acting parallel to the y axis. Determine the maximum shear stress.

Solution. If initially a slit is assumed at point 1 on the axis of symmetry so that $(\tau_{ho}t)_1 = 0$, then

$$(\tau_{ho}t)_{12} = \frac{-100 \times 10^3}{6 \times 10^6} \int_0^{s_1} s_1 \times 12 \times ds_1 = -0.1 s_1^2$$

$$(\tau_{ho}t)_2 = -0.1 \times 50^2 = -0.1 \times 2500 \, \text{N/mm}$$

$$(\tau_{ho}t)_{23} = -0.1 \times 2500 - \frac{100 \times 10^3}{6 \times 10^6} \int_0^{s_2} 50 \times 6 \times ds_2$$

$$= -0.1(2500 + 50 s_2)$$

$$(\tau_{ho}t)_3 = -0.1(2500 + 50 \times 150) = -0.1 \times 10\,000 \, \text{N/mm}$$

$$(\tau_{ho}t)_{34} = -0.1 \times 10\,000 - \frac{100 \times 10^3}{6 \times 10^6} \int_0^{s_3} (50 - s_3) \times 6 \times ds_3$$

$$= -0.1[10\,000 + (50 s_3 - s_3^2/2)]$$

$$(\tau_{ho}t)_4 = -0.1[10\,000 + (50 \times 100 - 100^2/2)] = -0.1 \times 10\,000 \, \text{N/mm}$$

$$= (\tau_{ho}t)_3, \text{ which is a symmetry check.}$$

The remainder of the shear stress distribution can therefore be obtained by symmetry.

The circulating shear flow $\tau_{hc}t$ can be determined by adapting equation 5.30. For this

$$\oint \tau_{ho}\mathrm{d}s = 2 \times (0.1 \times 50^3/3)/12$$

$$+ 2 \times (-0.1) \times (2500 \times 150 + 50 \times 150^2/2)/6$$

$$+ (-0.1) \times (10\,000 \times 100 + 50 \times 100^2/2 - 100^3/6)/6$$

$$= (-0.1) \times (6\,000\,000)/12 \text{ N/mm}$$

$$\oint \frac{1}{t}\mathrm{d}s = \frac{100}{12} + \frac{150}{6} + \frac{100}{6} + \frac{150}{6}$$

$$= 900/12.$$

Thus

$$\tau_{hc}t = (-0.1) \times \frac{(6\,000\,000)}{12} \times \frac{12}{900}$$

$$= 6666.7 \text{ N/mm}.$$

The maximum shear stress occurs at the centre of the segment 34, and adapting equation 5.29,

$$(\tau_h)_{max} = \{-0.1 \times (10\,000 + 50 \times 50 - 50^2/2) + 666.7\}/6 \text{ N/mm}^2$$

$$= -76.4 \text{ N/mm}^2$$

5.12.12   Example 12 – fully plastic moment of a plated UB

Problem. A 610×229 UB 125 beam (Figure 5.40a) of S275 steel is strengthened by welding a 300 mm $\times$ 20 mm plate of S275 steel to each flange. Compare the full plastic moments of the unplated and the plated sections.

Unplated section.

For $t_f = 19.6$ mm, $f_y = 265$ N/mm^2.   EN10025-2

$M_{pl,u} = f_y W_{pl,y} = 265 \times 3676 \times 10^3$ Nmm $= 974$ kNm.

Plated section.

For $t_p = 20$ mm, $f_y = 265$ N/mm^2.   EN10025-2

Because of the symmetry of the cross-section, the plastic neutral axis is at the centroid. Adapting equation 5.67,

$$M_{pl,p} = \{974 + 2 \times 265 \times (300 \times 20) \times (612.2/2 + 20/2)/10^6\} \text{ kNm}$$

$$= 1979 \text{ kNm} \approx 2 \times M_{pl,u}$$

5.12.13   Example 13 – fully plastic moment of a tee-section

Problem. The welded tee-section shown in Figure 5.40b is fabricated from two S275 steel plates. Compare the first yield and fully plastic moments.

First yield moment. The minimum elastic section modulus was determined in Section 5.12.2 as

$$W_{el,y} = 16.93 \text{ cm}^3.$$

For $t_{max} = 10$ mm, $f_y = 275$ N/mm^2.   EN10025-2

$$M_{el,y} = f_y W_{el,y} = 275 \times 16.93 \times 10^3 \text{ Nmm} = 4.656 \text{ kNm}.$$

Full plastic moment. For the plastic neutral axis to divide the cross-section into two equal areas, then using the thin-walled assumption leads to

$$(80 \times 5) + (z_p \times 10) = (82.5 - z_p) \times 10$$

so that $z_p = (82.5 \times 10 - 80 \times 5)/(2 \times 10) = 21.3$ mm.
  Using equation 5.67,

$$M_{pl,y} = 275 \times [(80 \times 5) \times 21.3 + (21.3 \times 10) \times 21.3/2$$
$$+ (82.5 - 21.3)^2 \times 10/2] \text{ Nmm}$$
$$= 8.117 \text{ kNm}.$$

Alternatively, the section properties $W_{el,y}$ and $W_{pl,y}$ can be calculated using Figure 5.6, and used to calculate $M_{el,y}$ and $M_{pl,y}$.

5.12.14   Example 14 – plastic collapse of a non-uniform beam

Problem. The two-span continuous beam shown in Figure 5.43a is a 610 × 229 UB 125 of S275 steel, with 2 flange plates 300 mm × 20 mm of S275 steel extending over a central length of 12 m. Determine the value of the applied loads Q at plastic collapse.

Solution. The full plastic moments of the beam (unplated and plated) were determined in Section 5.12.12 as $M_{pl,u} = 974$ kNm and $M_{pl,p} = 1979$ kNm.
  The bending moment diagram is shown in Figure 5.43b. Two plastic collapse mechanisms are possible, both with a plastic hinge $(-M_{pl,p})$ at the central support. For the first mechanism (Figure 5.43c), plastic hinges $(M_{pl,p})$ occur in the plated beam at the load points, but for the second mechanism (Figure 5.43d), plastic hinges $(M_{pl,u})$ occur in the unplated beam at the changes of cross-section.

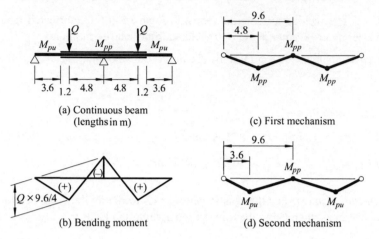

Figure 5.43 Worked example 14.

For the first mechanism shown in Figure 5.43c, inspection of the bending moment diagram shows that

$$M_{pl,p} + M_{pl,p}/2 = Q_1 \times 9.6/4$$

so that

$$Q_1 = (3 \times 1979/2) \times 4/9.6 = 1237 \text{ kN}.$$

For the second mechanism shown in Figure 5.43d, inspection of the bending moment diagram shows that

$$M_{pl,u} + 3.6M_{pl,p}/9.6 = (3.6/4.8) \times Q_2 \times 9.6/4$$

so that

$$Q_2 = [974 + (3.6/9.6) \times 1979] \times (4.8/3.6) \times 4/9.6 = 954 \text{ kN},$$

which is less than $Q_1 = 1237$ kN.

Thus plastic collapse occurs at $Q = 954$ kN by the second mechanism.

5.12.15 Example 15 – checking a simply supported beam

Problem. The simply supported 610 × 229 UB 125 of S275 steel shown in Figure 5.44a has a span of 6.0 m and is laterally braced at 1.5 m intervals. Check the adequacy of the beam for a nominal uniformly distributed dead load of 60 kNm together with a nominal uniformly distributed imposed load of 70 kNm.

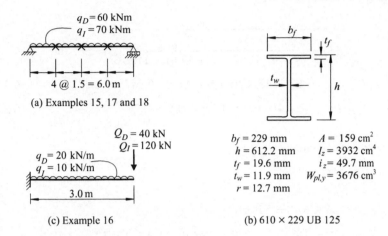

(a) Examples 15, 17 and 18

$q_D = 60$ kNm
$q_I = 70$ kNm

4 @ 1.5 = 6.0 m

$Q_D = 40$ kN
$Q_I = 120$ kN

$q_D = 20$ kN/m
$q_I = 10$ kN/m

3.0 m

(c) Example 16

$b_f = 229$ mm $A = 159$ cm^2
$h = 612.2$ mm $I_z = 3932$ cm^4
$t_f = 19.6$ mm $i_z = 49.7$ mm
$t_w = 11.9$ mm $W_{pl,y} = 3676$ cm^3
$r = 12.7$ mm

(b) 610×229 UB 125

Figure 5.44 Worked examples 15–18.

Classifying the section.

$$t_f = 19.6 \text{ mm}, \quad f_y = 265 \text{ N/mm}^2 \qquad \text{EN10025-2}$$

$$\varepsilon = \sqrt{(235/265)} = 0.942 \qquad \text{T5.2}$$

$$c_f/(t_f\varepsilon) = (229/2 - 11.9/2 - 12.7)/(19.6 \times 0.942) \qquad \text{T5.2}$$

$$= 5.19 < 9 \text{ and the flange is Class 1.} \qquad \text{T5.2}$$

$$c_w/(t_w\varepsilon) = (612.2 - 2 \times 19.6 - 2 \times 12.7)/(11.9 \times 0.942) \qquad \text{T5.2}$$

$$= 48.9 < 72 \text{ and the web is Class 1.} \qquad \text{T5.2}$$

(Note the general use of the minimum f_y obtained for the flange.)

Checking for moment.

$$q_{Ed} = (1.35 \times 60) + (1.5 \times 70) = 186 \text{ kNm}$$

$$M_{Ed} = 186 \times 6^2/8 = 837 \text{ kNm}$$

$$M_{c,Rd} = 3676 \times 10^3 \times 265/1.0 \text{ Nmm} \qquad 6.2.5$$

$$= 974 \text{ kNm} > 837 \text{ kNm} = M_{Ed}$$

which is satisfactory.

Checking for lateral bracing.

$$i_{f,z} = \sqrt{\frac{229^3 \times 19.6/12}{229 \times 19.6 + (612.2 - 2 \times 19.6) \times 11.9/6}} = 59.1 \text{ mm} \qquad 6.3.2.4$$

$$k_c L_c/(i_{f,z}\lambda_1) = 1.0 \times 1500/(59.1 \times 93.9 \times 0.942) = 0.287 \qquad 6.3.2.4$$

$$\bar{\lambda}_{c0} M_{c,Rd}/M_{Ed} = 0.4 \times 974/837 = 0.465 > 0.287 \qquad 6.3.2.4$$

and the bracing is satisfactory.

Checking for shear.

$$V_{Ed} = 186 \times 6/2 = 558 \text{ kN}$$

$$A_v = 159 \times 10^2 - 2 \times 229 \times 19.6 + (11.9 + 2 \times 12.7) \times 19.6$$

$$= 7654 \text{ mm}^2 \qquad 6.2.6(3)$$

$$V_{c,Rd} = 7654 \times (265/\sqrt{3})/1.0 \text{ N} = 1171 \text{ kN} > 558 \text{ kN} = V_{Ed} \qquad 6.2.6(2)$$

which is satisfactory.

Checking for bending and shear. The maximum M_{Ed} occurs at mid-span where $V_{Ed} = 0$, and the maximum V_{Ed} occurs at the support where $M_{Ed} = 0$, and so there is no need to check for combined bending and shear. (Note that in any case, 0.5 $V_{c,Rd} = 0.5 \times 1171 = 585.5$ kN > 558 kN $= V_{Ed}$ and so the combined bending and shear condition does not operate.)

Checking for bearing.

$$R_{Ed} = 186 \times 6/2 = 558 \text{ kN}$$

$$t_w = 11.9, f_{y,w} = 275 \text{ N/mm}^2 \qquad \text{EN10025-2}$$

$$h_w = 612.2 - 2 \times 19.6 = 573.0 \text{ mm}.$$

Assume $(s_s + c) = 100$ mm, which is not difficult to achieve.

$$k_F = 2 + 6 \times 100/573.0 = 3.047 \qquad \text{EC3-1-5, 6.1(4)(c)}$$

$$F_{cr} = 0.9 \times 3.047 \times 210\,000 \times 11.9^3/573.0 \text{ N} = 1694 \text{ kN}$$
$$\text{EC3-1-5, 6.4(1)}$$

$$m_1 = (265 \times 229)/(275 \times 11.9) = 18.54 \qquad \text{EC3-1-5, 6.5(1)}$$

$$m_2 = 0.02 \times (572.7/19.6)^2 = 17.09 \quad \text{if } \bar{\lambda}_F > 0.5 \qquad \text{EC3-1-5, 6.5(1)}$$

$$\ell_e = 3.047 \times 210\,000 \times 11.9^2/(2 \times 275 \times 573.0) = 287.5 > 100$$

EC3-1-5, 6.5(3)

$$\ell_{y1} = 100 + 19.6 \times \sqrt{(18.54/2 + (100/19.6)^2 + 17.09)} = 241.9$$

EC3-1-5, 6.5(3)

$$\ell_{y2} = 100 + 19.6 \times \sqrt{(18.54 + 17.09)} = 217.0 < 241.9 \quad \text{EC3-1-5, 6.5(3)}$$

$$\ell_y = 217.0\,\text{mm.} \qquad\qquad\qquad\qquad\qquad\quad \text{EC3-1-5, 6.5(3)}$$

$$\bar{\lambda}_F = \sqrt{\frac{217.0 \times 11.9 \times 275}{1694 \times 10^3}} = 0.648 > 0.5 \qquad \text{EC3-1-5, 6.4(1)}$$

$$\chi_F = 0.5/0.648 = 0.772 \qquad\qquad\qquad\qquad\qquad \text{EC3-1-5, 6.2}$$

$$L_{eff} = 0.772 \times 217.0 = 167.6\,\text{mm} \qquad\qquad\qquad \text{EC3-1-5, 6.2}$$

$$F_{Rd} = 275 \times 167.6 \times 11.9/1.0\,\text{N} = 548\,\text{kN} \qquad\qquad \text{EC3-1-5, 6.2}$$

$< 558\,\text{kN} = R_{Ed}$, and so the assumed value of $(s_s + c) = 100\,\text{mm}$

is inadequate.

5.12.16   Example 16 – designing a cantilever

Problem. A 3.0 m long cantilever has the nominal imposed and dead loads (which include allowances for self weight) shown in Figure 5.44c. Design a suitable UB section in S275 steel if the cantilever has sufficient bracing to prevent lateral buckling.

Selecting a trial section.

$$M_{Ed} = 1.35 \times (40 \times 3 + 20 \times 3^2/2) + 1.5 \times (120 \times 3 + 10 \times 3^2/2)$$
$$= 891\,\text{kNm.}$$

Assume that $f_y = 265$ MPa and that the section is Class 2.

$$W_{pl.y} \geq M_{Ed}/f_y = 891 \times 10^6/265\,\text{mm}^3 = 3362\,\text{cm}^3 \qquad\qquad 6.2.5$$

Choose a 610×229 UB 125 with $W_{pl.y} = 3676\,\text{cm}^3 > 3362\,\text{cm}^3$.

Checking the trial section. The trial section must now be checked by classifying the section and checking for moment, shear, moment and shear, and bearing. This process is similar to that used in Section 5.12.15 for checking a simply supported beam. If the section chosen fails any of the checks, then a new section must be chosen.

5.12.17 Example 17 – checking a plastically analysed beam

Problem. Check the adequacy of the non-uniform beam shown in Figure 5.43 and analysed plastically in Section 5.12.14. The beam and its flange plates are of S275 steel, and the concentrated loads have nominal dead load components of 250 kN (which include allowances for self weight), and imposed load components of 400 kN.

Classifying the sections. The 610×229 UB 125 was found to be Class 1 in Section 5.12.15.

For the 300×20 plate, $t_p = 20$ mm, $f_y = 265$ N/mm^2 EN10025-2

$$\varepsilon = \sqrt{(235/265)} = 0.942$$ T5.2

For the 300×20 plate as two outstands

$$c_p/(t_p\varepsilon) = (300/2)/(19.6 \times 0.942)$$ T5.2

$$= 8.13 < 9 \text{ and the outstands are Class 1.}$$ T5.2

For the internal element of the 300×20 plate,

$$c_p/(t_p\varepsilon) = 229/(20 \times 0.942)$$ T5.2

$$= 12.2 < 33 \text{ and the internal element is Class 1.}$$ T5.2

Checking for plastic collapse. The design loads are $Q_{Ed} = (1.35 \times 250) + (1.5 \times 400) = 937.5$ kN.

The plastic collapse load based on the nominal full plastic moments of the beam was calculated in Section 5.12.14 as

$$Q = 954 \text{ kN} > 937.5 \text{ kN} = Q_{Ed},$$

and so the resistance appears to be adequate.

Checking for lateral bracing. For the unplated segment,

$$L_{stable} = (60 - 40 \times 0) \times 0.924 \times 4.97 \times 10 = 2808 \text{ mm.}$$ 6.3.5.3

Thus a brace should be provided at the change of section (a plastic hinge location), and a second brace between this and the support.

For the plated segment,

$$i_z = \sqrt{\left\{ \frac{(3932 \times 10^4) + (2 \times 300^3 \times 20/12)}{(159 \times 10^2) + (2 \times 300 \times 20)} \right\}} = 68.1 \text{ mm}$$

and so,

$$L_{stable} = \{60 - 40 \times 974/(-1979)\} \times 0.924 \times 68.1 = 5109 \text{ mm.}$$ 6.3.5.3

Thus a single brace should be provided in the plated segment. A satisfactory location is at the load point.

Checking for shear. Under the collapse loads $Q = 954$ kN, the end reactions are

$$R_E = \{(954 \times 4.8) - 1979\}/9.6 = 270.6 \text{ kN}$$

and the central reaction is

$$R_C = 2 \times 954 - 2 \times 270.6 = 1366 \text{ kN}$$

so that the maximum shear at collapse is

$$V = 1366/2 = 683 \text{ kN}.$$

From Section 5.12.15, the design shear resistance is $V_{c,Rd} = 1171$ kN > 683 kN which is satisfactory.

Checking for bending and shear. At the central support, $V = 683$ kN and

$$0.5\, V_{c,Rd} = 0.5 \times 1171 = 586 \text{ kN} < 683 \text{ kN}$$

and so there is a reduction in the full plastic moment M_{pp}. This reduction is from 1979 kNm to 1925 kNm (after using Clause 6.2.8(3) of EC3). The plastic collapse load is reduced to 942 kN > 937.5 kN $= Q_{Ed}$, and so the resistance is adequate.

Checking for bearing. Web stiffeners are required for beams analysed plastically within $h/2$ of all plastic hinge points where the design shear exceeds 10% of the web capacity, or $0.1 \times 1171 = 117.1$ kN (Clause 5.6(2b) of EC3). Thus stiffeners are required at the hinge locations at the points of cross-section change and at the interior support. Load-bearing stiffeners should also be provided at the points of concentrated load, but are not required at the outer supports (see Section 5.12.15). An example of the design of load-bearing stiffeners is given in Section 4.9.9.

5.12.18 Example 18 – serviceability of a simply supported beam

Problem. Check the imposed load deflection of the 610×220 UB 125 of Figure 5.44a for a serviceability limit of $L/360$.

Solution. The central deflection w_c of a simply supported beam with uniformly distributed load q can be calculated using

$$w_c = \frac{5qL^4}{384EI_y} = \frac{5 \times 70 \times 6000^4}{384 \times 210\,000 \times 98\,610 \times 10^4} = 5.7 \text{ mm}. \qquad (5.73)$$

(The same result can be obtained using Figure 5.3.)
$L/360 = 6000/360 = 16.7$ mm > 5.7 mm $= w_c$ and so the beam is satisfactory.

5.13   Unworked examples

5.13.1   Example 19 – section properties

Determine the principal axis properties I_y, I_z of the half box section shown in Figure 5.45a.

5.13.2   Example 20 – elastic stresses and deflections

The box section beam shown in Figure 5.45b is to be used as a simply supported beam over a span of 20.0 m. The erection procedure proposed is to fabricate the beam as two half boxes (the right half as in Figure 5.45a), to erect them separately, and to pull them together before making the longitudinal connections between the flanges. The total construction load is estimated to be 10 kN/m × 20 m.

(a) Determine the maximum elastic bending stress and deflection of one half box.
(b) Determine the horizontal distributed force required to pull the two half boxes together.
(c) Determine the maximum elastic bending stress and deflection after the half boxes are pulled together.

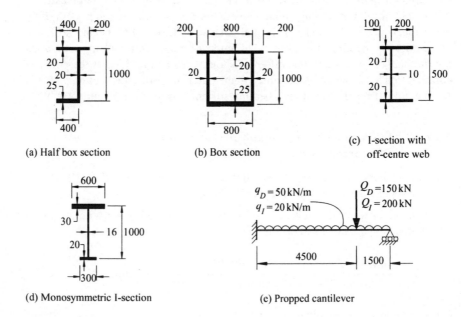

(a) Half box section   (b) Box section   (c) I-section with off-centre web

(d) Monosymmetric I-section   (e) Propped cantilever

Figure 5.45 Unworked examples.

5.13.3 Example 21 – shear stresses

Determine the shear stress distribution caused by a vertical shear force of 500 kN in the section shown in Figure 5.45c.

5.13.4 Example 22 – shear centre

Determine the position of the shear centre of the section shown in Figure 5.45c.

5.13.5 Example 23 – shear centre of a box section

Determine the shear flow distribution caused by a horizontal shear force of 2000 kN in the box section shown in Figure 5.45b, and determine the position of its shear centre.

5.13.6 Example 24 – plastic moment of a monosymmetric I-section

Determine the fully plastic moment and the shape factor of the monosymmetric I-beam shown in Figure 5.45d if the yield stress is 265 N/mm^2.

5.13.7 Example 25 – elastic design

Determine a suitable hot-rolled I-section member of S275 steel for the propped cantilever shown in Figure 5.45e by using the elastic method of design.

5.13.8 Example 26 – plastic collapse analysis

Determine the plastic collapse mechanism of the propped cantilever shown in Figure 5.45e.

5.13.9 Example 27 – plastic design

Use the plastic design method to determine a suitable hot-rolled I-section of S275 steel for the propped cantilever shown in Figure 5.45e.

References

1. Popov, E.P. (1968) *Introduction to Mechanics of Solids*, Prentice-Hall, Englewood Cliffs, New Jersey.
2. Hall, A.S. (1984) *An Introduction to the Mechanics of Solids*, 2nd edition, John Wiley, Sydney.
3. Pippard, A.J.S. and Baker, J.F. (1968) *The Analysis of Engineering Structures*, 4th edition, Edward Arnold, London.

4. Norris, C.H., Wilbur, J.B., and Utku, S. (1976) *Elementary Structural Analysis*, 3rd edition, McGraw-Hill, New York.
5. Harrison, H.B. (1973) *Computer Methods in Structural Analysis*, Prentice-Hall, Englewood Cliffs, New Jersey.
6. Harrison, H.B. (1990) *Structural Analysis and Design, Parts 1 and 2*, 2nd edition, Pergamon Press, Oxford.
7. Ghali, A., Neville, A.M., and Brown, T.G. (1997) *Structural Analysis – A Unified Classical and Matrix Approach*, 5th edition, Routledge, Oxford.
8. Coates, R.C., Coutie, M.G., and Kong, F.K. (1990) *Structural Analysis*, 3rd edition, Van Nostrand Reinhold (UK), Wokingham.
9. British Standards Institution (2005) *BS4-1:2005 Structural Steel Sections – Part1: Specification for hot-rolled sections*, BSI, London.
10. Bridge, R.Q. and Trahair, N.S. (1981) Bending, shear, and torsion of thin-walled beams, *Steel Construction*, Australian Institute of Steel Construction, **15**(1), pp. 2–18.
11. Hancock, G.J. and Harrison, H.B. (1972) A general method of analysis of stresses in thin-walled sections with open and closed parts, *Civil Engineering Transactions*, Institution of Engineers, Australia, **CE14**, No. 2, pp. 181–8.
12. Papangelis, J.P. and Hancock, G.J. (1995) *THIN-WALL – Cross-section Analysis and Finite Strip Buckling Analysis of Thin-Walled Structures*, Centre for Advanced Structural Engineering, University of Sydney.
13. Timoshenko, S.P. and Goodier, J.N. (1970) *Theory of Elasticity*, 3rd edition, McGraw-Hill, New York.
14. Winter, G. (1940) Stress distribution in and equivalent width of flanges of wide, thin-walled steel beams, *Technical Note 784*, National Advisory Committee for Aeronautics.
15. Abdel-Sayed, G. (1969) Effective width of steel deck-plates in bridges, *Journal of the Structural Division, ASCE*, **95**, No. ST7, pp. 1459–74.
16. Malcolm, D.J. and Redwood, R.G. (1970) Shear lag in stiffened box girders, *Journal of the Structural Division, ASCE*, **96**, No. ST7, pp. 1403–19.
17. Kristek, V. (1983) Shear lag in box girders, *Plated Structures. Stability and Strength*, (ed. R. Narayanan), Applied Science Publishers, London, pp. 165–94.
18. Baker, J.F. and Heyman, J. (1969) *Plastic Design of Frames – 1. Fundamentals*, Cambridge University Press, Cambridge.
19. Heyman, J. (1971) *Plastic Design of Frames – 2. Applications*, Cambridge University Press, Cambridge.
20. Horne, M.R. (1978) *Plastic Theory of Structures*, 2nd edition, Pergamon Press, Oxford.
21. Neal, B.G. (1977) *The Plastic Methods of Structural Analysis*, 3rd edition, Chapman and Hall, London.
22. Beedle, L.S. (1958) *Plastic Design of Steel Frames*, John Wiley, New York.
23. Horne, M.R. and Morris, L.J. (1981) *Plastic Design of Low-rise frames*, Granada, London.
24. Davies, J.M. and Brown, B.A. (1996) *Plastic Design to BS5950*, Blackwell Science Ltd, Oxford.
25. British Standards Institution (2006) Ευροχοδε 3 – *Design of Steel Structures – Part 1–5: Plated structural elements*, BSI, London.

Chapter 6

Lateral buckling of beams

6.1 Introduction

In the discussion given in Chapter 5 of the in-plane behaviour of beams, it was assumed that when a beam is loaded in its stiffer principal plane, it deflects only in that plane. If the beam does not have sufficient lateral stiffness or lateral support to ensure that this is so, then it may buckle out of the plane of loading, as shown in Figure 6.1. The load at which this buckling occurs may be substantially less than the beam's in-plane load resistance, as indicated in Figure 6.2.

For an idealised perfectly straight elastic beam, there are no out-of-plane deformations until the applied moment M reaches the elastic buckling moment M_{cr}, when the beam buckles by deflecting laterally and twisting, as shown in Figure 6.1. These two deformations are interdependent: when the beam deflects laterally, the applied moment has a component which exerts a torque about the deflected longitudinal axis which causes the beam to twist. This behaviour, which is important for long unrestrained I-beams whose resistances to lateral bending and torsion are low, is called elastic flexural–torsional buckling, (referred to as elastic lateral–torsional buckling in EC3). In this chapter, it will simply be referred to as lateral buckling.

The failure of a perfectly straight slender beam is initiated when the additional stresses induced by elastic buckling cause the first yield. However, a perfectly straight beam of intermediate slenderness may yield before the elastic buckling moment is reached, because of the combined effects of the in-plane bending stresses and any residual stresses, and may subsequently buckle inelastically, as indicated in Figure 6.2. For very stocky beams, the inelastic buckling moment may be higher than the in-plane plastic collapse moment M_p, in which case the moment resistance of the beam is not affected by lateral buckling.

In this chapter, the behaviour and design of beams which fail by lateral buckling and yielding are discussed. It is assumed that local buckling of the compression flange or of the web (which is dealt with in Chapter 4) does not occur. The behaviour and design of beams bent about both principal axes, and of beams with axial loads, are discussed in Chapter 7.

Figure 6.1 Lateral buckling of a cantilever.

6.2   Elastic beams

6.2.1   Buckling of straight beams

6.2.1.1   Simply supported beams with equal end moments

A perfectly straight elastic I-beam which is loaded by equal and opposite end
moments is shown in Figure 6.3. The beam is simply supported at its ends so
that lateral deflection and twist rotation are prevented, while the flange ends are
free to rotate in horizontal planes so that the beam ends are free to warp (see
Section 10.8.3). The beam will buckle at a moment M_{cr} when a deflected and
twisted equilibrium position, such as that shown in Figure 6.3, is possible. It is
shown in Section 6.12.1.1 that this position is given by

$$v = \frac{M_{cr}}{\pi^2 EI_z/L^2}\phi = \delta \sin \frac{\pi x}{L}, \tag{6.1}$$

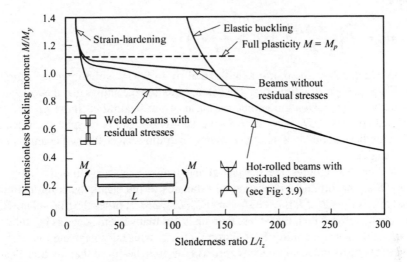

Figure 6.2 Lateral buckling strengths of simply supported I-beams.

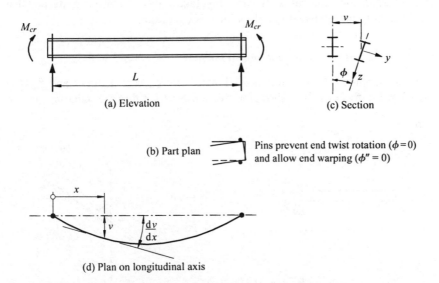

(a) Elevation

(c) Section

(b) Part plan

Pins prevent end twist rotation ($\phi = 0$)
and allow end warping ($\phi'' = 0$)

(d) Plan on longitudinal axis

Figure 6.3 Buckling of a simply supported beam.

where δ is the undetermined magnitude of the central deflection, and that the elastic buckling moment is given by

$$M_{cr} = M_{zx},$$

(6.2)

where

$$M_{zx} = \sqrt{\left\{\left(\frac{\pi^2 EI_z}{L^2}\right)\left(GI_t + \frac{\pi^2 EI_w}{L^2}\right)\right\}}, \tag{6.3}$$

and in which EI_z is the minor axis flexural rigidity, GI_t is the torsional rigidity, and EI_w is the warping rigidity of the beam. Equation 6.3 shows that the resistance to buckling depends on the geometric mean of the flexural stiffness EI_z and the torsional stiffness $(GI_t + \pi^2 EI_w/L^2)$. Equations 6.2 and 6.3 apply to all beams which are bent about an axis of symmetry, including equal flanged channels and equal angles.

Equation 6.3 ignores the effects of the major axis curvature $d^2w/dx^2 = -M_{cr}/EI_y$, and produces conservative estimates of the elastic buckling moment equal to $\sqrt{[(1 - EI_z/EI_y)\{1 - (GI_t + \pi^2 EI_w/L^2)/2EI_y\}]}$ times the true value. This correction factor, which is just less than unity for many beam sections but may be significantly less than unity for column sections, is usually neglected in design. Nevertheless, its value approaches zero as I_z approaches I_y so that the true elastic buckling moment approaches infinity. Thus an I-beam in uniform bending about its weak axis does not buckle, which is intuitively obvious. Research [1] has indicated that in some other cases the correction factor may be close to unity, and that it is prudent to ignore the effect of major axis curvature.

6.2.1.2   Beams with unequal end moments

A simply supported beam with unequal major axis end moments M and $\beta_m M$ is shown in Figure 6.4a. It is shown in Section 6.12.1.2 that the value of the end

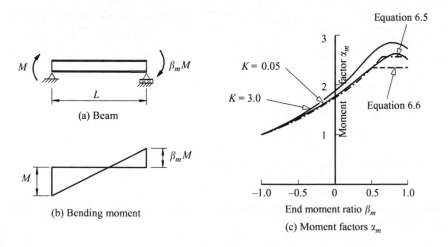

(a) Beam

(b) Bending moment

(c) Moment factors α_m

Figure 6.4 Buckling of beams with unequal end moments.

moment M_{cr} at elastic flexural–torsional buckling can be expressed in the form of

$$M_{cr} = \alpha_m M_{zx}, \tag{6.4}$$

in which the moment modification factor α_m which accounts for the effect of the non-uniform distribution of the major axis bending moment can be closely approximated by

$$\alpha_m = 1.75 + 1.05\beta_m + 0.3\beta_m^2 \leq 2.56, \tag{6.5}$$

or by

$$1/\alpha_m = 0.57 - 0.33\beta_m + 0.10\beta_m^2 \geq 0.43. \tag{6.6}$$

These approximations form the basis of a very simple method of predicting the buckling of the segments of a beam which is loaded only by concentrated loads applied through transverse members preventing local lateral deflection and twist rotation. In this case, each segment between load points may be treated as a beam with unequal end moments, and its elastic buckling moment may be estimated by using equation 6.4 and either equation 6.5 or 6.6 and by taking L as the segment length. Each buckling moment so calculated corresponds to a particular buckling load parameter for the complete load set, and the lowest of these parameters gives a conservative approximation of the actual buckling load parameter. This simple method ignores any buckling interactions between the segments. A more accurate method which accounts for these interactions is discussed in Section 6.8.2.

6.2.1.3 Beams with central concentrated loads

A simply supported beam with a central concentrated load Q acting at a distance $-z_Q$ above the centroidal axis of the beam is shown in Figure 6.5a. When the beam buckles by deflecting laterally and twisting, the line of action of the load moves with the central cross-section, but remains vertical, as shown in Figure 6.5c. The case when the load acts above the centroid is more dangerous than that of centroidal loading because of the additional torque $-Q z_Q \phi_{L/2}$ which increases the twisting of the beam and decreases its resistance to buckling.

It is shown in Section 6.12.1.3 that the dimensionless buckling load $QL^2 / \sqrt{(EI_z GI_t)}$ varies as shown in Figure 6.6 with the beam parameter

$$K = \sqrt{(\pi^2 EI_w / GI_t L^2)} \tag{6.7}$$

and the dimensionless height ε of the point of application of the load given by

$$\varepsilon = \frac{z_Q}{L} \sqrt{\left(\frac{EI_z}{GI_t}\right)}. \tag{6.8}$$

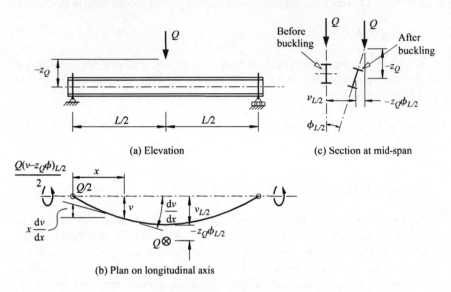

(a) Elevation

(c) Section at mid-span

(b) Plan on longitudinal axis

Figure 6.5 I-beam with a central concentrated load.

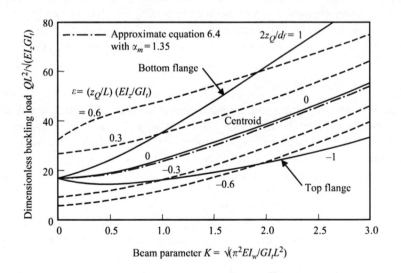

Figure 6.6 Elastic buckling of beams with central concentrated loads.

For centroidal loading ($\varepsilon = 0$), the elastic buckling load Q increases with the beam parameter K in much the same way as does the buckling moment of beams with equal and opposite end moments (see equation 6.3). The elastic buckling load

$$Q_{cr} = 4M_{cr}/L \qquad (6.9)$$

can be approximated by using equation 6.4 with the moment modification factor α_m (which accounts for the effect of the non-uniform distribution of major axis bending moment) equal to 1.35.

The elastic buckling load also varies with the load height parameter ε, and although the resistance to buckling is high when the load acts below the centroidal axis, it decreases significantly as the point of application rises, as shown in Figure 6.6. For equal flanged I-beams, the parameter ε can be transformed into

$$\frac{2z_Q}{d_f} = \frac{\pi}{K}\varepsilon \qquad (6.10)$$

where d_f is the distance between flange centroids. The variation of the buckling load with $2z_Q/d_f$ is shown by the solid lines in Figure 6.6, and it can be seen that the differences between top ($2z_Q/d_f = -1$) and bottom ($2z_Q/d_f = 1$) flange loading increase with the beam parameter K. This effect is therefore more important for deep beam-type sections of short span than for shallow column-type sections of long span. Approximate expressions for the variations of the moment modification factor α_m with the beam parameter K which account for the dimensionless load height $2z_Q/d_f$ for equal flanged I-beams are given in [2]. Alternatively, the maximum moment at elastic buckling $M_{cr} = QL/4$ may be approximated by using

$$\frac{M_{cr}}{M_{zx}} = \alpha_m \left\{ \sqrt{\left[1 + \left(\frac{0.4\alpha_m z_Q N_{cr,z}}{M_{zx}}\right)^2\right]} + \frac{0.4\alpha_m z_Q N_{cr,z}}{M_{zx}} \right\} \qquad (6.11)$$

and $\alpha_m \approx 1.35$, in which

$$N_{cr,z} = \pi^2 EI_z/L^2. \qquad (6.12)$$

6.2.1.4   Other loading conditions

The effect of the distribution of the applied load along the length of a simply supported beam on its elastic buckling strength has been investigated numerically by many methods, including those discussed in [3–5]. A particularly powerful computer method is the finite element method [6–10], while the finite integral method [11, 12], which allows accurate numerical solutions of the coupled minor axis bending and torsion equations to be obtained, has been used extensively. Many particular cases have been studied [13–16], and tabulations of elastic buckling loads are available [2, 3, 5, 13, 15, 17], as is a user-friendly computer program [18] for analysing elastic flexural–torsional buckling.

Some approximate solutions for the maximum moments M_{cr} at elastic buckling of simply supported beams which are loaded along their centroidal axes can be obtained from equation 6.4 by using the moment modification factors α_m given in Figure 6.7. It can be seen that the more dangerous loadings are those which

Beam	Moment distribution	α_m	Range
$M\left(\overleftrightarrow{}\right)\beta_m M$	$M\cdot$ ⬐ $\beta_m M$	$1.75+1.05\,\beta_m+0.3\beta_m^2$ 2.5	$-1\leq\beta_m\leq0.6$ $0.6\leq\beta_m\leq1$
$Q\downarrow \quad \downarrow Q$ ⬐ $\overset{}{\underset{\mid 2a\mid}{}}$	$-\dfrac{QL}{2}\left(1-\dfrac{2a}{L}\right)$	$1.0+0.35(1-2a/L)^2$	$0\leq2a/L\leq1$
$Q\downarrow$ ⬐ $\overset{}{\underset{\mid a\mid}{}}$	$-\dfrac{QL}{4}\{1-(2a/L)^2\}$	$1.35+0.4(2a/L)^2$	$0\leq2a/L\leq1$
$Q\downarrow$ $\left(\overset{}{\underset{\mid L/2\mid L/2\mid}{}}\right)\dfrac{3\beta_m QL}{16}$	$\dfrac{3\beta_m QL/16}{}$ $-\dfrac{QL}{4}(1-3\beta_m/8)$	$1.35+0.15\beta_m$ $-1.2+3.0\beta_m$	$0\leq\beta_m\leq0.9$ $0.9\leq\beta_m\leq1$
$\dfrac{\beta_m QL}{8}\left(\overset{Q\downarrow}{\underset{\mid L/2\mid L/2\mid}{}}\right)\dfrac{\beta_m QL}{8}$	$\dfrac{\beta_m QL/8}{}$ $-\dfrac{QL}{4}(1-\beta_m/2)$	$1.35+0.36\beta_m$	$0\leq\beta_m\leq1$
q $\left(\overset{}{}\right)\dfrac{\beta_m qL}{8}$	$\dfrac{\beta_m qL^2}{8}$ $-\dfrac{qL^2}{8}(1-\beta_m/4)^2$	$1.13+0.10\beta_m$ $-1.25+3.5\beta_m$	$0\leq\beta_m\leq0.7$ $0.7\leq\beta_m\leq1$
$\dfrac{\beta_m qL^2}{12}\left(\overset{q}{}\right)\dfrac{\beta_m qL}{12}$	$\dfrac{\beta_m qL^2/12}{}$ $-\dfrac{qL^2}{8}(1-2\beta_m/3)$	$1.13+0.12\beta_m$ $-2.38+4.8\beta_m$	$0\leq\beta_m\leq0.75$ $0.75\leq\beta_m\leq1$

Figure 6.7 Moment modification factors for simply supported beams.

produce more nearly constant distributions of major axis bending moment, and that the worst case is that of equal and opposite end moments for which $\alpha_m = 1.0$.

For other beam loadings than those shown in Figure 6.7, the moment modification factor α_m may be approximated by using

$$\alpha_m = \frac{1.75M_{max}}{\sqrt{(M_2^2 + M_3^2 + M_4^2)}} \leq 2.5 \tag{6.13}$$

in which M_{max} is the maximum moment, M_2, M_4 are the moments at the quarter points, and M_3 is the moment at the mid-point of the beam.

The effect of load height on the elastic buckling moment M_{cr} may generally be approximated by using equation 6.11 with α_m obtained from Figure 6.7 or equation 6.13.

6.2.2   Bending and twisting of crooked beams

Real beams are not perfectly straight, but have small initial crookednesses and twists which cause them to bend and twist at the beginning of loading. If a simply supported beam with equal and opposite end moments M has an initial crookedness and twist rotation which are given by

$$\frac{v_0}{\delta_0} = \frac{\phi_0}{\theta_0} = \sin\frac{\pi x}{L}, \tag{6.14}$$

in which the central initial crookedness δ_0 and twist rotation θ_0 are related by

$$\frac{\delta_0}{\theta_0} = \frac{M_{zx}}{\pi^2 E I_z / L^2},\tag{6.15}$$

then the deformations of the beam are given by

$$\frac{v}{\delta} = \frac{\phi}{\theta} = \sin\frac{\pi x}{L},\tag{6.16}$$

in which

$$\frac{\delta}{\delta_0} = \frac{\theta}{\theta_0} = \frac{M/M_{zx}}{1 - M/M_{zx}},\tag{6.17}$$

as shown in Section 6.12.2. The variations of the dimensionless central deflection δ/δ_0 and twist rotation θ/θ_0 are shown in Figure 6.8, and it can be seen that deformation begins at the commencement of loading, and increases rapidly as the elastic buckling moment M_{zx} is approached.

The simple load–deformation relationships of equations 6.16 and 6.17 are of the same forms as those of equations 3.8 and 3.9 for compression members with sinusoidal initial crookedness. It follows that the Southwell plot technique for extrapolating the elastic buckling loads of compression members from experimental measurements (see Section 3.2.2) may also be used for beams.

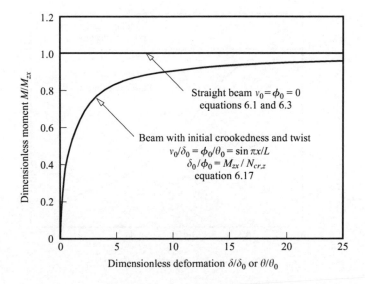

Figure 6.8 Lateral deflection and twist of a beam with equal end moments.

As the deformations increase with the applied moments M, so do the stresses. It is shown in Section 6.12.2 that the limiting moment M_L at which a beam without residual stresses first yields is given by

$$\frac{M_L}{M_y} = \frac{1}{\Phi + \sqrt{\Phi^2 - \overline{\lambda}^2}} \tag{6.18}$$

in which

$$\Phi = (1 + \eta + \overline{\lambda}^2)/2, \tag{6.19}$$

$$\overline{\lambda} = \sqrt{(M_y/M_{zx})} \tag{6.20}$$

is a generalised slenderness, $M_y = W_{el,y} f_y$ is the nominal first yield moment, and η is a factor defining the imperfection magnitudes, when the central crookedness δ_0 is given by

$$\frac{\delta_0 N_{cr,z}}{M_{zx}} = \theta_0 = \frac{W_{el,z}/W_{el,y}}{1 + (d_f/2)\,(N_{cr,z}/M_{zx})}\,\eta, \tag{6.21}$$

in which $N_{cr,z}$ is given by equation 6.12. Equations 6.18–6.20 are similar to equations 3.11, 3.12, and 3.5 for the limiting axial force at first yield of a

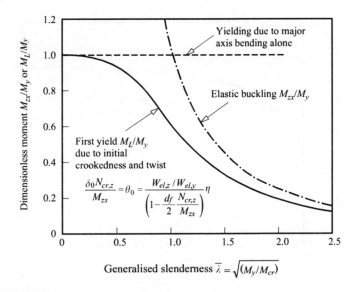

Figure 6.9 Buckling and yielding of beams.

compression member. The variation of the dimensionless limiting moment M_L/M_y for $\eta = \overline{\lambda}^2/4$ is shown in Figure 6.9, in which $\overline{\lambda} = \sqrt{(M_y/M_{zx})}$ plotted along the horizontal axis is equivalent to the generalised slenderness ratio used in Figure 3.4 for an elastic compression member. Figure 6.9 shows that the limiting moments of short beams approach the yield moment M_y, while for long beams the limiting moments approach the elastic buckling moment M_{zx}.

6.3   Inelastic beams

The solution for the buckling moment M_{zx} of a perfectly straight simply supported I-beam with equal end moments given by equations 6.2 and 6.3 is only valid while the beam remains elastic. In a short-span beam, yielding occurs before the ultimate moment is reached, and significant portions of the beams are inelastic when buckling commences. The effective rigidities of these inelastic portions are reduced by yielding, and consequently, the buckling moment is also reduced.

For beams with equal and opposite end moments ($\beta_m = -1$), the distribution of yield across the section does not vary along the beam, and when there are no residual stresses, the inelastic buckling moment can be calculated from a modified form of equation 6.3 as

$$M_I = \sqrt{\left\{ \left[\frac{\pi^2 (EI_z)_t}{L^2} \right] \left[(GI_t)_t + \frac{\pi^2 (EI_w)_t}{L^2} \right] \right\}} \tag{6.22}$$

in which the subscripted quantities $(\ )_t$ are the reduced inelastic rigidities which are effective at buckling. Estimates of these rigidities can be obtained by using the tangent moduli of elasticity (see Section 3.3.1) which are appropriate to the varying stress levels throughout the section. Thus the values of E and G are used in the elastic areas, while the strain-hardening moduli E_{st} and G_{st} are used in the yielded and strain-hardened areas (see Section 3.3.4). When the effective rigidities calculated in this way are used in equation 6.22, a lower bound estimate of the buckling moment is determined (Section 3.3.3). The variation of the dimensionless buckling moment M/M_y with the geometrical slenderness ratio L/i_z of a typical rolled-steel section which has been stress-relieved is shown in Figure 6.2. In the inelastic range, the buckling moment increases almost linearly with decreasing slenderness from the first yield moment $M_y = W_{el,y}f_y$ to the full plastic moment $M_p = W_{pl,y}f_y$, which is reached soon after the flanges are fully yielded, beyond which buckling is controlled by the strain-hardening moduli E_{st}, G_{st}.

The inelastic buckling moment of a beam with residual stresses can be obtained in a similar manner, except that the pattern of yielding is not symmetrical about the section major axis, so that a modified form of equation 6.76 for a monosymmetric I-beam must be used instead of equation 6.22. The inelastic buckling moment varies markedly with both the magnitude and the distribution of the residual stresses. The moment at which inelastic buckling initiates depends mainly on the magnitude of the residual compressive stresses at the compression flange tips, where yielding causes significant reductions in the effective rigidities $(EI_z)_t$ and $(EI_w)_t$. The flange-tip residual stresses are comparatively high in hot-rolled beams, especially those with high ratios of flange to web area, and so the inelastic buckling is initiated comparatively early in these beams, as shown in Figure 6.2. The residual stresses in hot-rolled beams decrease away from the flange tips (see Figure 3.9 for example), and so the extent of yielding increases and the effective rigidities steadily decrease as the applied moment increases. Because of this, the inelastic buckling moment decreases in an approximately linear fashion as the slenderness increases, as shown in Figure 6.2.

In beams fabricated by welding flange plates to web plates, the compressive residual stresses at the flange tips, which increase with the welding heat input, are usually somewhat smaller than those in hot-rolled beams, and so the initiation of inelastic buckling is delayed, as shown in Figure 6.2. However, the variations of the residual stresses across the flanges are more nearly uniform in welded beams, and so, once flange yielding is initiated, it spreads quickly through the flange with little increase in moment. This causes large reductions in the inelastic buckling moments of stocky beams, as indicated in Figure 6.2.

When a beam has a more general loading than that of equal and opposite end moments, the in-plane bending moment varies along the beam, and so when yielding occurs its distribution also varies. Because of this the beam acts as if non-uniform, and the torsion equilibrium equation becomes more complicated. Nevertheless, numerical solutions have been obtained for some hot-rolled beams with a number of different loading arrangements [19, 20], and some of these (for unequal end moments M and $\beta_m M$) are shown in Figure 6.10, together with approximate solutions given by

$$\frac{M_I}{M_p} = 0.7 + \frac{0.3(1 - 0.7M_p/M_{cr})}{(0.61 - 0.3\beta_m + 0.07\beta_m^2)} \tag{6.23}$$

in which M_{cr} is given by equations 6.4 and 6.5.

In this equation, the effects of the bending moment distribution are included in both the elastic buckling resistance $M_{cr} = \alpha_m M_{zx}$ through the use of the end moment ratio β_m in the moment modification factor α_m, and also through the direct use of β_m in equation 6.23. This latter use causes the inelastic buckling moments M_I to approach the elastic buckling moment M_{cr} as the end moment ratio increases towards $\beta_m = 1$.

The most severe case is that of equal and opposite end moments ($\beta_m = -1$), for which yielding is constant along the beam so that the resistance to lateral buckling is reduced everywhere. Less severe cases are those of beams with unequal end moments M and $\beta_m M$ with $\beta_m > 0$, where yielding is confined to short regions near the supports, for which the reductions in the section properties are comparatively unimportant. The least severe case is that of equal end moments that bend the beam

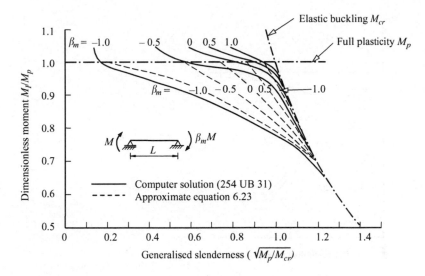

Figure 6.10 Inelastic buckling of beams with unequal end moments.

in double curvature ($\beta_m = 1$), for which the moment gradient is steepest and the regions of yielding are most limited.

The range of modified slenderness $\sqrt{(M_p/M_{cr})}$ for which a beam can reach the full plastic moment M_p depends very much on the loading arrangement. An approximate expression for the limit of this range for beams with end moments M and $\beta_m M$ can be obtained from equation 6.23 as

$$\sqrt{\left(\frac{M_p}{M_{cr}}\right)_p} = \sqrt{\left(\frac{0.39 + 0.30\beta_m - 0.07\beta_m^2}{0.70}\right)}. \tag{6.24}$$

In the case of a simply supported beam with an unbraced central concentrated load, yielding is confined to a small central portion of the beam, so that any reductions in the section properties are limited to this region. Inelastic buckling can be approximated by using equation 6.23 with $\beta_m = -0.7$ and $\alpha_m = 1.35$.

6.4 Real beams

Real beams differ from the ideal beams analysed in Section 6.2.1 in much the same way as do real compression members (see Section 3.4.1). Thus any small imperfections such as initial crookedness, twist, eccentricity of load, or horizontal load components cause the beam to behave as if it had an equivalent initial crookedness and twist (see Section 6.2.2), as shown by curve A in Figure 6.11. On the other hand, imperfections such as residual stresses or variations in material

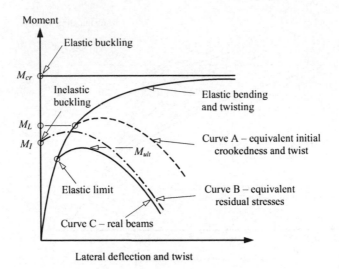

Figure 6.11 Behaviour of real beams.

properties cause the beam to behave as shown by curve B in Figure 6.11. The behaviour of real beams having both types of imperfection is indicated by curve C in Figure 6.11, which shows a transition from the elastic behaviour of a beam with curvature and twist to the inelastic post-buckling behaviour of a beam with residual stresses.

6.5   Design against lateral buckling

6.5.1   General

It is possible to develop a refined analysis of the behaviour of real beams which includes the effects of all types of imperfection. However, the use of such an analysis is not warranted because the magnitudes of the imperfections are uncertain. Instead, design rules are often based on a simple analysis for one type of equivalent imperfection which allows approximately for all imperfections, or on approximations of experimental results such as those shown in Figure 6.12.

For the EC3 method of designing against lateral buckling, the maximum moment in the beam at elastic lateral buckling M_{cr} and the beam section resistance $W_y f_y$ are used to define a generalised slenderness

$$\overline{\lambda}_{LT} = \sqrt{(W_y f_y / M_{cr})} \tag{6.25}$$

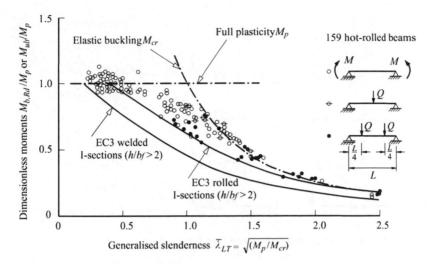

Figure 6.12 Moment resistances of beams in near-uniform bending.

and this is used in a modification of the first yield equation 6.18 which approximates the experimental resistances of many beams in near-uniform bending, such as those shown in Figure 6.12.

6.5.2 Elastic buckling moment

The EC3 method for designing against lateral buckling is an example of a more general approach to the analysis and the design of structures whose strengths are governed by the interaction between yielding and buckling. Another example of this approach was developed in Section 3.7, in which the relationship between the ultimate strength of a simply supported uniform member in uniform compression and its squash and elastic buckling loads was used to determine the in-plane ulti-mate strengths of other compression members. This method has been called the method of design by buckling analysis, because the maximum moment at elastic buckling M_{cr} must be used, rather than the approximations of somewhat variable accuracy which have been used in the past.

The first step in the method of design by buckling analysis is to determine the load at which the elastic lateral buckling takes place, so that the elastic buckling moment M_{cr} can be calculated. This varies with the beam geometry, its loading, and its restraints, but unfortunately there is no simple general method of finding it.

However, there are a number of general computer programs which can be used for finding M_{cr} [6, 8, 10, 21–23], including the user-friendly computer program PRFELB [18] which can calculate the elastic buckling moment for any beam or cantilever under any loading or restraint conditions. While such programs have

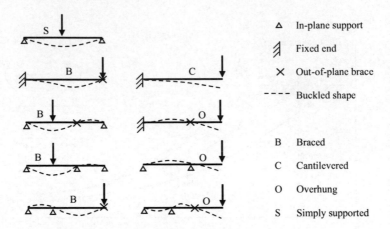

△	In-plane support
⫯	Fixed end
×	Out-of-plane brace
----	Buckled shape
B	Braced
C	Cantilevered
O	Overhung
S	Simply supported

Figure 6.13 Beams, cantilevers, and overhanging beams.

not been widely used in the past, it is expected that the use of EC3 will result in their much greater application.

Alternatively, many approximations for the elastic buckling moments M_{cr} under a wide range of loading conditions are given in [16]. Later sections summarise the methods of obtaining M_{cr} for restrained beams, cantilevers, and overhanging beams, braced and continuous beams (Figure 6.13), rigid frames, and monosymmetric and non-uniform beams under common loading conditions which cause bending about an axis of symmetry.

6.5.3 EC3 design buckling moment resistances

The EC3 design buckling moment resistance $M_{b,Rd}$ is defined by

$$\frac{M_{b,Rd}\gamma_{M1}}{W_y f_y} = \frac{1}{\Phi_{LT} + \sqrt{\Phi_{LT}^2 - \beta\overline{\lambda}_{LT}^2}} \tag{6.26}$$

and

$$\Phi_{LT} = 0.5 \left\{ 1 + \alpha_{LT}(\overline{\lambda}_{LT} - \overline{\lambda}_{LT,0}) + \beta\overline{\lambda}_{LT}^2 \right\}, \tag{6.27}$$

in which γ_{M1} is the partial factor for member instability which has a recommended value of 1.0 in EC3, $W_y f_y$ is the section moment resistance, $\overline{\lambda}_{LT}$ is the modified slenderness given by equation 6.25, and the values of α_{LT}, β and $\overline{\lambda}_{LT,0}$ depend on the type of beam section. Equations 6.25–6.27 are similar to equations 6.20, 6.18, and 6.19 for the first yield of a beam with initial crookedness and twist, except that the section moment resistance $W_y f_y$ replaces the yield moment M_y, the term $\alpha_{LT}(\overline{\lambda}_{LT} - \overline{\lambda}_{LT,0})$ replaces η, and the term β is introduced.

The EC3 uniform bending design buckling resistances $M_{b,Rd}$ for $\beta = 0.75$, $\overline{\lambda}_{LT,0} = 0.4$, and $\alpha_{LT} = 0.49$ (rolled I-sections with $h/b > 2$) are compared with experimental results for beams in near-uniform bending in Figure 6.12. For very slender beams with high values of the modified slenderness $\overline{\lambda}_{LT}$, the design buckling moment resistance $M_{b,Rd}$ shown in Figure 6.12 approaches the elastic buckling moment M_{cr}, while for stocky beams the moment resistance $M_{b,Rd}$ reaches the section resistance $W_y f_y$, and so is governed by yielding or local buckling, as discussed in Section 4.7.2. For beams of intermediate slenderness, equations 6.25–6.27 provide a transition between these limits, which is close to the lower bound of the experimental results shown in Figure 6.12. Also shown in Figure 6.12 are the EC3 design buckling moment resistances for $\alpha_{LT} = 0.76$ (welded I-sections with $h/b > 2$).

The EC3 provides two methods of design, a simple but conservative method which may be applied to any type of beam section, and a less conservative limited method.

For the simple general method, $\beta = 1.0, \overline{\lambda}_{LT,0} = 0.2$, and the imperfection factor α_{LT} depends on the type of beam section, as set out in Tables 6.4 and 6.3 of EC3. This method is conservative because it uses a very low threshold modified slenderness of $\overline{\lambda}_{LT,0} = 0.2$, above which the buckling resistance is reduced below the section resistance $W_y f_y$, and because it ignores the increased inelastic buckling resistances of beams in non-uniform bending shown for example in Figure 6.10.

For the less conservative limited method for uniform beams of rolled I-section, $\beta = 0.75$ and $\overline{\lambda}_{LT,0} = 0.4$, and a modified design buckling moment resistance

$$M_{b,Rd,mod} = M_{b,Rd}/f \le M_{b,Rd} \tag{6.28}$$

is determined using

$$f = 1 - 0.5(1 - k_c)\{1 - 2(\overline{\lambda}_{LT} - 0.8)^2\} \tag{6.29}$$

in which the values of the correction factor k_c depend on the bending moment distribution, and are given by

$$k_c = 1/\sqrt{C_1} = 1/\sqrt{\alpha_m} \tag{6.30}$$

in which α_m is the moment modification factor obtained from Figure 6.7 or equation 6.5, 6.6, or 6.13. The EC3 design buckling moment resistances $M_{b,Rd,mod}$ for $\beta = 0.75$, $\overline{\lambda}_{LT,0} = 0.4$, and $\alpha_{LT} = 0.49$ (rolled I-sections with $h/b > 2$) are compared with the inelastic buckling moment approximations of equation 6.23 in Figure 6.14.

6.5.4 Lateral buckling design procedures

For the EC3 strength design of a beam against lateral buckling, the distribution of the bending moment and the value of the maximum moment M_{Ed} are determined

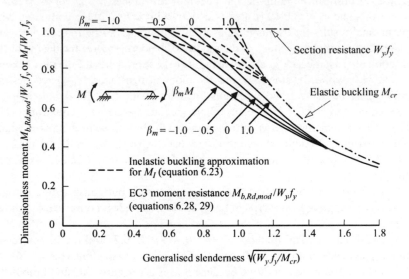

Figure 6.14 EC3 design buckling moment resistances.

by an elastic analysis (if the beam is statically indeterminate), or by statics (if the beam is statically determinate). The strength design loads are obtained by summing the nominal loads multiplied by the appropriate partial load factors γ_F (see Section 1.5.6).

In both the EC3 simple general and less conservative methods of checking a uniform equal-flanged beam, the section moment resistance $W_y f_y / \gamma_{M0}$ is checked first, the elastic buckling moment M_{cr} is determined, and the modified slenderness $\bar{\lambda}_{LT}$ is calculated using equation 6.25. The appropriate EC3 values of α_{LT}, β, and $\bar{\lambda}_{LT,0}$ (the values of these differ according to which of the two methods of design is used) are then selected and the design buckling moment resistance $M_{b,Rd}$ calculated using equations 6.26 and 6.27. In the simple general method, the beam is satisfactory when

$$M_{Ed} \leq M_{b,Rd} \leq W_y f_y / \gamma_{M0} \tag{6.31}$$

In the less conservative method, $M_{b,Rd}$ in equation 6.31 is replaced by $M_{b,Rd,mod}$ obtained using equations 6.28–6.30.

The beam must also be checked for shear, shear and bending, and bearing as discussed in Sections 5.6.1.5–5.6.1.7, and for serviceability, as discussed in Section 5.7.

When a beam is to be designed, the beam section is not known, and so a trial section must be chosen. An iterative process is then used, in which the trial section is evaluated and a new trial section chosen, until a satisfactory section is found.

One method of finding a first-trial section is to make initial guesses for f_y (usually the nominal value) and $M_{b,Rd}/W_y f_y$ (say 0.5), to use these to calculate a target section plastic modulus

$$W_{pl,y} \geq \frac{M_{Ed}/f_y}{M_{b,Rd}/(W_y f_y)} \tag{6.32}$$

and then to select a suitable trial section.

After the trial section has been selected, its elastic buckling moment M_{cr}, yield stress f_y, and design moment resistance $M_{b,Rd}$ can be found and used to calculate a new value of $W_{pl,y}$, and to select a new trial section. The iterative process usually converges within a few cycles, but convergence can be hastened by using the mean of the previous and current values of $M_{b,Rd}/W_y f_y$ in the calculation of the target section modulus.

Worked examples of checking and designing beams against lateral buckling are given in Sections 6.15.1–6.15.7

6.5.5 Checking beams supported at both ends

6.5.5.1 Section moment resistance

The classification of a specified beam cross-section as Class 1, 2, 3, or 4 is described in Sections 4.7.2 and 5.6.1.2, and the determination of the design section moment resistance $M_{c,Rd} = W_y f_y / \gamma_{M0}$ in Sections 4.7.2 and 5.6.1.3.

6.5.5.2 Elastic buckling moment

The elastic buckling moment M_{cr} of a simply supported beam depends on its geometry, loading, and restraints. It may be obtained by calculating M_{zx} (equation 6.3), $N_{cr,z}$ (equation 6.12), and α_m (equation 6.13), and substituting these into equation 6.11, or by using a computer program such as those referred to in Section 6.5.2.

6.5.5.3 Design buckling moment resistance

The design buckling moment resistance $M_{b,Rd}$ can be obtained as described in Section 6.5.3.

6.6 Restrained beams

6.6.1 Simple supports and rigid restraints

In the previous sections it was assumed that the beam was supported laterally only at its ends. When a beam with equal and opposite end moments ($\beta_m = -1.0$) has an additional rigid restraint at its centre which prevents lateral deflection and twist rotation, then its buckled shape is given by

$$\frac{\phi}{(\phi)_{L/4}} = \frac{v}{(v)_{L/4}} = \sin\frac{\pi x}{L/2},$$ (6.33)

and its elastic buckling moment is given by

$$M_{cr} = \sqrt{\left\{\left(\frac{\pi^2 EI_z}{(L/2)^2}\right)\left(GI_t + \frac{\pi^2 EI_w}{(L/2)^2}\right)\right\}}.$$ (6.34)

The end supports of a beam may also differ from simple supports. For example, both ends of the beam may be rigidly built-in against lateral rotation about the minor axis and against end warping. If the beam has equal and opposite end moments, then its buckled shape is given by

$$\frac{\phi}{(\phi)_{L/2}} = \frac{v}{(v)_{L/2}} = \frac{1}{2}\left(1 - \cos\frac{\pi x}{L/2}\right),$$ (6.35)

and its elastic buckling moment is given by equation 6.34.

In general, the elastic buckling moment M_{cr} of a restrained beam with equal and opposite end moments ($\beta_m = -1.0$) can be expressed as

$$M_{cr} = \sqrt{\left\{\left(\frac{\pi^2 EI_z}{L_{cr}^2}\right)\left(GI_t + \frac{\pi^2 EI_w}{L_{cr}^2}\right)\right\}},$$ (6.36)

in which

$$L_{cr} = k_{cr}L$$ (6.37)

is the effective length and k_{cr} is an effective length factor.

When a beam has several rigid restraints which prevent local lateral deflection and twist rotation, then the beam is divided into a series of segments. The elastic buckling of each segment may be approximated by using its length as the effective length L_{cr}. One segment will be the most critical, and the elastic buckling moment of this segment will provide a conservative estimate of the elastic buckling resistance of the whole beam. This method ignores the interactions between adjacent segments which increase the elastic buckling resistance of the beam. Approximate methods of allowing for these interactions are given in Section 6.8.

The use of the effective length concept can be extended to beams with loading conditions other than equal and opposite end moments. In general, the maximum moment M_{cr} at elastic buckling depends on the beam section, its loading, and its restraints, so that

$$\frac{M_{cr}L}{\sqrt{(EI_z GI_t)}} = \text{fn}\left(\frac{\pi^2 EI_w}{GI_t L^2}, \text{ loading}, \frac{2z_Q}{d_f}, \text{ restraints}\right).$$ (6.38)

A partial separation of the effect of the loading from that of the restraint conditions may be achieved for beams with *centroidal* loading ($z_Q = 0$) by approximating the maximum moment M_{cr} at elastic buckling by using

$$M_{cr} = \alpha_m \sqrt{\left\{ \left(\frac{\pi^2 EI_z}{L_{cr}^2} \right) \left(GI_t + \frac{\pi^2 EI_w}{L_{cr}^2} \right) \right\}}, \tag{6.39}$$

for which it is assumed that the factor α_m depends only on the in-plane bending moment distribution, and that the effective length L_{cr} depends only on the restraint conditions. This approximation allows the α_m factors calculated for simply supported beams (see Section 6.2.1) and the effective lengths L_{cr} determined for restrained beams with equal and opposite end moments ($\alpha_m = 1$) to be used more generally.

Unfortunately, the form of equation 6.39 is not suitable for beams with loads acting *away from the centroid*. It has been suggested that approximate solutions for M_{cr} may be obtained by using equation 6.11 with L_{cr} substituted for L in equations 6.3 and 6.12, but it has been reported [16] that predictions obtained in this way are sometimes of somewhat variable accuracy.

6.6.2 Intermediate restraints

In the previous sub-section it was shown that the elastic buckling moment of a simply supported I-beam is substantially increased when a restraint is provided which prevents the centre of the beam from deflecting laterally and twisting. This restraint need not be completely rigid, but may be elastic, provided its translational and rotational stiffnesses exceed certain minimum values.

The case of the beam with equal and opposite end moments M shown in Figure 6.15a is analysed in [24]. This beam has a central translational restraint of stiffness α_t (where α_t is the ratio of the lateral force exerted by the restraint to the lateral deflection of the beam measured at the height z_t of the restraint), and a central torsional restraint of stiffness α_r (where α_r is the ratio of the torque exerted by the restraint to the twist rotation of the beam). It is shown in [24] that the elastic buckling moment M_{cr} can be expressed in the standard form of equation 6.39 when the stiffnesses α_t, α_r are related to each other and the effective length factor k_{cr} through

$$\frac{\alpha_t L^3}{16 EI_z} \left(1 + \frac{2z_t/d_f}{2z_c/d_f} \right) = \frac{\left(\frac{\pi}{2k_{cr}} \right)^3 \cot \frac{\pi}{2k_{cr}}}{\frac{\pi}{2k_{cr}} \cot \frac{\pi}{2k_{cr}} - 1}, \tag{6.40}$$

$$\frac{\alpha_r L^3}{16 EI_w} \bigg/ \left(1 - \frac{2z_t 2z_c}{d_f d_f} \right) = \frac{\left(\frac{\pi}{2k_{cr}} \right)^3 \cot \frac{\pi}{2k_{cr}}}{\frac{\pi}{2k_{cr}} \cot \frac{\pi}{2k_{cr}} - 1}, \tag{6.41}$$

in which

$$z_c = \frac{M_{cr}}{\pi^2 EI_z/(k_{cr}L)^2} = \frac{d_f}{2} \sqrt{(1 + k_{cr}^2/K^2)} \tag{6.42}$$

and $K = \sqrt{(\pi^2 EI_w/GI_t L^2)}$.

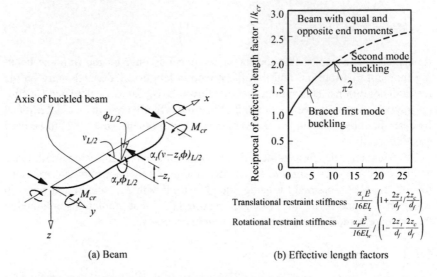

(a) Beam   (b) Effective length factors

Figure 6.15 Beam with elastic intermediate restraints.

These relationships are shown graphically in Figure 6.15b, and are similar to that shown in Figure 3.16c for compression members with intermediate restraints. It can be seen that the effective length factor k_{cr} varies from 1 when the restraints are of zero stiffness to 0.5 when

$$\frac{\alpha_t L^3}{16EI_z}\left(1+\frac{2z_t/d_f}{2z_c/d_f}\right) = \frac{\alpha_r L^3}{16EI_w}\bigg/\left(1-\frac{2z_t}{d_f}\frac{2z_c}{d_f}\right) = \pi^2. \tag{6.43}$$

If the restraint stiffnesses exceed these values, the beam buckles in the second mode with zero central deflection and twist at a moment which corresponds to $k_{cr} = 0.5$. When the height $-z_t$ of the translational restraint above the centroid is equal to $-d_f^2/4z_c$, the required rotational stiffness α_r given by equation 6.43 is zero. Since z_c is never less than $d_f/2$ (see equation 6.42), it follows that a top flange translational restraint of stiffness

$$\alpha_L = \frac{16\pi^2 EI_z/L^3}{1+d_f/2z_c} \tag{6.44}$$

is always sufficient to brace the beam into the second mode. This minimum stiffness can be expressed as

$$\alpha_L = \frac{8M_{cr}}{Ld_f}\frac{1}{\{1+\sqrt{(1+4K^2)}\}}, \tag{6.45}$$

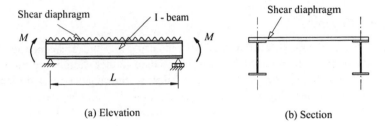

(a) Elevation   (b) Section

Figure 6.16 Diaphragm-braced I-beams.

and the greatest value of this is

$$\alpha_L = \frac{4M_{cr}}{L\,d_f}.\tag{6.46}$$

The flange force Q_f at elastic buckling can be approximated by

$$Q_f = M_{cr}/d_f,\tag{6.47}$$

and so the minimum top-flange translational stiffness can be approximated by

$$\alpha_L = 4Q_f/L.\tag{6.48}$$

This is of the same form as equation 3.33 for the minimum stiffness of intermediate restraints for compression members.

There are no restraint stiffness requirements in EC3. This follows the finding [25, 26] that compression member restraints which are capable of transmitting 2.5% of the force in the compression member invariably are stiff enough to ensure the second mode buckling. Thus EC3 generally requires any restraining element at a plastic hinge location to be capable of transmitting 2.5% of the flange force in the beam being restrained.

The influence of intermediate restraints on beams with central concentrated and uniformly distributed loads has also been studied, and many values of the minimum restraint stiffnesses required to cause the beams to buckle as if rigidly braced have been determined [16, 24, 27–30].

The effects of diaphragm bracing on the lateral buckling of simply supported beams with equal and opposite end moments (see Figure 6.16) have also been investigated [16, 31, 32], and a simple method of determining whether a diaphragm is capable of providing full bracing has been developed [31].

6.6.3　Elastic end restraints

6.6.3.1　General

When a beam forms part of a rigid-jointed structure, the adjacent members elastically restrain the ends of the beam (i.e. they induce restraining moments which are proportional to the end rotations). These restraining actions significantly modify the elastic buckling moment of the beam. Four different types of restraining moment may act at each end of a beam, as shown in Figure 6.17. They are

(a) the major axis end moment M which provides restraint about the major axis,
(b) the bottom flange end moment M_B, and
(c) the top flange end moment M_T, which provide restraints about the minor axis and against end warping, and
(d) the axial torque T_0 which provides restraint against end twisting.

6.6.3.2　Major axis end moments

The major axis end restraining moments M vary directly with the applied loads, and can be determined by a conventional in-plane bending analysis. The degree of restraint experienced at one end of the beam depends on the major axis stiffness α_y of the adjacent member (which is defined as the ratio of the end moment to the end rotation). This may be expressed by the ratio R_1 of the actual restraining moment to the maximum moment required to prevent major axis end rotation. Thus R_1 varies from 0 when there is no restraining moment to 1 when there is no end rotation. For beams which are symmetrically loaded and restrained, the restraint parameter

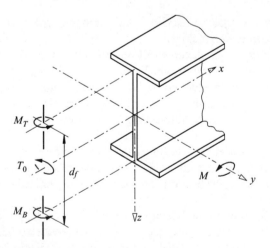

Figure 6.17 End-restraining moments.

R_1 is related to the stiffness α_y of each adjacent member by

$$\alpha_y = \frac{EI_y}{L} \frac{2R_1}{1 - R_1}. \tag{6.49}$$

Although the major axis end moments are independent of the buckling deformations, they do affect the buckling load of the beam because of their effect on the in-plane bending moment distribution (see Section 6.2.1.4). Many particular cases have been studied, and tabulations of buckling loads are available [5, 12, 13, 33–35].

6.6.3.3 Minor axis end restraints

On the other hand, the flange end moments M_B and M_T remain zero until the buckling load of the beam is reached, and then increase in proportion to the flange end rotations. Again, the degree of end restraint can be expressed by the ratio of the actual restraining moment to the maximum value required to prevent end rotation. Thus, the minor axis end restraint parameter R_2 (which describes the relative magnitude of the restraining moment $M_B + M_T$) varies between 0 and 1, and the end warping restraint parameter R_4 (which describes the relative magnitude of the differential flange end moments $(M_T - M_B)/2$) varies from 0 when the ends are free to warp to 1 when end warping is prevented. The particular case of symmetrically restrained beams with equal and opposite end moments (Figure 6.18) is analysed in Section 6.13.1. It is assumed for this that the minor axis and end warping restraints take the form of equal rotational restraints which act at each flange end and whose

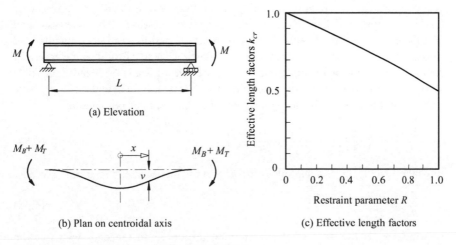

(a) Elevation

(b) Plan on centroidal axis

(c) Effective length factors

Figure 6.18 Elastic buckling of end-restrained beams.

stiffnesses are such that

$$\frac{\text{Flange end moment}}{\text{Flange end rotation}} = \frac{-EI_z}{L} \frac{R}{1-R}, \tag{6.50}$$

in which case

$$R_2 = R_4 = R. \tag{6.51}$$

It is shown in Section 6.13.1 that the moment M_{cr} at which the restrained beam buckles elastically is given by equations 6.36 and 6.37 when the effective length factor k_{cr} is the solution of

$$\frac{R}{1-R} = \frac{-\pi}{2k_{cr}} \cot \frac{\pi}{2k_{cr}}. \tag{6.52}$$

It can be seen from the solutions of this equation shown in Figure 6.18c that the effective length factor k_{cr} decreases from 1 to 0.5 as the restraint parameter R increases from 0 to 1. These solutions are exactly the same as those obtained from the compression member effective length chart of Figure 3.21a when

$$k_1 = k_2 = \frac{1-R}{1-0.5R}, \tag{6.53}$$

which suggests that the effective length factors k_{cr} for beams with unequal end restraints may be approximated by using the values given by Figure 3.21a.

The elastic buckling of symmetrically restrained beams with unequal end moments has also been analysed [36], while solutions have been obtained for many other minor axis and end warping restraint conditions [16, 27, 32–34].

6.6.3.4 Torsional end restraints

The end torques T_0 which resist end twist rotations also remain zero until elastic buckling occurs, and then increase with the end twist rotations. It has been assumed that the ends of all the beams discussed so far are rigidly restrained against end twist rotations. When the end restraints are elastic instead of rigid, some end twist rotation occurs during buckling and the elastic buckling load is reduced. Analytical studies [21] of beams in uniform bending with elastic torsional end restraints have shown that the reduced buckling moment $M_{zx,r}$ can be approximated by

$$\frac{M_{zx,r}}{M_{zx}} = \sqrt{\left\{ \frac{1}{(4.9 + 4.5 K^2)R_3 + 1} \right\}} \tag{6.54}$$

in which $1/R_3$ is the dimensionless stiffness of the torsional end restraints given by

$$\frac{1}{R_3} = \frac{\alpha_x L}{GI_t} \tag{6.55}$$

in which α_x is the ratio of the restraining torque T_0 to the end twist rotation $(\phi)_0$. Reductions for other loading conditions can be determined from the elastic

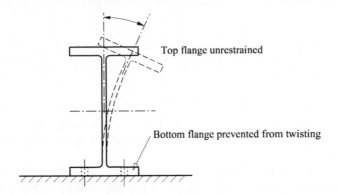

Figure 6.19 End distortion of a beam with an unrestrained top flange.

buckling solutions in [16, 34, 37] or by using a computer program [18] to carry out an elastic buckling analysis.

Another situation for which end twist rotation is not prevented is illustrated in Figure 6.19, where the bottom flange of a beam is simply supported at its end and prevented from twisting but the top flange is unrestrained. In this case, beam buckling may be accentuated by distortion of the cross-section which results in the web bending shown in Figure 6.19. Studies of the distortional buckling [38–40] of beams such as that shown in Figure 6.19 have suggested that the reduction in the buckling capacity can be allowed for approximately by using an effective length factor

$$k_{cr} = 1 + \left(d_f/6L\right)\left(t_f/t_w\right)^3\left(1 + b_f/d_f\right)/2, \qquad (6.56)$$

in which t_f and t_w are the flange and web thicknesses and b_f is the flange width.

The preceding discussion has dealt with the effects of each type of end restraint, but most beams in rigid-jointed structures have all types of elastic restraint acting simultaneously. Many such cases of combined restraints have been analysed, and tabular, graphical, or approximate solutions are given in [12, 16, 33, 35, 41].

6.7   Cantilevers and overhanging beams

6.7.1   Cantilevers

The support conditions of cantilevers differ from those of beams in that a cantilever is usually assumed to be completely fixed at one end (so that lateral deflection, lateral rotation, and warping are prevented) and completely free (to deflect, rotate, and warp) at the other. The elastic buckling solution for such a cantilever in uniform bending caused by an end moment M which rotates ϕ_L with the end of the cantilever [16] can be obtained from the solution given by equations 6.2 and 6.3 for simply supported beams by replacing the beam length L by twice the cantilever length $2L$, whence

$$M_{cr} = \sqrt{\left\{ \left(\frac{\pi^2 EI_z}{4L^2} \right) \left(GI_t + \frac{\pi^2 EI_w}{4L^2} \right) \right\}}. \tag{6.57}$$

This procedure is similar to the effective length method used to obtain the buckling load of a cantilever column (see Figure 3.15e). This loading condition is very unusual and probably never occurs, and so the solution of equation 6.57 has no practical relevance.

Cantilevers with other loading conditions are not so easily analysed, but numerical solutions are available [16, 37, 42, 43]. The particular case of a cantilever with an end concentrated load Q is discussed in Section 6.12.1.4, and plots of the dimensionless elastic buckling moments $QL^2 / \sqrt{(EI_z GI_t)}$ for bottom flange, centroidal, and top flange loading are given in Figure 6.20, together with plots of the dimensionless elastic buckling moments $(qL^3/2) / \sqrt{(EI_z GI_t)}$ of cantilevers with uniformly distributed loads q.

More accurate approximations may be obtained by using

$$\frac{QL^2}{\sqrt{(EI_z GI_t)}} = 11 \left\{ 1 + \frac{1.2\varepsilon}{\sqrt{(1 + 1.2^2 \varepsilon^2)}} \right\}$$

$$+ 4(K - 2) \left\{ 1 + \frac{1.2(\varepsilon - 0.1)}{\sqrt{(1 + 1.2^2(\varepsilon - 0.1)^2)}} \right\} \tag{6.58}$$

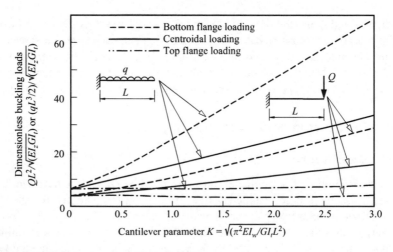

Figure 6.20 Elastic buckling loads of cantilevers.

or

$$\frac{qL^3}{2\sqrt{(EI_zGI_t)}} = 27\left\{1 + \frac{1.4(\varepsilon - 0.1)}{\sqrt{(1 + 1.4^2(\varepsilon - 0.1)^2)}}\right\}$$

$$+ 10(K - 2)\left\{1 + \frac{1.3(\varepsilon - 0.1)}{\sqrt{(1 + 1.3^2(\varepsilon - 0.1)^2)}}\right\} \tag{6.59}$$

Cantilevers which have restraints at the unsupported end which prevent lateral deflection and twist rotation (Figure 6.13) may be treated as beams which are supported laterally at both ends, as in Section 6.6.

Intermediate restraints which prevent lateral deflection and twist rotation divide a cantilever into segments (Figure 6.13). The elastic buckling of each segment can be approximated by assuming that there are no interactions between adjacent segments. The segment with the free end can then be treated as an overhanging beam as in Section 6.7.2 following, while the interior segments may be treated as beams supported laterally at both ends, as in Section 6.6.

6.7.2 Overhanging beams

Overhanging beams are similar to cantilevers in that they are free to deflect, rotate, and warp at one end (Figure 6.13). However, lateral rotation and warping are not completely prevented at the supported end, but are elastically restrained by the continuity of the overhanging beam with its continuation beyond the support. The buckling moments of some examples are reported in [16, 37].

It is usually very difficult to predict the degrees of lateral rotation and warping restraint, in which case they should be assumed to be zero. The elastic buckling moments of such overhanging beams with either concentrated end load or uniformly distributed load can be predicted by using [16]

$$\frac{QL^2}{\sqrt{(EI_zGI_t)}} = 6\left\{1 + \frac{1.5(\varepsilon - 0.1)}{\sqrt{(1 + 1.5^2(\varepsilon - 0.1)^2)}}\right\}$$

$$+ 1.5(K - 2)\left\{1 + \frac{3(\varepsilon - 0.3)}{\sqrt{(1 + 3^2(\varepsilon - 0.3)^2)}}\right\} \tag{6.60}$$

or

$$\frac{qL^3}{2\sqrt{(EI_zGI_t)}} = 15\left\{1 + \frac{1.8(\varepsilon - 0.3)}{\sqrt{(1 + 1.8^2(\varepsilon - 0.3)^2)}}\right\}$$

$$+ 40(K - 2)\left\{1 + \frac{2.8(\varepsilon - 0.4)}{\sqrt{(1 + 2.8^2(\varepsilon - 0.4)^2)}}\right\}. \tag{6.61}$$

Overhanging beams which have restraints at the unsupported end which prevent lateral deflection and twist rotation (Figure 6.13) may be treated as beams which are supported laterally at both ends, as in Section 6.6.

Intermediate restraints which prevent lateral deflection and twist rotation divide an overhanging beam into segments (Figure 6.13). The elastic buckling of each segment can be approximated by assuming that there are no interactions between adjacent segments. The segment with the free end is an overhanging beam, while the interior segments may be treated as beams supported laterally at both ends, as in Section 6.6.

6.8   Braced and continuous beams

Perhaps the simplest types of rigid-jointed structure are the braced beam and the continuous beam (Figure 6.13), which can be regarded as a series of segments which are rigidly connected together at points where lateral deflection and twist are prevented. In general, the interaction between the segments during buckling depends on the loading pattern for the whole beam, and so also do the magnitudes of the buckling loads. This behaviour is similar to the in-plane buckling behaviour of rigid frames discussed in Section 8.3.5.3.

6.8.1   Beams with only one segment loaded

When only one segment of a braced or continuous beam is loaded, its elastic buckling load may be evaluated approximately when tabulations for segments with elastic end restraints are available [12, 33, 35]. To do this it is first necessary to determine the end restraint parameters R_1, R_2, R_4, ($R_3 = 0$ because it is assumed that twisting is prevented at the supports and brace points) from the stiffnesses of the adjacent unloaded segments. For example, when only the centre span of the symmetrical three span continuous beam shown in Figure 6.21a is loaded ($Q_1 = 0$), then the end spans provide elastic restraints which depend on their stiffnesses. By analysing the major axis bending, minor axis bending, and differential flange bending of the end spans, it can be shown [34] that

$$R_1 = R_2 = R_4 = \frac{1}{1 + 2L_1/3L_2} \tag{6.62}$$

approximately. With these values, the elastic buckling load for the centre span can be determined from the tabulations in [33, 35]. Some elastic buckling loads (for beams of narrow rectangular section for which $K = 0$) determined in this way are shown in a non-dimensional form in Figure 6.22.

A similar procedure can be followed when only the outer spans of the symmetrical three span continuous beam shown in Figure 6.21 are loaded ($Q_2 = 0$). In this

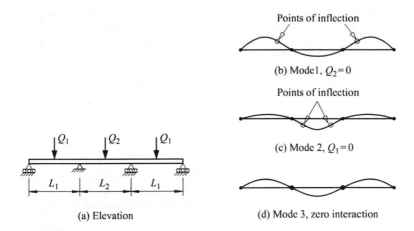

Figure 6.21 Buckling modes for a symmetrical three span continuous beam.

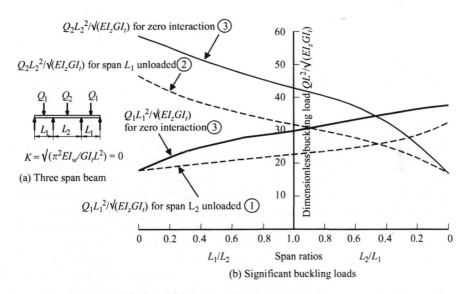

Figure 6.22 Significant buckling loads of symmetrical three span beams.

case the restraint parameters [12] are given by

$$R_1 = R_2 = R_4 = \frac{1}{1 + 3L_2/2L_1} \tag{6.63}$$

approximately. Some dimensionless buckling loads (for beams of narrow rectangular section) determined by using these restraint parameters in the tabulations of

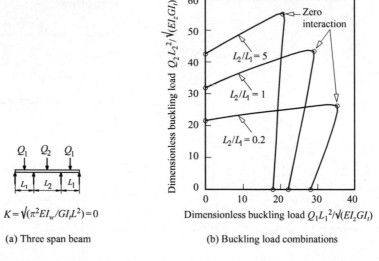

$K = \sqrt{(\pi^2 E I_w / G I_t L^2)} = 0$

(a) Three span beam

(b) Buckling load combinations

Figure 6.23 Elastic buckling load combinations of symmetrical three span beams.

[12] are shown in Figure 6.22. Similar diagrams have been produced [44] for two span beams of narrow rectangular section.

6.8.2 Beams with general loading

When more than one segment of a braced or continuous beam is loaded, the buckling loads can be determined by analysing the interaction between the segments. This has been done for a number of continuous beams of narrow rectangular section [44], and the results for some symmetrical three span beams are shown in Figure 6.23. These indicate that as the loads Q_1 on the end spans increase from zero, so does the buckling load Q_2 of the centre span until a maximum value is reached, and that a similar effect occurs as the centre span load Q_2 increases from zero.

The results shown in Figure 6.23 suggest that the elastic buckling load interaction diagram can be closely and safely approximated by drawing straight lines as shown in Figure 6.24 between the following three significant load combinations:

1   When only the end spans are loaded ($Q_2 = 0$), they are restrained during buckling by the centre span, and the buckled shape has inflection points in the end spans, as shown in Figure 6.21b. In this case the buckling loads Q_1 can be determined by using $R_1 = R_2 = R_4 = 1/(1 + 3L_2/2L_1)$ in the tabulations of [12].
2   When only the centre span is loaded ($Q_1 = 0$), it is restrained by the end spans, and the buckled shape has inflection points in the centre span, as shown in

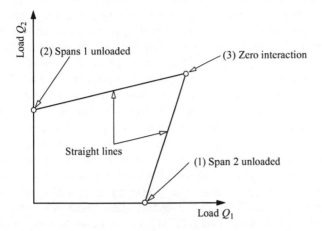

Figure 6.24 Straight line approximation.

Figure 6.21c. In this case the buckling load Q_2 can be determined by using $R_1 = R_2 = R_4 = 1/(1 + 2L_1/3L_2)$ in the tabulations of [33, 35].

3   Between these two extremes exists the zero interaction load combination for which the buckled shape has inflection points at the internal supports, as shown in Figure 6.21d, and each span buckles as if unrestrained in the buckling plane. In this case the buckling loads can be determined by using

$$R_2 = R_4 = 0, \tag{6.64}$$

$$R_{11} = \frac{1 + Q_2 L_2^2 / Q_1 L_1^2}{1 + 3L_2/2L_1} \tag{6.65}$$

$$R_{12} = \frac{1 + Q_1 L_1^2 / Q_2 L_2^2}{1 + 2L_1/3L_2}, \tag{6.66}$$

in the tabulations of [12, 33, 35]. Some zero interaction load combinations (for three span beams of narrow rectangular section) determined in this way are shown in Figure 6.22, while other zero interaction combinations are given in [44].

Unfortunately, this comparatively simple approximate method has proved too complex for use in routine design, possibly because the available tabulations of elastic buckling loads are not only insufficient to cover all the required loading and restraint conditions, but also too detailed to enable them to be easily used. Instead, an approximate method of analysis [45] is often used, in which the effects of lateral continuity between adjacent segments are ignored and each segment is regarded as being simply supported laterally. Thus the elastic buckling of each segment is analysed for its in-plane moment distribution (the moment modification factors of equation 6.13 or Figure 6.7 may be used) and for an effective length L_{cr} equal to the segment length L. The so-determined elastic buckling moment of each segment is then used to evaluate a corresponding beam load set, and the lowest of these is taken as the elastic buckling load set. This method produces a lower bound estimate which is sometimes remarkably close to the true buckling load set.

However, this is not always the case, and so a much more accurate but still reasonably simple method has been developed [36]. In this method, the accuracy of the lower bound estimate (obtained as described above) is improved by allowing for the interactions between the critical segment and the adjacent segments at buckling. This is done by using a simple approximation for the destabilising effects of the in-plane bending moments on the stiffnesses of the adjacent segments, and by approximating the restraining effects of these segments on the critical segment by using the effective length chart of Figure 3.21a for braced compression members to estimate the effective length of the critical beam segment. A step-by-step summary [36] is as follows:

(1) Determine the properties EI_z, GI_t, EI_w, L of each segment.
(2) Analyse the in-plane bending moment distribution through the beam, and determine the moment modification factors α_m for each segment from equation 6.13 or Figure 6.7.
(3) Assume all effective length factors k_{cr} are equal to unity.
(4) Calculate the maximum moment M_{cr} in each segment at elastic buckling from

$$M_{cr} = \alpha_m \sqrt{\left\{\left(\frac{\pi^2 EI_z}{L_{cr}^2}\right)\left(GI_t + \frac{\pi^2 EI_w}{L_{cr}^2}\right)\right\}} \qquad (6.67)$$

with $L_{cr} = L$, and the corresponding beam buckling loads Q_s.
(5) Determine a lower-bound estimate of the beam buckling load as the lowest value Q_{ms} of the loads Q_s, and identify the segment associated with this as the critical segment 12. (This is the approximate method [45] described in the preceding paragraph.)
(6) If a more accurate estimate of the beam buckling load is required, use the values Q_{ms} and Q_{rs1}, Q_{rs2} calculated in step 5 together with Figure 6.25 (which is similar to Figure 3.19 for braced compression members) to approximate the stiffnesses α_{r1}, α_{r2} of the segments adjacent to the critical segment 12.
(7) Calculate the stiffness of the critical segment 12 from $2EI_{zm}/L_m$.
(8) Calculate the stiffness ratios k_1, k_2 from

$$k_{1,2} = \frac{2EI_{zm}/L_m}{0.5\alpha_{r1,r2} + 2EI_{zm}/L_m}. \qquad (6.68)$$

(9) Determine the effective length factor k_{cr} for the critical segment 12 from Figure 3.21a, and the effective length $L_{cr} = k_{cr}L$.

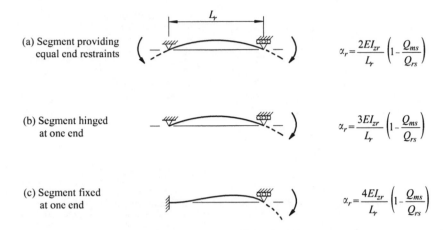

(a) Segment providing
    equal end restraints
$$\alpha_r = \frac{2EI_{zr}}{L_r}\left(1 - \frac{Q_{ms}}{Q_{rs}}\right)$$

(b) Segment hinged
    at one end
$$\alpha_r = \frac{3EI_{zr}}{L_r}\left(1 - \frac{Q_{ms}}{Q_{rs}}\right)$$

(c) Segment fixed
    at one end
$$\alpha_r = \frac{4EI_{zr}}{L_r}\left(1 - \frac{Q_{ms}}{Q_{rs}}\right)$$

Figure 6.25 Stiffness approximations for restraining segments.

(10) Calculate the elastic buckling moment M_{cr} of the critical segment 12 using L_{cr} in equation 6.67, and from this the corresponding improved approximation of the elastic buckling load Q_{cr} of the beam.

It should be noted that while the calculations for the lower bound (the first five steps) are made for all segments, those for the improved estimate are only made for the critical segment, and so comparatively little extra effort is involved. The development and application of this method is further described in [36]. The use of user-friendly computer programs such as in [18] eliminates the need for these approximate procedures.

6.9   Rigid frames

Under some conditions, the elastic buckling loads of rigid frames with only one member loaded can be determined from the available tabulations [12, 33, 35] in a similar manner to that described in Section 6.8.1 for braced and continuous beams. For example, consider a symmetrically loaded beam which is rigidly connected to two equal cross-beams as shown in Figure 6.26a. In this case the comparatively large major axis bending stiffnesses of the cross-beams ensure that end twisting of the loaded beam is effectively prevented, and it may therefore be assumed that $R_3 = 0$. If the cross-beams are of open section so that their torsional stiffness is comparatively small, then it is not unduly conservative to assume that they do not restrain the loaded beam about its major axis, and so

$$R_1 = 0. \tag{6.69}$$

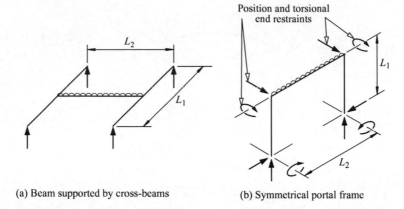

(a) Beam supported by cross-beams   (b) Symmetrical portal frame

Figure 6.26 Simple rigid-jointed structures.

By analysing the minor axis and differential flange bending of the cross-beams, it can be shown that

$$R_2 = R_4 = 1/(1 + L_1 I_{z2}/6L_2 I_{z1}). \tag{6.70}$$

This result is based on the assumption that there is no likelihood of the cross-beams themselves buckling, so that their effective minor axis rigidity can be taken as EI_{z1}. With these values for the restraint parameters, the elastic buckling load can be determined from the tabulations in [33, 35].

The buckling loads of the symmetrical portal frame shown in Figure 6.26b can be determined in a similar manner provided there are sufficient external restraints to position-fix the joints of the frame. The columns of portal frames of this type are usually placed so that the plane of greatest bending stiffness is that of the frame. In this case, the minor axis stiffness of each column provides only a small resistance against end twisting of the beam, and this resistance is reduced by the axial force transmitted by the column, so that it may be necessary to provide additional torsional end restraints to the beam. If these additional restraints are stiff enough (and the information given in [16, 34] will give some guidance), then it may be assumed without serious error that $R_3 = 0$. The major axis stiffness of a pinned base column is $3EI_{y1}/L_1$ and of a fixed base column is $4EI_{y1}/L_1$, and it can be shown by analysing the in-plane flexure of a portal frame that for pinned base portals

$$R_1 = 1/(1 + 2L_1 I_{y2}/3L_2 I_{y1}), \tag{6.71}$$

and for fixed base portals,

$$R_1 = 1/(1 + L_1 I_{y2}/2L_2 I_{y1}). \tag{6.72}$$

If the columns are of open cross-section, their torsional stiffness is comparatively small, and it is not unduly conservative to assume that the columns do not restrain the beam element or its flanges against minor axis flexure, and so

$$R_2 = R_4 = 0. \tag{6.73}$$

Using these values of the restraint parameters, the elastic buckling load can be determined from the tabulations in [33, 35].

When more than one member of a rigid frame is loaded, the buckling restraint parameters cannot be easily determined because of the interactions between members which take place during buckling. The typical member of such a frame acts as a beam-column which is subjected to a combination of axial and transverse loads and end moments. The buckling behaviour of beam-columns and of rigid frames is treated in detail in Chapters 7 and 8.

6.10 Monosymmetric beams

6.10.1 Elastic buckling resistance

When a monosymmetric I-beam (see Figure 6.27) which is loaded in its plane of symmetry twists during buckling, the longitudinal bending stresses $M_y z/I_y$ exert

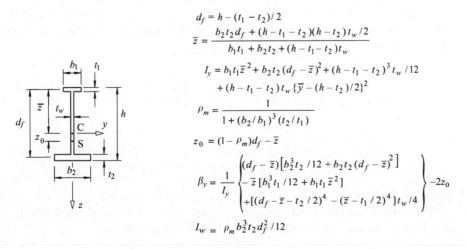

$$d_f = h - (t_1 - t_2)/2$$

$$\bar{z} = \frac{b_2 t_2 d_f + (h - t_1 - t_2)(h - t_2) t_w/2}{b_1 t_1 + b_2 t_2 + (h - t_1 - t_2) t_w}$$

$$I_y = b_1 t_1 \bar{z}^2 + b_2 t_2 (d_f - \bar{z})^2 + (h - t_1 - t_2)^3 t_w/12$$
$$+ (h - t_1 - t_2) t_w \{\bar{y} - (h - t_2)/2\}^2$$

$$\rho_m = \frac{1}{1 + (b_2/b_1)^3 (t_2/t_1)}$$

$$z_0 = (1 - \rho_m) d_f - \bar{z}$$

$$\beta_y = \frac{1}{I_y} \left\{ \begin{array}{l} (d_f - \bar{z})[b_2^3 t_2/12 + b_2 t_2 (d_f - \bar{z})^2] \\ -\bar{z}[b_1^3 t_1/12 + b_1 t_1 \bar{z}^2] \\ +[(d_f - \bar{z} - t_2/2)^4 - (\bar{z} - t_1/2)^4] t_w/4 \end{array} \right\} - 2z_0$$

$$I_w = \rho_m b_2^3 t_2 d_f^2/12$$

Figure 6.27 Properties of monosymmetric sections.

a torque (see Section 6.14)

$$T_M = M_y \beta_y \frac{d\phi}{dx} \tag{6.74}$$

in which

$$\beta_y = \frac{1}{I_y} \int_A (y^2 z + z^3) dA - 2z_0 \tag{6.75}$$

is the monosymmetry property of the cross-section. An explicit expression for β_y for an I-section is given in Figure 6.27.

The action of the torque T_M can be thought of as changing the effective torsional rigidity of the section from GI_t to $(GI_t + M_y \beta_y)$, and is related to the effect which causes some short concentrically loaded compression members to buckle torsionally (see Section 3.7.5). In that case the compressive stresses exert a disturbing torque so that there is a reduction in the effective torsional rigidity. In doubly symmetric beams, the disturbing torque exerted by the compressive bending stresses is exactly balanced by the restoring torque due to the tensile stresses, and β_y is zero. In monosymmetric beams, however, there is an imbalance which is dominated by the stresses in the smaller flange which is further from the shear centre. Thus, when the smaller flange is in compression there is a reduction in the effective torsional rigidity ($M_y \beta_y$ is negative), while the reverse is true ($M_y \beta_y$ is positive) when the smaller flange is in tension. Consequently, the resistance to buckling is increased when the larger flange is in compression, and decreased when the smaller flange is in compression.

The elastic buckling moment M_{cr} of a simply supported monosymmetric beam with equal and opposite end moments can be obtained by substituting M_{cr} for M_{zx} and the effective rigidity $(GI_t + M_{cr}\beta_y)$ for GI_t in equation 6.3 for a doubly symmetric beam and rearranging, whence

$$M_{cr} = \sqrt{\left(\frac{\pi^2 EI_z}{L^2}\right)} \left\{ \sqrt{\left[GI_t + \frac{\pi^2 EI_w}{L^2} + \left\{\frac{\beta_y}{2}\sqrt{\left(\frac{\pi^2 EI_z}{L^2}\right)}\right\}^2\right]} \right.$$
$$\left. + \frac{\beta_y}{2}\sqrt{\left(\frac{\pi^2 EI_z}{L^2}\right)} \right\}, \tag{6.76}$$

in which the warping section constant I_w is as given in Figure 6.27.

The evaluation of the monosymmetry property β_y is not straightforward, and it has been suggested that the more easily calculated parameter

$$\rho_m = I_{cz}/I_z \tag{6.77}$$

should be used instead, where I_{cz} is the section minor axis second moment of area of the compression flange. The monosymmetry property β_y may then be approximated [46] by

$$\beta_y = 0.9 d_f (2\rho_m - 1)(1 - I_z^2/I_y^2) \tag{6.78}$$

and the warping section constant I_w by

$$I_w = \rho_m (1 - \rho_m) I_z d_f^2. \tag{6.79}$$

The variations of the dimensionless elastic buckling moment $M_{cr}L/\sqrt{(EI_zGI_t)}$ with the values of ρ_m and $K_m = \sqrt{(\pi^2 EI_z d_f^2/4GI_t L^2)}$ are shown in Figure 6.28. The dimensionless buckling resistance for a T-beam with the flange in compression ($\rho_m = 1.0$) is significantly higher than for an equal flanged I-beam ($\rho_m = 0.5$) with the same value of K_m, but the resistance is greatly reduced for a T-beam with the flange in tension ($\rho_m = 0.0$).

The elastic flexural–torsional buckling of simply supported monosymmetric beams with other loading conditions has been investigated numerically, and tabulated solutions and approximating equations are available [5, 13, 16, 42, 47–49] for beams under moment gradient or with central concentrated loads or uniformly distributed loads. Solutions are also available [16, 42, 50] for cantilevers with concentrated end loads or uniformly distributed loads. These solutions can be used to find the maximum moment M_{cr} in the beam or cantilever at elastic buckling.

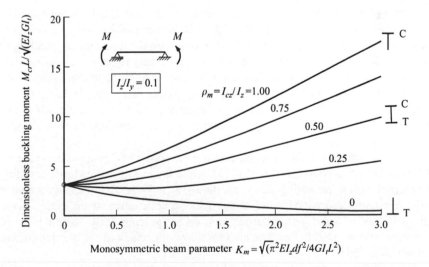

Figure 6.28 Monosymmetric I-beams in uniform bending.

6.10.2 Design rules

EC3 has no specific rules for designing monosymmetric beams against lateral buckling, because it regards them as being the same as doubly symmetric beams. Thus it requires the value of the elastic buckling moment resistance M_{cr} of the monosymmetric beam to be used in the simple general design method described in Section 6.5.3. However, the logic of this approach has been questioned [49], and in some cases may lead to less safe results.

A worked example of checking a monosymmetric T-beam is given in Section 6.15.6.

6.11 Non-uniform beams

6.11.1 Elastic buckling resistance

Non-uniform beams are often more efficient than beams of constant section, and are frequently used in situations where the major axis bending moment varies along the length of the beam. Non-uniform beams of narrow rectangular section are usually tapered in their depth. Non-uniform I-beams may be tapered in their depth, or less commonly in their flange width, and rarely in their flange thickness, while steps in flange width or thickness are common.

Depth reductions in narrow rectangular beams produce significant reductions in their minor axis flexural rigidities EI_z and torsional rigidities GI_t. Because of this, there are also significant reductions in their resistances to lateral buckling. Closed form solutions for the elastic buckling loads of many tapered beams and cantilevers are given in the papers cited in [13, 16, 51, 52].

Depth reductions in I-beams have no effect on the minor axis flexural rigidity EI_z, and little effect on the torsional rigidity GI_t, although they produce significant reductions in the warping rigidity EI_w. It follows that the resistance to buckling of a beam which does not depend primarily on its warping rigidity is comparatively insensitive to depth tapering. On the other hand, reductions in the flange width cause significant reductions in GI_t and even greater reductions in EI_z and EI_w, while reductions in flange thickness cause corresponding reductions in EI_z and EI_w and in GI_t. Thus the resistance to buckling varies significantly with changes in the flange geometry.

General numerical methods of calculating the elastic buckling loads of tapered I-beams have been developed in [51–53], while the elastic and inelastic buckling of tapered monosymmetric I-beams are discussed in [53–55]. Solutions for beams with constant flanges and linearly tapered depths under unequal end moments are given in [56, 57], and more general solutions are given in [58]. The buckling of I-beams with stepped flanges has also been investigated, and many solutions are tabulated in [59].

Approximations for the elastic buckling moment M_{cr} of a simply supported non-uniform beam can be obtained by reducing the value calculated for a uniform beam having the properties of the maximum section of the beam by multiplying it by

$$\alpha_{st} = 1.0 - 1.2 \left(\frac{L_r}{L}\right) \left\{1 - \left(0.6 + 0.4\frac{d_{min}}{d_{max}}\right)\frac{A_{f,min}}{A_{f,max}}\right\}$$ (6.80)

in which $A_{f,min}, A_{f,max}$ are the flange areas and d_{min}, d_{max} are the section depths at the minimum and maximum cross-sections, and L_r is either the portion of the beam length which is reduced in section, or is $0.5L$ for a tapered beam. This equation agrees well with the buckling solutions shown in Figure 6.29 for central concentrated loads on stepped [59] and tapered [51] beams whose minimum cross-sections are at their simply supported ends.

6.11.2 Design rules

EC3 requires non-uniform beams to be designed against lateral buckling by using the methods described in Section 6.5.3 with the section moment resistance $M_{c,Rd} = W_y f_y$ and the elastic buckling moment M_{cr} being the values for the most critical section where the ratio of $M_{Ed}/M_{c,Rd}$ of the design bending moment to the section moment capacity is greatest.

A worked example of checking a stepped beam is given in Section 6.15.7.

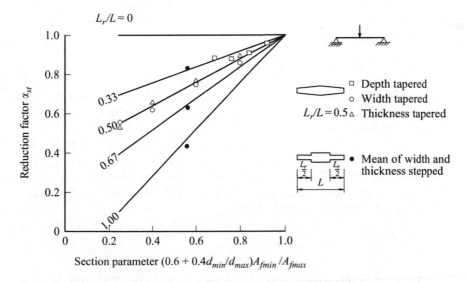

Figure 6.29 Reduction factors for stepped and tapered I-beams.

6.12   Appendix – elastic beams

6.12.1   Buckling of straight beams

6.12.1.1   Beams with equal end moments

The elastic buckling moment M_{cr} of the beam shown in Figure 6.3 can be determined by finding a deflected and twisted position which is one of equilibrium. The differential equilibrium equation of bending of the beam is

$$EI_z \frac{d^2 v}{dx^2} = -M_{cr} \phi \qquad (6.81)$$

which states that the internal minor axis moment of resistance $EI_z d^2 v / dx^2$ must exactly balance the disturbing component $-M_{cr} \phi$ of the applied bending moment M_{cr} at every point along the length of the beam. The differential equation of torsion of the beam is

$$GI_t \frac{d\phi}{dx} - EI_w \frac{d^3 \phi}{dx^3} = M_{cr} \frac{dv}{dx} \qquad (6.82)$$

which states that the sum of the internal resistance to uniform torsion $GI_t d\phi / dx$ and the internal resistance to warping torsion $-EI_w d^3 \phi / dx^3$ must exactly balance the disturbing torque $M_{cr} dv / dx$ caused by the applied moment M_{cr} at every point along the length of the beam.

The derivation of the left-hand side of equation 6.82 is fully discussed in Sections 10.2 and 10.3. The torsional rigidity GI_t in the first term determines the beam's resistance to uniform torsion, for which the rate of twist $d\phi / dx$ is constant, as shown in Figure 10.1a. For thin-walled open sections, the torsion constant I_t is approximately given by the summation

$$I_t \approx \sum bt^3 / 3$$

in which b is the length and t the thickness of each rectangular element of the cross-section. Accurate expressions for I_t are given in [3: Chapter 10] from which the values for hot-rolled I-sections have been calculated [4: Chapter 10].

The warping rigidity EI_w in the second term of equation 6.82 determines the additional resistance to non-uniform torsion, for which the flanges bend in opposite directions, as shown in Figure 10.1b and c. When this flange bending varies along the length of the beam, flange shear forces are induced which exert a torque $-EI_w d^3 \phi / dx^3$. For equal flanged I-beams,

$$I_w = \frac{I_z d_f^2}{4},$$

in which d_f is the distance between flange centroids. An expression for I_w for a monosymmetric I-beam is given in Figure 6.27.

When equations 6.81 and 6.82 are both satisfied at all points along the beam, then the deflected and twisted position is one of equilibrium. Such a position is defined by the buckled shape

$$v = \frac{M_{cr}}{\pi^2 E I_z / L^2} \phi = \delta \sin \frac{\pi x}{L}, \qquad (6.1)$$

in which the maximum deflection δ is indeterminate. This buckled shape satisfies the boundary conditions at the supports of lateral deflection prevented,

$$(v)_0 = (v)_L = 0, \qquad (6.83)$$

twist rotation prevented,

$$(\phi)_0 = (\phi)_L = 0, \qquad (6.84)$$

and ends free to warp (see Section 10.8.3),

$$\left(\frac{d^2 \phi}{dx^2} \right)_0 = \left(\frac{d^2 \phi}{dx^2} \right)_L = 0. \qquad (6.85)$$

Equation 6.1 also satisfies the differential equilibrium equations (equations 6.81 and 6.82) when $M_{cr} = M_{zx}$, where

$$M_{zx} = \sqrt{\left\{ \left(\frac{\pi^2 E I_z}{L^2} \right) \left(G I_t + \frac{\pi^2 E I_w}{L^2} \right) \right\}} \qquad (6.3)$$

which defines the moment at elastic lateral buckling.

6.12.1.2   Beams with unequal end moments

The major axis bending moment M_y and shear V_z in the beam with unequal end moments M and $\beta_m M$ shown in Figure 6.4a are given by

$$M_y = M - (1 + \beta_m) Mx/L,$$

and

$$V_z = -(1 + \beta_m) M/L.$$

When the beam buckles, the minor axis bending equation is

$$E I_z \frac{d^2 v}{dx^2} = -M_y \phi,$$

and the torsion equation is

$$GI_t \frac{d\phi}{dx} - EI_w \frac{d^3\phi}{dx^3} = M_y \frac{dv}{dx} - V_z v.$$

These equations reduce to equations 6.81 and 6.82 for the case of equal and opposite end moments ($\beta_m = -1$).

Closed form solutions of these equations are not available, but numerical methods [3–11] have been used. The numerical solutions can be conveniently expressed in the form of

$$M_{cr} = \alpha_m M_{zx} \tag{6.4}$$

in which the factor α_m accounts for the effect of the non-uniform distribution of the bending moment M_y on elastic lateral buckling. The variation of α_m with the end moment ratio β_m is shown in Figure 6.4c for the two extreme values of the beam parameter

$$K = \sqrt{\left(\frac{\pi^2 EI_w}{GJL^2} \right)} \tag{6.7}$$

of 0.05 and 3. Also shown in Figure 6.4c is the approximation

$$\alpha_m = 1.75 + 1.05\beta_m + 0.3\beta_m^2 \leq 2.56. \tag{6.5}$$

6.12.1.3   Beams with central concentrated loads

The major axis bending moment M_y and the shear V_z in the beam with a central concentrated load Q shown in Figure 6.5 are given by

$$M_y = Qx/2 - Q \langle x - L/2 \rangle$$
$$V_z = Q/2 - Q \langle x - L/2 \rangle^0,$$

in which the values of the second terms are taken as zero when the values inside the Macaulay brackets $\langle \rangle$ are negative. When the beam buckles, the minor axis bending equation is

$$EI_z \frac{d^2v}{dx^2} = -M_y \phi,$$

and the torsion equation is

$$GI_t \frac{d\phi}{dx} - EI_w \frac{d^3\phi}{dx^3} = \frac{Q}{2}(v - z_Q \phi)_{L/2}(1 - 2\langle x - L/2 \rangle^0) + M_y \frac{dv}{dx} - V_z v$$

in which z_Q is the distance of the point of application of the load below the centroid, and $(Q/2)(v - z_Q\phi)_{L/2}$ is the end torque.

Numerical solutions of these equations for the dimensionless buckling load $QL^2/\sqrt{(EI_zGI_t)}$ are available [3, 13, 16, 42], and some of these are shown in Figure 6.6, in which the dimensionless height ε of the point of application of the load is generally given by

$$\varepsilon = \frac{z_Q}{L}\sqrt{\left(\frac{EI_z}{GI_t}\right)}, \qquad (6.8)$$

or for the particular case of equal flanged I-beams, by

$$\varepsilon = \frac{K}{\pi}\frac{2z_Q}{d_f}.$$

6.12.1.4   Cantilevers with concentrated end loads

The elastic buckling of a cantilever with a concentrated end load Q applied at a distance z_Q below the centroid can be predicted from the solutions of the differential equations of minor axis bending

$$EI_z\frac{d^2v}{dx^2} = -M_y\phi,$$

and of torsion

$$GI_t\frac{d\phi}{dx} - EI_w\frac{d^3\phi}{dx^3} = Q(v - z_Q\phi)_L + M_y\frac{dv}{dx} - V_zv,$$

in which

$$M_y = -Q(L - x)$$

and

$$V_z = Q.$$

These solutions must satisfy the fixed end ($x = 0$) boundary conditions of

$$(v)_0 = (\phi)_0 = (dv/dx)_0 = (d\phi/dx)_0 = 0$$

and the condition that the end $x = L$ is free to warp, whence

$$(d^2\phi/dx^2)_L = 0.$$

Numerical solutions of these equations are available [16, 37, 42, 43], and some of these are shown in Figure 6.20.

6.12.2 Deformations of beams with initial crookedness and twist

The deformations of a simply supported beam with initial crookedness and twist caused by equal and opposite end moments M can be analysed by considering the minor axis bending and torsion equations

$$EI_z \frac{\mathrm{d}^2 v}{\mathrm{d}x^2} = -M(\phi + \phi_0), \tag{6.86}$$

$$GI_t \frac{\mathrm{d}\phi}{\mathrm{d}x} - EI_w \frac{\mathrm{d}^3 \phi}{\mathrm{d}x^3} = M\left(\frac{\mathrm{d}v}{\mathrm{d}x} - \frac{\mathrm{d}v_0}{\mathrm{d}x}\right), \tag{6.87}$$

which are obtained from equations 6.81 and 6.82 by adding the additional moment $M\phi_0$ and torque $M\mathrm{d}v_0/\mathrm{d}x$ induced by the initial twist and crookedness.

If the initial crookedness and twist rotation are such that

$$\frac{v_0}{\delta_0} = \frac{\phi_0}{\theta_0} = \sin\frac{\pi x}{L}, \tag{6.14}$$

in which the central initial crookedness δ_0 and twist rotation θ_0 are related by

$$\frac{\delta_0}{\theta_0} = \frac{M_{zx}}{\pi^2 EI_z/L^2}, \tag{6.15}$$

then the solution of equations 6.86 and 6.87 which satisfies the boundary conditions (equations 6.83–6.85) is given by

$$\frac{v}{\delta} = \frac{\phi}{\theta} = \sin\frac{\pi x}{L}, \tag{6.16}$$

in which

$$\frac{\delta}{\delta_0} = \frac{\theta}{\theta_0} = \frac{M/M_{zx}}{1 - M/M_{zx}}. \tag{6.17}$$

The maximum longitudinal stress in the beam is the sum of the stresses due to major axis bending, minor axis bending, and warping, and is equal to

$$\sigma_{max} = \frac{M}{W_{el,y}} - \frac{EI_z}{W_{el,z}} \left(\frac{\mathrm{d}^2(v + d_f\phi/2)}{\mathrm{d}x^2}\right)_{L/2}.$$

If the elastic limit is taken as the yield stress f_y, then the limiting nominal stress σ_L for which this elastic analysis is valid is given by

$$\sigma_L = f_y - \frac{\delta_0 N_{cr,z}}{M_{zx}} \left(1 + \frac{d_f}{2}\frac{N_{cr,z}}{M_{zx}}\right) \frac{1}{W_{el,z}} \frac{M_L}{1 - M_L/M_{zx}},$$

or

$$M_L = M_y - \frac{\delta_0 N_{cr,z}}{M_{zx}} \left(1 + \frac{d_f}{2} \frac{N_{cr,z}}{M_{zx}}\right) \frac{W_{el,y}}{W_{el,z}} \frac{M_L}{1 - M_L/M_{zx}}$$

in which $N_{cr,z} = \pi^2 E I_z / L^2, M_L = f_L W_{el,y}$ is the limiting moment at first yield, and $M_y = f_y W_{el,y}$ is the nominal first yield moment. This can be solved for the dimensionless limiting moment M_L/M_y. In the case where the central crookedness δ_0 is given by

$$\frac{\delta_0 N_{cr,z}}{M_{zx}} = \theta_0 = \frac{W_{el,z}/W_{el,y}}{1 + (d_f/2)(N_{cr,z}/M_{zx})} \; \eta, \tag{6.21}$$

in which η defines the magnitudes δ_0, θ_0, then the dimensionless limiting moment simplifies to

$$\frac{M_L}{M_y} = \frac{1}{\Phi + \sqrt{\Phi^2 - \bar{\lambda}^2}} \tag{6.18}$$

in which

$$\Phi = (1 + \eta + \bar{\lambda}^2)/2, \tag{6.19}$$

and

$$\bar{\lambda} = \sqrt{(M_y/M_{zx})} \tag{6.20}$$

is a generalised slenderness.

6.13 Appendix – effective lengths of beams

6.13.1 Beams with elastic end restraints

The beam shown in Figure 6.18 is restrained at its ends against minor axis rotations dv/dx and against warping rotations $(d_f/2)d\phi/dx$, and the boundary conditions at the end $x = L/2$ can be expressed in the form of

$$\frac{M_B + M_T}{(dv/dx)_{L/2}} = \frac{-EI_z}{L} \frac{2R_2}{1 - R_2},$$

and

$$\frac{M_T - M_B}{(d_f/2)(d\phi/dx)_{L/2}} = \frac{-EI_z}{L} \frac{2R_4}{1 - R_4},$$

in which M_T and M_B are the flange minor axis end restraining moments

$$M_T = 1/2EI_z(\mathrm{d}^2v/\mathrm{d}x^2)_{L/2} + (d_f/4)EI_z(\mathrm{d}^2\phi/\mathrm{d}x^2)_{L/2}$$
$$M_B = 1/2EI_z(\mathrm{d}^2v/\mathrm{d}x^2)_{L/2} - (d_f/4)EI_z(\mathrm{d}^2\phi/\mathrm{d}x^2)_{L/2},$$

and R_2 and R_4 are the dimensionless minor axis bending and warping end restraint parameters. If the beam is symmetrically restrained, then similar conditions apply at the end $x = -L/2$. The other support conditions are

$$(v)_{\pm L/2} = (\phi)_{\pm L/2} = 0.$$

The particular case for which the minor axis and warping end restraints are equal, so that

$$R_2 = R_4 = R \qquad\qquad\qquad (6.51)$$

may be analysed. The differential equilibrium equations for a buckled position v, ϕ of the beam are

$$EI_z\frac{\mathrm{d}^2v}{\mathrm{d}x^2} = -M_{cr}\phi + (M_B + M_T)$$

and

$$GI_t\frac{\mathrm{d}\phi}{\mathrm{d}x} - EI_w\frac{\mathrm{d}^3\phi}{\mathrm{d}x^3} = M_{cr}\frac{\mathrm{d}v}{\mathrm{d}x}.$$

These differential equations and the boundary conditions are satisfied by the buckled shape

$$v = \frac{M_{cr}\phi}{\pi^2EI_z/k_{cr}^2L^2} = A\left(\cos\frac{\pi x}{k_{cr}L} - \cos\frac{\pi}{2k_{cr}}\right),$$

in which the effective length factor k_{cr} satisfies

$$\frac{R}{1-R} = -\frac{\pi}{2k_{cr}}\cot\frac{\pi}{2k_{cr}}. \qquad\qquad (6.52)$$

The solutions of this equation are shown in Figure 6.18c.

6.14 Appendix – monosymmetric beams

When a monosymmetric beam (see Figure 6.27) is bent in its plane of symmetry and twisted, the longitudinal bending stresses σ exert a torque which is similar to that which causes some short concentrically loaded compression members to buckle torsionally (see Sections 3.7.5 and 3.11). The longitudinal bending force

$$\sigma \, \delta A = \frac{M_y z}{I_y} \delta A$$

acting on an element δA of the cross-section rotates $a_0 \mathrm{d}\phi/\mathrm{d}x$ (where a_0 is the distance to the axis of twist through the shear centre $y_0 = 0, z_0$), and so its transverse component $\sigma \cdot \delta A \cdot a_0 \mathrm{d}\phi/\mathrm{d}x$ exerts a torque $\sigma \cdot \delta A \cdot a_0 \mathrm{d}\phi/\mathrm{d}x \cdot a_0$ about the axis of twist. The total torque T_M exerted is

$$T_M = \frac{\mathrm{d}\phi}{\mathrm{d}x} \int_A a_0^2 \frac{M_y z}{I_y} \mathrm{d}A$$

in which

$$a_0^2 = y^2 + (z - z_0)^2.$$

Thus

$$T_M = M_y \beta_y \frac{\mathrm{d}\phi}{\mathrm{d}x}, \tag{6.74}$$

in which the monosymmetry property β_y of the cross-section is given by

$$\beta_y = \frac{1}{I_y} \left\{ \int_A z^3 \, \mathrm{d}A + \int_A y^2 z \mathrm{d}A \right\} - 2z_0. \tag{6.75}$$

An explicit expression for β_y for a monosymmetric I-section is given in Figure 6.27, and this can also be used for tee-sections by putting the flange thickness t_1 or t_2 equal to zero. Also given in Figure 6.27 is an explicit expression for the warping section constant I_w of a monosymmetric I-section. For a tee-section, I_w is zero.

6.15   Worked examples

6.15.1   Example I – checking a beam supported at both ends

Problem. The 7.5 m long 610×229 UB 125 of S275 steel shown in Figure 6.30 is simply supported at both ends where lateral deflections v are effectively prevented and twist rotations ϕ are partially restrained. Check the adequacy of the beam for a central concentrated top flange load caused by an unfactored dead load of 60 kN (which includes an allowance for self-weight) and an unfactored imposed load of 100 kN.

Design bending moment.

$$M_{Ed} = \{(1.35 \times 60) + (1.5 \times 100)\} \times 7.5/4 = 433 \text{ kNm}.$$

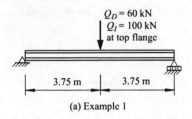

(a) Example 1

Quantity	610×229 UB 125	457×191 UB 82	Units
b_f	229.0	191.3	mm
t_f	19.6	16.0	mm
h	612.2	460.0	mm
t_w	11.9	9.9	mm
r	12.7	10.2	mm
$W_{pl.y}$	3676	1831	cm³
I_z	3932	1871	cm⁴
I_t	154	69.2	cm⁴
I_w	3.45	0.922	dm⁶

(b) Section properties

Figure 6.30 Examples 1 and 4.

Section resistance.
As in Section 5.12.15, $f_y = 265$ N/mm², the section is Class 1, and the section resistance is $M_{c,Rd} = 974$ kNm > 433 kNm $= M_{Ed}$ and the section resistance is adequate.

Elastic buckling moment.
Using equation 6.56 to allow for the partial torsional end restraints

$$k_{cr} = 1 + \{(612.2 - 19.6)/(6 \times 7500)\}(19.6/11.9)^3$$
$$\times \{1 + 229.0/(612.2 - 19.6)\}/2$$
$$= 1.041$$

so that $L_{cr} = 1.041 \times 7500 = 7806$ mm.
   Adapting equation 6.36,

$$M_{zx0} = \sqrt{\left\{\frac{\pi^2 \times 210\,000 \times 3932 \times 10^4}{7806^2} \times \left(81\,000 \times 154 \times 10^4 + \frac{\pi^2 \times 210\,000 \times 3.45 \times 10^{12}}{7806^2}\right)\right\}} \text{ Nmm}$$

$$= 569 \text{ kNm}.$$

Using equation 6.12, $N_{cr,z} = \pi^{\cdot 2} \times 210\,000 \times 3932 \times 10^4/7500^2$ N $= 1449$ kN.

Using Figure 6.7, $\alpha_m = 1.35$

$$\frac{0.4\alpha_m z_Q N_{cr,z}}{M_{zx0}} = \frac{0.4 \times 1.35 \times (-612.2/2) \times 1449 \times 10^3}{569 \times 10^6} = -0.421$$

Adapting equation 6.11,

$$M_{cr} = 1.35 \times 569\{\sqrt{[1 + (-0.421)^2]} + (-0.421)\} \text{ Nmm} = 510 \text{ kNm}.$$

(Using the computer program PRFELB [18] with $L = 7806$ mm leads to $M_{cr} = 522$ kNm.)

Member resistance.
Using equation 6.25, $\overline{\lambda}_{LT} = \sqrt{(974/510)} = 1.382$

$$h/b = 612.2/229.0 = 2.67 > 2$$

Using the EC3 simple general method with $\beta = 1.0, \overline{\lambda}_{LT,0} = 0.2$ (Clause 6.3.2.2) and $\alpha_{LT} = 0.34$ (Tables 6.4 and 6.3), and equation 6.27,

$$\Phi_{LT} = 0.5\{1 + 0.34(1.382 - 0.2) + 1.0 \times 1.382^2\} = 1.656$$

Using equation 6.26,

$$M_{b,Rd} = 974/\{1.656 + \sqrt{(1.656^2 - 1.382^2)}\} \text{ kNm}$$
$$= 379 \text{ kNm} < 433 \text{ kNm} = M_{Ed}$$

and the beam appears to be inadequate.
Using the EC3 less conservative method with $\beta = 0.75, \overline{\lambda}_{LT,0} = 0.4$ (Clause 6.3.2.3) and $\alpha_{LT} = 0.49$ (Tables 6.5 and 6.3), and equation 6.27,

$$\Phi_{LT} = 0.5\{1 + 0.49(1.382 - 0.4) + 0.75 \times 1.382^2\} = 1.457$$

Using equation 6.26,

$$M_{b,Rd} = 974/\{1.457 + \sqrt{(1.457^2 - 0.75 \times 1.382^2)}\} \text{ kNm}$$
$$= 426 \text{ kNm} < 433 \text{ kNm} = M_{Ed}$$

and the design moment resistance still appears to be inadequate.
The design moment resistance is further increased by using equations 6.29, 6.30, and 6.28 to find

$$f = 1 - 0.5 \times (1 - 1/\sqrt{1.35})\{1 - 2 \times (1.382 - 0.8)^2\} = 0.978, \text{ and}$$
$$M_{b,Rd,mod} = 426/0.978 = 436 \text{ kNm} > 433 \text{ kNm} = M_{Ed}$$

and the design moment resistance is adequate after all.

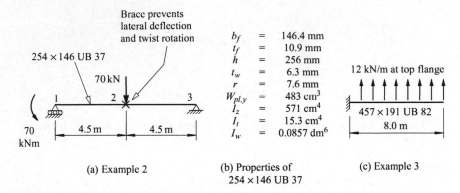

(a) Example 2

(b) Properties of
254 × 146 UB 37

(c) Example 3

Figure 6.31 Examples 2 and 3.

6.15.2   Example 2 – checking a braced beam

Problem. The 9 m long 254 × 146 UB 37 braced beam of S275 steel shown in Figure 6.31a and b has a central concentrated design load of 70 kN (which includes an allowance for self-weight) and a design end moment of 70 kNm. Lateral deflections v and twist rotations ϕ are effectively prevented at both ends and by a brace at mid-span. Check the adequacy of the braced beam.

Design bending moment.

$$R_3 = \{(70 \times 4.5) - 70\}/9 = 27.22 \, \text{kN}, M_{Ed,2} = 27.22 \times 4.5 = 122.5 \, \text{kNm}.$$

Section resistance.

$$t_f = 10.9 \, \text{mm}, \, f_y = 275 \, \text{N/mm}^2 \qquad\qquad \text{EN10025-2}$$

$$\varepsilon = \sqrt{(235/275)} = 0.924 \qquad\qquad \text{T5.2}$$

$$c_f/(t_f\varepsilon) = (146.4/2 - 6.3/2 - 7.6)/(10.9 \times 0.924) \qquad\qquad \text{T5.2}$$

$$= 6.20 < 9 \text{ and the flange is Class 1.} \qquad\qquad \text{T5.2}$$

$$c_w/(t_w\varepsilon) = (256 - 2 \times 10.9 - 2 \times 7.6)/(6.3 \times 0.924) \qquad\qquad \text{T5.2}$$

$$= 37.6 < 72 \text{ and the web is Class 1.} \qquad\qquad \text{T5.2}$$

$$M_{c,Rd} = 275 \times 483 \times 10^3 \, \text{Nmm} \qquad\qquad 6.2.5$$

$$= 132.8 \, \text{kNm} > 122.5 \, \text{kNm} = M_{Ed}$$

and the section resistance is adequate.

Elastic buckling moment.
The beam is fully restrained at mid-span, and so consists of two equal length segments 12 and 23. By inspection, the check will be controlled by segment 23 which has the lower moment gradient, and therefore the lower value of α_m. Using Figure 6.7, $\alpha_m = 1.75$.

Using equation 6.3,

$$M_{zx} = \sqrt{\left\{ \frac{\pi^2 \times 210\,000 \times 571 \times 10^4}{4500^2} \times \left(81\,000 \times 15.3 \times 10^4 + \frac{\pi^2 \times 210\,000 \times 0.0857 \times 10^{12}}{4500^2} \right) \right\}} \quad \text{Nmm}$$

$$= 111.2 \, \text{kNm}.$$

Using equation 6.4,

$$M_{cr} = 1.75 \times 111.2 = 194.6 \, \text{kNm}.$$

(Using the computer program PRFELB [18] on the segment 23 alone leads to $M_{cr} = 204.5 \, \text{kNm}$. Using PRFELB on the complete beam leads to $M_{cr} = 237.9 \, \text{kNm}$, which indicates the increase caused by the restraint offered by segment 12 to segment 23. The increased elastic buckling moment may be approximated by using the improved method given in Section 6.8.2, as demonstrated in Section 6.15.5.)

Member resistance.
Using equation 6.25, $\overline{\lambda}_{LT} = \sqrt{(132.8/194.6)} = 0.826$

$$h/b = 256/146.4 = 1.75 < 2$$

Using the EC3 simple general method with $\beta = 1.0, \overline{\lambda}_{LT,0} = 0.2$ (Clause 6.3.2.2) and $\alpha_{LT} = 0.21$ (Tables 6.4 and 6.3), and equation 6.27,

$$\Phi_{LT} = 0.5\{1 + 0.21(0.826 - 0.2) + 1.0 \times 0.826^2\} = 0.907$$

Using equation 6.26,

$$M_{b,Rd} = 132.8/\{0.907 + \sqrt{(0.907^2 - 0.826^2)}\} \, \text{kNm}$$

$$= 103.7 \, \text{kNm} < 122.5 \, \text{kNm} = M_{Ed}$$

and the beam appears to be inadequate.

Using the EC3 less conservative method with $\beta = 0.75, \bar{\lambda}_{LT,0} = 0.4$ (Clause 6.3.2.3) and $\alpha_{LT} = 0.34$ (Tables 6.5 and 6.3), and equation 6.27,

$$\Phi_{LT} = 0.5\{1 + 0.34(0.826 - 0.4) + 0.75 \times 0.826^2\} = 0.828$$

Using equation 6.26,

$$M_{b,Rd} = 132.8/\{0.828 + \sqrt{(0.828^2 - 0.75 \times 0.826^2)}\} \, kNm = 106.6 \, kNm.$$

The design moment resistance is further increased by using equations 6.29, 6.30 and 6.28 to find

$$f = 1 - 0.5 \times (1 - 1/\sqrt{1.75})\{1 - 2 \times (0.826 - 0.8)^2\} = 0.878, \text{ and}$$
$$M_{b,Rd,mod} = 106.6/0.878 = 121.4 \, kNm < 122.5 \, kNm = M_{Ed}$$

and the design moment resistance is just inadequate.

6.15.3   Example 3 – checking a cantilever

Problem. The 8.0 m long 457 × 191 UB 82 cantilever of S275 steel shown in Figure 6.31c has the section properties shown in Figure 6.30b. The cantilever has lateral, torsional, and warping restraints at the support, is free at the tip, and has a factored upwards design uniformly distributed load of 12 kN/m (which includes an allowance for self-weight) acting at the top flange. Check the adequacy of the cantilever.

Design bending moment.

$$M_{Ed} = 12 \times 8^2/2 = 384 \, kNm.$$

Section resistance.

$$t_f = 16.0 \, mm, f_y = 275 \, N/mm^2 \qquad\qquad\qquad \text{EN10025-2}$$
$$\varepsilon = \sqrt{(235/275)} = 0.924 \qquad\qquad\qquad\qquad\quad \text{T5.2}$$
$$c_f/(t_f\varepsilon) = (191.3/2 - 9.9/2 - 10.2)/(16.0 \times 0.924) \qquad \text{T5.2}$$
$$= 5.44 < 9 \text{ and the flange is Class 1.} \qquad\qquad \text{T5.2}$$

$$c_w/(t_w\varepsilon) = (460.2 - 2 \times 16.0 - 2 \times 10.2)/(9.9 \times 0.924) \qquad \text{T5.2}$$

$$= 44.6 < 72 \text{ and the web is Class 1.} \qquad \text{T5.2}$$

$$M_{c,Rd} = 275 \times 1831 \times 10^3 \text{ Nmm} \qquad \text{6.2.5}$$

$$= 503.5 \text{ kNm} > 384 \text{ kNm} = M_{Ed}$$

and the section resistance is adequate.

Elastic buckling moment.
The upwards load at the top flange is equivalent to a downwards load at the bottom flange, so that $z_Q = 460.0/2 = 230.0$ mm. Using equation 6.8,

$$\varepsilon = \frac{230.0}{8000} \sqrt{\left(\frac{210\,000 \times 1871 \times 10^4}{81\,000 \times 69.2 \times 10^4}\right)} = 0.241$$

Using equation 6.7,

$$K = \sqrt{\{(\pi^2 \times 210\,000 \times 0.922 \times 10^{12})/(81\,000 \times 69.2 \times 10^4 \times 8000^2)\}}$$

$$= 0.730$$

Using equation 6.59,

$$\frac{qL^3}{2\sqrt{EI_zGI_t}} = 27\left\{1 + \frac{1.4(0.241 - 0.1)}{\sqrt{\{1 + 1.4^2(0.241 - 0.1)^2\}}}\right\}$$

$$+ 10(0.730 - 2)\left\{1 + \frac{1.3(0.241 - 0.1)}{\sqrt{\{1 + 1.3^2(0.241 - 0.1)^2\}}}\right\}$$

$$= 17.23$$

$$M_{cr} = qL^2/2 = 17.23 \times \sqrt{(210\,000 \times 1871 \times 10^4}$$

$$\times\ 81\,000 \times 69.2 \times 10^4)/8000 \text{ Nmm}$$

$$= 1011 \text{ kNm.}$$

(Using the computer program PRFELB [18] leads to $M_{cr} = 1051$ kNm.)

Member resistance.
Using equation 6.25, $\bar{\lambda}_{LT} = \sqrt{(503.5/1011)} = 0.706$

$$h/b = 460.0/191.3 = 2.40 > 2$$

Using the EC3 simple general method with $\beta = 1.0$, $\bar{\lambda}_{LT,0} = 0.2$ (Clause 6.3.2.2) and $\alpha_{LT} = 0.34$ (Tables 6.4 and 6.3), and equation 6.27,

$$\Phi_{LT} = 0.5\{1 + 0.34(0.706 - 0.2) + 1.0 \times 0.706^2\} = 0.835$$

Using equation 6.26,

$$M_{b,Rd} = 503.5/\{0.835 + \sqrt{(0.835^2 - 0.706^2)}\}$$
$$= 393.0 \, \text{kNm} > 384 \, \text{kNm} = M_{Ed}$$

and the design moment resistance is adequate.

6.15.4   Example 4 – designing a braced beam

Problem. Determine a suitable UB of S275 steel for the simply supported beam of Section 6.15.1 if twist rotations are effectively prevented at the ends and if a brace is added which effectively prevents lateral deflection v and twist rotation ϕ at mid-span.

Design bending moment.
Using Section 6.15.1, $M_{Ed} = 433 \, \text{kNm}$.

Selecting a trial section.
The central brace divides the beam into two identical segments, each of length $L = 3750 \, \text{mm}$, and eliminates the effect of the load height.
   Guess $f_y = 275 \, \text{N/mm}^2$ and $M_{b,Rd}/W_y f_y = 0.9$.
   Using equation 6.32, $W_{pl,y} \geq (433 \times 10^6/275)/0.9 \, \text{mm}^3 = 1750 \, \text{cm}^3$.
   Try a 457×191 UB 82 with $W_{pl,y} = 1831 \, \text{cm}^3 > 1750 \, \text{cm}^3$.

Section resistance.
As in Section 6.15.3, $f_y = 275 \, \text{N/mm}^2$, $M_{c,Rd} = 503.5 \, \text{kNm} > 433 \, \text{kNm} = M_{Ed}$ and the section resistance is adequate.

Elastic buckling moment.
Using Figure 6.7, $\alpha_m = 1.75$.
   Using equation 6.3,

$$M_{zx} = \sqrt{\left\{ \frac{\pi^2 \times 210\,000 \times 1871 \times 10^4}{3750^2} \times \left(81\,000 \times 69.2 \times 10^4 + \frac{\pi^2 \times 210\,000 \times 0.922 \times 10^{12}}{3750^2} \right) \right\}} \quad \text{Nmm}$$

$$= 727.5 \, \text{kNm}$$

Using equation 6.4,

$$M_{cr} = 1.75 \times 727.5 = 1273 \, \text{kNm}$$

(Using the computer program PRFELB [18] leads to $M_{cr} = 1345 \, \text{kNm}$.)

Member resistance.
Using equation 6.25, $\overline{\lambda}_{LT} = \sqrt{(503.5/1273)} = 0.629$

$$h/b = 460.0/191.3 = 2.40 > 2$$

Using the EC3 simple general method with $\beta = 1.0, \overline{\lambda}_{LT,0} = 0.2$ (Clause 6.3.2.2) and $\alpha_{LT} = 0.34$ (Tables 6.4 and 6.3), and equation 6.27,

$$\Phi_{LT} = 0.5\{1 + 0.34(0.629 - 0.2) + 1.0 \times 0.629^2\} = 0.771$$

Using equation 6.26,

$$M_{b,Rd} = 503.5/\{0.771 + \sqrt{(0.771^2 - 0.629^2)}\} \text{ kNm}$$
$$= 414 \text{ kNm} < 433 \text{ kNm} = M_{Ed}$$

and the beam appears to be inadequate.

Using the EC3 less conservative method with $\beta = 0.75, \overline{\lambda}_{LT,0} = 0.4$ (Clause 6.3.2.3) and $\alpha_{LT} = 0.49$ (Tables 6.5 and 6.3), and equation 6.27,

$$\Phi_{LT} = 0.5\{1 + 0.49(0.629 - 0.4) + 0.75 \times 0.629^2\} = 0.704$$

Using equation 6.26,

$$M_{b,Rd} = 503.5/\{0.704 + \sqrt{(0.704^2 - 0.75 \times 0.629^2)}\} \text{ kNm}$$
$$= 437 \text{ kNm} > 433 \text{ kNm} = M_{Ed}$$

and so the design moment resistance is adequate after all.

6.15.5   Example 5 – checking a braced beam by buckling analysis

Problem. Use the method of elastic buckling analysis given in Section 6.8.2 to check the braced beam of Section 6.15.2.

Elastic buckling analysis.
The application of the method of elastic buckling analysis given in Section 6.8.2 is summarised below, using the same step numbering as in Section 6.8.2.

(2) The beam is fully restrained at mid-span, and so consists of two segments 12 and 23.
For segment 12, $M_1 = -70$ kNm and $M_2 = 122.5$ kNm (as in Section 6.15.2).
$\beta_{m12} = 70/122.5 = 0.571$
Using equation 6.5, $\alpha_{m12} = 1.75 + 1.05 \times 0.571 + 0.3 \times 0.571^2$
$= 2.448 < 2.56$
For segment 23, $\beta_{m23} = 0/122.5 = 0$ and $\alpha_{m23} = 1.75$ as in Section 6.15.2
(3) $L_{cr12} = L_{cr23} = 4500$ mm.

(4) Using $M_{zx12} = M_{zx23} = 111.2\,\text{kNm}$ from Section 6.15.2,

$$M_{cr12} = 2.448 \times 111.2 = 272.3\,\text{kNm}.$$

$$Q_{s12} = 70 \times 272.3/122.5 = 155.6\,\text{kN}.$$

$$M_{cr23} = 1.75 \times 111.2 = 194.6\,\text{kNm}.$$

$$Q_{s23} = 70 \times 194.6/122.5 = 111.2\,\text{kN}.$$

(5) $Q_{s23} < Q_{s12}$ and so 23 is the critical segment.
(6) $\alpha_{r12} = (3 \times 210\,000 \times 571 \times 10^4/4500) \times (1 - 111.2/155.6)$
     $= 227.9 \times 10^6\,\text{Nmm}.$
(7) $\alpha_{23} = 2 \times 210\,000 \times 571 \times 10^4/4500 = 532.9 \times 10^6\,\text{Nmm}.$
(8) $k_2 = 532.9 \times 10^6/(0.5 \times 227.9 \times 10^6 + 532.9 \times 10^6) = 0.824$
     $k_3 = 1.0$ (zero restraint at 3).
(9) Using Figure 3.21a, $k_{cr23} = 0.93, L_{cr23} = 0.93 \times 4500 = 4185\,\text{mm}.$
(10) Using equation 6.39,

$$M_{zx} = \sqrt{\left\{ \dfrac{\pi^2 \times 210\,000 \times 571 \times 10^4}{4185^2} \times \left(81\,000 \times 15.3 \times 10^4 + \dfrac{\pi^2 \times 210\,000 \times 0.0857 \times 10^{12}}{4185^2} \right) \right\}}\ \text{Nmm}$$

$$= 123.4\,\text{kNm}$$

and $M_{cr23} = 1.75 \times 123.4 = 215.9\,\text{kNm}.$

(Using the computer program PRFELB [18] leads to $M_{cr23} = 237.9\,\text{kNm}.$)

Member resistance.
$M_{c,Rd} = 132.8\,\text{kNm}$ as in Section 6.15.2
   Using equation 6.25, $\bar{\lambda}_{LT} = \sqrt{(132.8/215.9)} = 0.784$

$$h/b = 256/146.4 = 1.75 < 2$$

Using the EC3 less conservative method with $\beta = 0.75, \bar{\lambda}_{LT,0} = 0.4$ (Clause 6.3.2.3) and $\alpha_{LT} = 0.34$ (Tables 6.5 and 6.3), and equation 6.27,

$$\Phi_{LT} = 0.5\{1 + 0.34(0.784 - 0.4) + 0.75 \times 0.784^2\} = 0.796$$

Using equation 6.26,

$$M_{b,Rd} = 132.8/\{0.796 + \sqrt{(0.796^2 - 0.75 \times 0.784^2)}\}\,\text{kNm} = 105.6\,\text{kNm}.$$

Using equations 6.29, 6.30, and 6.28,

$$f = 1 - 0.5 \times (1 - 1/\sqrt{1.75})\{1 - 2 \times (0.784 - 0.8)^2\} = 0.878, \text{ and}$$

$$M_{b,Rd,mod} = 105.6/0.878 = 120.2 \, \text{kNm} < 122.5 \, \text{kNm} = M_{Ed}$$

and the design moment resistance is just inadequate.

6.15.6   Example 6 – checking a T-beam

Problem. The 5.0 m long simply supported monosymmetric 229×305 BT 63 T-beam of S275 steel shown in Figure 6.32 has the section properties shown in Figure 6.32c. Determine the uniform bending design moment resistances when the flange is in either compression or tension.

Section resistance (flange in compression).

$$t_f = 19.6 \, \text{mm}, \quad f_y = 265 \, \text{N/mm}^2 \qquad\qquad \text{EN10025-2}$$

$$\varepsilon = \sqrt{(235/265)} = 0.942 \qquad\qquad \text{T5.2}$$

$$c_f/(t_f \varepsilon) = (229.0/2 - 11.9/2 - 12.7)/(19.6 \times 0.942) \qquad \text{T5.2}$$

$$= 5.19 < 9 \text{ and the flange is Class 1.} \qquad\qquad \text{T5.2}$$

The plastic neutral axis bisects the area, and in this case lies in the flange (at $z_p = \{229.0 \times 19.6 + (306.0 - 19.6) \times 11.9\}/(2 \times 229.0) = 17.2$ mm).
    Thus the whole web is in tension, and is automatically Class 1.

$$M_{c,Rd} = 265 \times 531 \times 10^3 \, \text{Nmm} = 140.7 \, \text{kNm}. \qquad\qquad 6.2.5$$

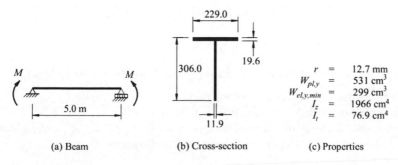

r	=	12.7 mm
$W_{pl,y}$	=	531 cm^3
$W_{el,y,min}$	=	299 cm^3
I_z	=	1966 cm^4
I_t	=	76.9 cm^4

(a) Beam   (b) Cross-section   (c) Properties

Figure 6.32 Example 6.

Elastic buckling (flange in compression).
Using Figure 6.27,

$$d_f = 306.0 - (19.6 + 0)/2 = 296.2 \text{ mm}.$$

$$\bar{z} = \frac{0 + (306.0 - 19.6 - 0) \times (306.0 - 0) \times 11.9/2}{(229.0 \times 19.6) + 0 + (306.0 - 19.6 - 0) \times 11.9} = 66.0 \text{ mm}.$$

$$I_y = (229.0 \times 19.6 \times 66.0^2) + 0 + (306.0 - 19.6 - 0)^3 \times 11.9/12$$
$$+ (306.0 - 19.6 - 0) \times 11.9 \times \{66.0 - (306.0 - 0)/2\}^2$$
$$= 68.64 \times 10^6 \text{ mm}^4$$

$$1 - \rho_m = 0$$

$$z_0 = 0 \times 296.2 - 66.0 = -66.0 \text{ mm}.$$

$$\beta_y = \frac{1}{68.64 \times 10^6} \left\{ \begin{array}{l} (296.2 - 66.0) \times (0 + 0) - 66.0 \\ \times (229^3 \times 19.6/12 + 229 \times 19.6 \times 66.0^2) \\ + [(296.2 - 66.0 - 0)^4 \\ - (66.0 - 19.6/2)^4] \times 11.9/4 \end{array} \right\}$$

$$- 2 \times (-66.0) = 215.6 \text{ mm}.$$

$$I_w = 0$$

$$\frac{\pi^2 E I_z}{L^2} = \frac{\pi^2 \times 210\,000 \times 1966 \times 10^4}{5000^2} = 1.630 \times 10^6 \text{N}$$

$$\frac{\beta_y}{2} \sqrt{\frac{\pi^2 E I_z}{L^2}} = \frac{215.6}{2} \times \sqrt{1.630 \times 10^6} = 137.6 \times 10^3$$

Using equation 6.76,

$$M_{cr} = \sqrt{(1.633 \times 10^6)} \times \left\{ \sqrt{[81\,000 \times 76.9 \times 10^4 + 0 + (137.6 \times 10^3)^2]} \right.$$
$$\left. + 137.6 \times 10^3 \right\} = 540.1 \text{ kNm}.$$

(Using the computer program PRFELB [18] leads to $M_{cr} = 540.1$ kNm.)

Member resistance (flange in compression).
Using equation 6.25, $\bar{\lambda}_{LT} = \sqrt{(140.7/540.3)} = 0.510$
    Using the EC3 simple general method with $\beta = 1.0, \bar{\lambda}_{LT,0} = 0.2$ (Clause 6.3.2.2) and $\alpha_{LT} = 0.76$ (Tables 6.4 and 6.3), and equation 6.27,

$$\Phi_{LT} = 0.5\{1 + 0.76(0.510 - 0.2) + 1.0 \times 0.510^2\} = 0.748$$

Using equation 6.26,

$$M_{b,Rd} = 140.7/\{0.748 + \sqrt{(0.748^2 - 0.510^2)}\} \text{ kNm} = 108.6 \text{ kNm}.$$

Section resistance (flange in tension).
Using Table 4.2 of EN1993-1-5 [60] and the ratio of the web outstand tip compression to tension of

$$\psi = \sigma_2/\sigma_1 = -(66.0 - 19.6/2 - 12.7)/(306.0 - 19.6/2 - 66.0)$$
$$= -0.189,$$

$$k_\sigma = 0.57 - 0.21 \times (-0.189) + 0.07 \times (-0.189)^2 = 0.612, \text{ and}$$

$$21\sqrt{k_\sigma} = 21 \times \sqrt{(0.612)} = 16.4$$

$$c_w/(t_w\varepsilon) = (306.0 - 19.6 - 12.7)/(11.9 \times 0.942) \qquad\qquad \text{T5.2}$$
$$= 24.4 > 16.4 = 21\sqrt{k_\sigma} \text{ and the web is Class 4.} \qquad \text{T5.2}$$

The section may be treated as Class 3 if the effective value of ε is increased so that the Class 3 limit is just satisfied, in which case

$$\varepsilon = \frac{c_w/t_w}{21\sqrt{k_\sigma}} = \frac{(306.0 - 19.6 - 12.7)/11.9}{16.4} = 1.400 \qquad\qquad 5.5.2(9)$$

in which case the maximum design compressive stress is

$$\sigma_{com,Rd} = \frac{f_y/\gamma_{M0}}{\varepsilon^2} = \frac{265/1.0}{1.400^2} = 135 \text{ N/mm}^2 \qquad\qquad 5.5.2(9)$$

and so the section resistance is

$$M_{c,Rd} = 135 \times 299 \times 10^3 \text{ Nmm} = 40.4 \text{ kNm.} \qquad\qquad 6.2.5$$

(If the effective section method of Clause 4.3(4) of EN1993-1-5 [60] is used, then an increased section resistance of $M_{c,Rd} = W_{el,y,min}f_y = 53.2 \text{ kNm}$ is obtained.)

Elastic buckling (flange in tension).
Using Figure 6.27 leads to $\beta_y = -215.6$ mm.
 Using equation 6.76 leads to $M_{cr} = 188.3$ kNm.
 (Using the computer program PRFELB [18] leads to $M_{cr} = 188.3$ kNm.)

Member resistance (flange in tension).
Using equation 6.25, $\overline{\lambda}_{LT} = \sqrt{(40.4/188.3)} = 0.463$
 Using the EC3 simple general method as before and equation 6.27,

$$\Phi_{LT} = 0.5\{1 + 0.76(0.463 - 0.2) + 1.0 \times 0.463^2\} = 0.708$$

Using equation 6.26,

$$M_{b,Rd} = 40.4/\{0.708 + \sqrt{(0.708^2 - 0.463^2)}\} \text{ kNm} = 32.6 \text{ kNm.}$$

6.15.7   Example 7 – checking a stepped beam

Problem. The simply supported non-uniform welded beam of S275 steel shown in Figure 6.33 has a central concentrated design load Q acting at the bottom flange. The beam has a 960×16 web, and its equal flanges are 300×32 in the central 2.0 m region and 300×20 in the outer 3.0 m regions. Determine the maximum value of Q.

Section properties.

$$W_{pl,yi} = 300 \times 32 \times (960 + 32) + 960^2 \times 16/4 = 13.21 \times 10^6 \text{ mm}^3.$$

$$W_{pl,yo} = 300 \times 20 \times (960 + 20) + 960^2 \times 16/4 = 9.566 \times 10^6 \text{ mm}^3.$$

$$I_{zi} = 2 \times 300^3 \times 32/12 = 144 \times 10^6 \text{ mm}^4.$$

$$I_{ti} = 2 \times 300 \times 32^3/3 + 960 \times 16^3/3 = 7.864 \times 10^6 \text{ mm}^4.$$

$$I_{wi} = 144 \times 10^6 \times (960 + 32)^2/4 = 35.43 \times 10^{12} \text{ mm}^6.$$

Section resistances.

$t_{fi} = 32 \text{ mm}, f_y = 265 \text{ N/mm}^2.$	EN10025-2
$t_{fo} = 20 \text{ mm}, f_y = 265 \text{ N/mm}^2.$	EN10025-2
$\varepsilon = \sqrt{(235/265)} = 0.942$	T5.2
$c_{fo}/(t_{fo}\varepsilon) = (300 - 16)/2/(20 \times 0.942)$	T5.2
$\quad = 7.54 < 9$ and the flange is Class 1.	T5.2
$c_w/(t_w\varepsilon) = 960/(16 \times 0.942)$	T5.2
$\quad = 63.7 < 72$ and the web is Class 1.	T5.2
$M_{c,Rd,o} = 265 \times 9.566 \times 10^6 \text{ Nmm} = 2535 \text{ kNm}.$	6.2.5
$M_{c,Rd,i} = 265 \times 13.21 \times 10^6 \text{ Nmm} = 3501 \text{ kNm}.$	6.2.5

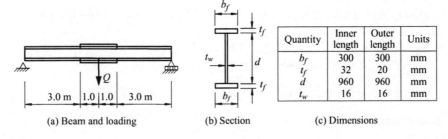

Quantity	Inner length	Outer length	Units
b_f	300	300	mm
t_f	32	20	mm
d	960	960	mm
t_w	16	16	mm

(a) Beam and loading   (b) Section   (c) Dimensions

Figure 6.33 Example 7.

Critical section.
At mid-span, $M_{Ed,4} = Q \times 8/4 = 2.0\,Q$ kNm.
At the change of section, $M_{Ed,3} = (Q/2) \times 3 = 1.5\,Q$ kNm.
$M_{Ed,4}/M_{c,Rd,i} = 2.0 \times Q/3501 = 0.000571\,Q$
$M_{Ed,3}/M_{c,Rd,o} = 1.5 \times Q/2535 = 0.000592\,Q > 0.000571\,Q$
and so the critical section is at the change of section.

Elastic buckling.
For a uniform beam having the properties of the central cross-section, using
equation 6.3,

$$M_{zxi} = \sqrt{\left\{ \frac{\pi^2 \times 210\,000 \times 144 \times 10^6}{8000^2} \times \left(81\,000 \times 7.864 \times 10^6 + \frac{\pi^2 \times 210\,000 \times 35.43 \times 10^{12}}{8000^2} \right) \right\}}\ \text{Nmm}$$

$$= 2885\ \text{kNm}.$$

Using equation 6.12, $N_{cr,z} = \pi^{\cdot 2} \times 210000 \times 144 \times 10^6/8000^2$ N $= 4663$ kN.
    Using Figure 6.7, $\alpha_m = 1.35$

$$\frac{0.4\alpha_m z_Q N_{cr,z}}{M_{zxi}} = \frac{0.4 \times 1.35 \times (960/2) \times 4663 \times 10^3}{2885 \times 10^6} = 0.491$$

Using equation 6.11,

$$M_{cri} = 1.35 \times 2885\{\sqrt{[1 + 0.491^2]} + 0.491\}\ \text{Nmm} = 5854\ \text{kNm}.$$

(Using the computer program PRFELB [18] leads to $M_{cri} = 5972$ kNm.)
    For the non-uniform beam,

$$L_r/L = 6.0/8.0 = 0.75$$

$$\left(0.6 + 0.4\frac{d_{min}}{d_{max}} \right) \frac{A_{f\,min}}{A_{f\,max}} = \left(0.6 + 0.4 \times \frac{960 + 2 \times 20}{960 + 2 \times 32} \right) \frac{300 \times 20}{300 \times 32} = 0.619$$

Using equation 6.80,

$$\alpha_{st} = 1.0 - 1.2 \times \left(\frac{6000}{8000} \right) \times (1 - 0.619) = 0.657$$

$$M_{cr} = 0.657 \times 5854 = 3847\ \text{kNm}.$$

(Using the computer program PRFELB [18] leads to $M_{cr} = 4420$ kNm.)

Moment resistance.
At the critical section, $M_{cro} = 3847 \times 3000/4000 = 2885$ kNm.

Using the values for the critical section in equation 6.25, $\bar{\lambda}_{LT} = \sqrt{(2535/2885)} = 0.937$

$$h/b = (960 + 2 \times 20)/300 = 3.33 > 2.$$

Using the EC3 simple general method with $\beta = 1.0$ and $\bar{\lambda}_{LT,0} = 0.2$ (Clause 6.3.2.2) and $\alpha_{LT} = 0.34$ (Tables 6.4 and 6.3) and equation 6.27,

$$\Phi_{LT} = 0.5\{1 + 0.34(0.937 - 0.2) + 0.937^2\} = 1.065$$

Using equation 6.26,

$$M_{b,Rd} = 2535/\{1.065 + \sqrt{(1.065^2 - 0.937^2)}\} \text{ kNm}$$
$$= 1615 \text{ kNm} = (Q_{b,Rd}/2) \times 3.0$$
$$Q_{b,Rd} = 1615 \times 2/3.0 = 1077 \text{ kN}.$$

6.16   Unworked examples

6.16.1   Example 8 – checking a continuous beam

A continuous 533×210 UB 92 beam of S275 steel has two equal spans of 7.0 m. A uniformly distributed design load q is applied along the centroidal axis. Determine the maximum value of q.

6.16.2   Example 9 – checking an overhanging beam

The overhanging beam shown in Figure 6.34a is prevented from twisting and deflecting laterally at its end and supports. The supported segment is a $250 \times 150 \times 10$ RHS of S275 steel and is rigidly connected to the overhanging segment which is a 254×146 UB 37 of S275 steel. Determine the maximum design load Q that can be applied at the end of the UB.

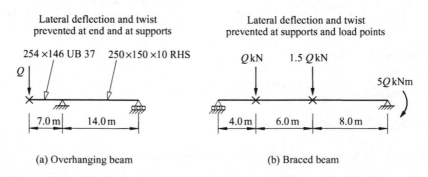

(a) Overhanging beam   (b) Braced beam

Figure 6.34 Unworked examples 9 and 10.

6.16.3   Example 10 – checking a braced beam

Determine the maximum moment at elastic buckling of the braced 533×210 UB 92 of S275 steel shown in Figure 6.34b.

6.16.4   Example 11 – checking a monosymmetric beam

A simply supported 6.0 m long 530×210 UB 92 beam of S275 steel has equal and opposite end moments. The moment resistance is to be increased by welding a full length 25×16 mm plate of S275 steel to one flange. Determine the increases in the moment resistances of the beam when the plate is welded to

(a) the compression flange, or
(b) the tension flange.

6.16.5   Example 12 – checking a non-uniform beam

A simply supported 530×210 UB 92 beam of S275 steel has a concentrated shear centre load at the centre of the 10.0 m span. The moment resistance is to be increased by welding 250×16 plates of S275 steel to the top and bottom flanges over a central length of 4.0 m. Determine the increase in the moment resistance of the beam.

References

1. Vacharajittiphan P., Woolcock, S.T., and Trahair, N.S (1974) Effect of in-plane deformation on lateral buckling, *Journal of Structural Mechanics,* **3**, pp. 29–60.
2. Nethercot, D.A. and Rockey, K.C. (1971) A unified approach to the elastic lateral buckling of beams, *The Structural Engineer,* **49**, pp. 321–30.
3. Timoshenko, S.P. and Gere, J.M. (1961) *Theory of Elastic Stability,* 2nd edition, McGraw-Hill, New York.
4. Bleich, F. (1952) *Buckling Strength of Metal Structures,* McGraw-Hill, New York.
5. Structural Stability Research Council (1998) *Guide to Design Criteria for Metal Compression Members,* 5th edition, (ed. T.V. Galambos) John Wiley, New York.
6. Barsoum, R.S. and Gallagher, R.H. (1970) Finite element analysis of torsional and torsional-flexural stability problems, *International Journal of Numerical Methods in Engineering,* **2**, pp. 335–52.
7. Krajcinovic, D. (1969) A consistent discrete elements technique for thin-walled assemblages, *International Journal of Solids and Structures,* **5**, pp. 639–62.
8. Powell, G. and Klingner, R. (1970) Elastic lateral buckling of steel beams, *Journal of the Structural Division, ASCE,* **96**, No. ST9, pp. 1919–32.
9. Nethercot, D.A. and Rockey, K.C. (1971) Finite element solutions for the buckling of columns and beams, *International Journal of Mechanical Sciences,* **13**, pp. 945–9.
10. Hancock, G.J. and Trahair, N.S. (1978) Finite element analysis of the lateral buckling of continuously restrained beam-columns, *Civil Engineering Transactions, Institution of Engineers, Australia,* **CE20**, No. 2, pp. 120 7.

11. Brown, P.T. and Trahair, N.S. (1968) Finite integral solution of differential equations, *Civil Engineering Transactions, Institution of Engineers, Australia*, **CE10**, pp. 193–6.
12. Trahair, N.S. (1968) Elastic stability of propped cantilevers, *Civil Engineering Transactions, Institution of Engineers, Australia*, **CE10**, pp. 94–100.
13. Column Research Committee of Japan (1971) *Handbook of Structural Stability*, Corona, Tokyo.
14. Lee, G.C. (1960) A survey of literature on the lateral instability of beams, *Welding Research Council Bulletin*, No. 63, August.
15. Clark, J.W. and Hill, H.N. (1960) Lateral buckling of beams, *Journal of the Structural Division, ASCE*, **86**, No. ST7, pp. 175–96.
16. Trahair, N.S. (1993) *Flexural-Torsional Buckling of Structures*, E and FN Spon, London.
17. Galambos, T.V. (1968) *Structural Members and Frames*, Prentice-Hall, Englewood Cliffs, New Jersey.
18. Papangelis, J.P., Trahair, N.S., and Hancock, G.J. (1998) Elastic flexural-torsional buckling of structures by computer, *Computers and Structures*, **68**, pp. 125–37.
19. Nethercot, D.A. and Trahair, N.S. (1976) Inelastic lateral buckling of determinate beams, *Journal of the Structural Division, ASCE*, **102**, No. ST4, pp. 701–17.
20. Trahair, N.S. (1983) Inelastic lateral buckling of beams, Chapter 2 in *Beams and Beam-Columns. Stability and Strength*, (ed. R. Narayanan), Applied Science Publishers, London, pp. 35–69.
21. Nethercot, D.A. (1972) Recent progress in the application of the finite element method to problems of the lateral buckling of beams, *Proceedings of the EIC Conference on Finite Element Methods in Civil Engineering*, Montreal, June, pp. 367–91.
22. Vacharajittiphan, P. and Trahair, N.S. (1975) Analysis of lateral buckling in plane frames, *Journal of the Structural Division, ASCE*, **101**, No. ST7, pp. 1497–516.
23. Vacharajittiphan, P. and Trahair, N.S. (1974) Direct stiffness analysis of lateral buckling, *Journal of Structural Mechanics*, **3**, pp. 107–37.
24. Mutton, B.R. and Trahair, N.S. (1973) Stiffness requirements for lateral bracing, *Journal of the Structural Division, ASCE*, **99**, No. ST10, pp. 2167–82.
25. Mutton, B.R. and Trahair, N.S. (1975) Design requirements for column braces, *Civil Engineering Transactions, Institution of Engineers, Australia*, **CE17**, No. 1, pp. 30–5.
26. Trahair, N.S. (1999) Column bracing forces, *Australian Journal of Structural Engineering, Institution of Engineers, Australia*, **SE2**, Nos 2,3, pp. 163–8.
27. Nethercot, D.A. (1973) Buckling of laterally or torsionally restrained beams, *Journal of the Engineering Mechanics Division, ASCE*, **99**, No. EM4, pp. 773–91.
28. Trahair, N.S. and Nethercot, D.A. (1984) Bracing requirements in thin-walled structures, Chapter 3 of *Developments in Thin-Walled Structures – 2*, (eds J. Rhodes and A.C. Walker), Elsevier Applied Science Publishers, Barking, pp. 93–130.
29. Valentino, J., Pi, Y.-L., and Trahair, N.S. (1997) Inelastic buckling of steel beams with central torsional restraints, *Journal of Structural Engineering, ASCE*, **123**, No. 9, pp. 1180–6.
30. Valentino, J. and Trahair, N.S. (1998) Torsional restraint against elastic lateral buckling, *Journal of Structural Engineering, ASCE*, **124**, No. 10, pp. 1217–25.
31. Nethercot, D.A. and Trahair, N.S. (1975) Design of diaphragm braced I-beams, *Journal of the Structural Division, ASCE*, **101**, No. ST10, pp. 2045–61.

32. Trahair, N.S. (1979) Elastic lateral buckling of continuously restrained beam-columns, *The Profession of a Civil Engineer,* (eds D. Campbell-Allen and E.H. Davis), Sydney University Press, Sydney, pp. 61–73.

33. Austin, W.J., Yegian, S., and Tung, T.P. (1955) Lateral buckling of elastically end-restrained beams, *Proceedings of the ASCE,* **81**, No. 673, April, pp. 1–25.

34. Trahair, N.S. (1965) Stability of I-beams with elastic end restraints, *Journal of the Institution of Engineers, Australia,* **37**, pp. 157–68.

35. Trahair, N.S. (1966) Elastic stability of I-beam elements in rigid-jointed structures, *Journal of the Institution of Engineers, Australia,* **38**, pp. 171–80.

36. Nethercot, D.A. and Trahair, N.S. (1976) Lateral buckling approximations for elastic beams, *The Structural Engineer,* **54**, pp. 197–204.

37. Trahair, N.S. (1983) Lateral buckling of overhanging beams, *Instability and Plastic Collapse of Steel Structures,* (ed. L.J. Morris), Granada, London, pp. 503–18.

38. Bradford, M.A. and Trahair, N.S. (1981) Distortional buckling of I-beams, *Journal of the Structural Division, ASCE,* **107**, No. ST2, pp. 355–70.

39. Bradford, M.A. and Trahair, N.S. (1983) Lateral stability of beams on seats, *Journal of Structural Engineering, ASCE,* **109**, No. ST9, pp. 2212–15.

40. Bradford, M.A. (1992) Lateral-distortional buckling of steel I-section members, *Journal of Constructional Steel Research,* **23**, Nos 1–3, pp. 97–116.

41. Trahair, N.S. (1966) The bending stress rules of the draft ASCA1, *Journal of the Institution of Engineers, Australia,* **38**, pp. 131–41.

42. Anderson, J.M. and Trahair, N.S. (1972) Stability of monosymmetric beams and cantilevers, *Journal of the Structural Division, ASCE,* **98**, No. ST1, pp. 269–86.

43. Nethercot, D.A. (1973) The effective lengths of cantilevers as governed by lateral buckling, *The Structural Engineer,* **51**, pp. 161–8.

44. Trahair, N.S. (1968) Interaction buckling of narrow rectangular continuous beams, *Civil Engineering Transactions, Institution of Engineers, Australia,* **CE10**, No. 2, pp. 167–72.

45. Salvadori, M.G. (1951) Lateral buckling of beams of rectangular cross-section under bending and shear, *Proceedings of the First US National Congress of Applied Mechanics,* pp. 403–5.

46. Kitipornchai, S., and Trahair, N.S. (1980) Buckling properties of monosymmetric I-beams, *Journal of the Structural Division, ASCE,* **106**, No. ST5, pp. 941–57.

47. Kitipornchai, S., Wang, C.M., and Trahair, N.S. (1986) Buckling of monosymmetric I-beams under moment gradient, *Journal of Structural Engineering, ASCE,* **112**, No. 4, pp. 781–99.

48. Wang, C.M. and Kitipornchai, S. (1986) Buckling capacities of monosymmetric I-beams, *Journal of Structural Engineering, ASCE,* **112**, No. 11, pp. 2373–91.

49. Woolcock, S.T. Kitipornchai, S., and Bradford, M.A. (1999) *Design of Portal Frame Buildings,* 3rd edn., Australian Institute of Steel Construction, Sydney.

50. Wang, C.M. and Kitipornchai, S. (1986) On stability of monosymmetric cantilevers, *Engineering Structures,* **8**, No. 3, pp. 169–80.

51. Kitipornchai, S. and Trahair, N.S. (1972) Elastic stability of tapered I-beams, *Journal of the Structural Division, ASCE,* **98**, No. ST3, pp. 713–28.

52. Nethercot, D.A. (1973) Lateral buckling of tapered beams, *Publications, IABSE,* **33**, pp. 173–92.

53. Bradford, M.A. and Cuk, P.E. (1988) Elastic buckling of tapered monosymmetric I-beams, *Journal of Structural Engineering, ASCE,* **114**, No. 5, pp. 977–96.

54. Kitipornchai, S., and Trahair, N.S. (1975) Elastic behaviour of tapered monosymmetric I-beams, *Journal of the Structural Division, ASCE,* **101**, No. ST8, pp. 1661–78.
55. Bradford, M.A. (1989) Inelastic buckling of tapered monosymmetric I-beams, *Engineering Structures*, **11**, No. 2, pp. 119–126.
56. Lee, G.C., Morrell, M.L., and Ketter, R.L. (1972) Design of tapered members, *Bulletin* 173, Welding Research Council, June.
57. Bradford, M.A. (1988) Stability of tapered I-beams, *Journal of Constructional Steel Research*, **9**, pp. 195–216.
58. Bradford, M.A. (2006) Lateral buckling of tapered steel members, Chapter 1 in *Analysis and Design of Plated Structures. Vol.1:Stability*, Woodhead Publishing Ltd, Cambridge, pp. 1–25.
59. Trahair, N.S. and Kitipornchai, S. (1971) Elastic lateral buckling of stepped I-beams, *Journal of the Structural Division, ASCE,* **97**, No. ST10, pp. 2535–48.
60. British Standards Institution (2006) Eurocode 3: Design of Steel Structures – Part 1–5: Plated Structural Elements, BSI, London.

Chapter 7

Beam-columns

7.1 Introduction

Beam-columns are structural members which combine the beam function of transmitting transverse forces or moments with the compression (or tension) member function of transmitting axial forces. Theoretically, all structural members may be regarded as beam-columns, since the common classifications of tension members, compression members, and beams are merely limiting examples of beam-columns. However, the treatment of beam-columns in this chapter is generally limited to members in axial compression. The behaviour and design of members with moments and axial tension are treated in Chapter 2.

Beam-columns may act as if isolated, as in the case of eccentrically loaded compression members with simple end connections, or they may form part of a rigid frame. In this chapter, the behaviour and design of isolated beam-columns are treated. The ultimate resistance and design of beam-columns in frames are discussed in Chapter 8.

It is convenient to discuss the behaviour of isolated beam-columns under the three separate headings of In-Plane Behaviour, Flexural–Torsional Buckling, and Biaxial-Bending, as is done in Sections 7.2–7.4. When a beam-column is bent about its weaker principal axis, or when it is prevented from deflecting laterally while being bent about its stronger principal axis (as shown in Figure 7.1a), its action is confined to the plane of bending. This in-plane behaviour is related to the bending of beams discussed in Chapter 5 and to the buckling of compression members discussed in Chapter 3. When a beam-column which is bent about its stronger principal axis is not restrained laterally (as shown in Figure 7.1b), it may buckle prematurely out of the plane of bending by deflecting laterally and twisting. This action is related to the flexural–torsional buckling of beams (referred to as lateral–torsional buckling in EC3) discussed in Chapter 6. More generally, however, a beam-column may be bent about both principal axes, as shown in Figure 7.1c. This biaxial bending, which commonly occurs in three-dimensional rigid frames, involves interactions of beam bending and twisting with beam and column buckling. Sections 7.2–7.4 discuss the in-plane behaviour, flexural–torsional buckling, and biaxial bending of isolated beam-columns.

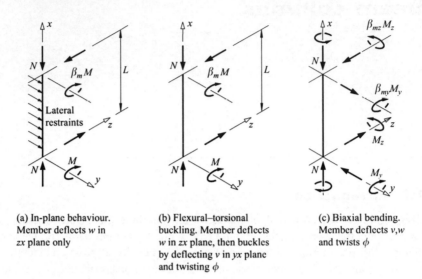

(a) In-plane behaviour. Member deflects w in zx plane only

(b) Flexural–torsional buckling. Member deflects w in zx plane, then buckles by deflecting v in yx plane and twisting ϕ

(c) Biaxial bending. Member deflects v,w and twists ϕ

Figure 7.1 Beam-column behaviour.

7.2 In-plane behaviour of isolated beam-columns

When the deformations of an isolated beam-column are confined to the plane of bending (see Figure 7.1a), its behaviour shows an interaction between beam bending and compression member buckling, as indicated in Figure 7.2. Curve 1 of this figure shows the linear behaviour of an elastic beam, while curve 6 shows the limiting behaviour of a rigid-plastic beam at the full plastic moment M_{pl}. Curve 2 shows the transition of a real elastic – plastic beam from curve 1 to curve 6. The elastic buckling of a concentrically loaded compression member at its elastic buckling load $N_{cr,y}$ is shown by curve 4. Curve 3 shows the interaction between bending and buckling in an elastic member, and allows for the additional moment $N\delta$ exerted by the axial load. Curve 7 shows the interaction between the bending moment and axial force which causes the member to become fully plastic. This curve allows for reduction from the full plastic moment M_{pl} to $M_{pl,r}$ caused by the axial load, and for the additional moment $N\delta$. The actual behaviour of a beam-column is shown by curve 5 which provides a transition from curve 3 for an elastic member to curve 7 for full plasticity.

7.2.1 Elastic beam-columns

A beam-column is shown in Figures 7.1a and 7.3a which is acted on by axial forces N and end moments M and $\beta_m M$, where β_m can have any value between -1 (single curvature bending) and $+1$ (double curvature bending). For isolated

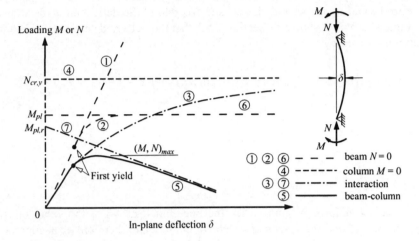

Figure 7.2 In-plane behaviour of a beam-column.

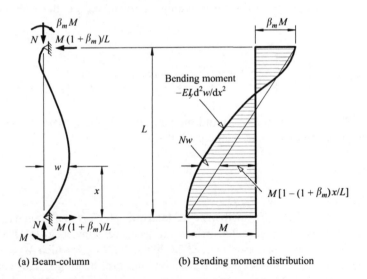

(a) Beam-column   (b) Bending moment distribution

Figure 7.3 In-plane bending moment distribution in a beam-column.

beam-columns, these end moments are independent of the deflections, and have no effect on the elastic buckling load $N_{cr,y} = \pi^2 EI_y/L^2$. (They are therefore quite different from the end restraining moments discussed in Section 3.6.3 which increase with the end rotations and profoundly affect the elastic buckling load.) The beam-column is prevented from bending about the minor axis out of the plane of the end moments, and is assumed to be straight before loading.

The bending moment in the beam-column is the sum of $M - M(1 + \beta_m)x/L$ due to the end moments and shears, and Nw due to the deflection w, as shown in Figure 7.3b. It is shown in Section 7.5.1 that the deflected shape of the beam-column is given by

$$w = \frac{M}{N}\left[\cos \mu x - (\beta_m \text{cosec } \mu L + \cot \mu L) \sin \mu x - 1 + (1 + \beta_m)\frac{x}{L}\right],$$

$$(7.1)$$

where

$$\mu^2 = \frac{N}{EI_y} = \frac{\pi^2}{L^2}\frac{N}{N_{cr,y}}.$$

$$(7.2)$$

As the axial force N approaches the buckling load $N_{cr,y}$, the value of μL approaches π, and the values of $\text{cosec}\mu L$, $\cot \mu L$, and therefore of w approach infinity, as indicated by curve 3 in Figure 7.2. This behaviour is similar to that of a compression member with geometrical imperfections (see Section 3.2.2). It is also shown in Section 7.5.1 that the maximum moment M_{max} in the beam-column is given by

$$M_{max} = M\left[1 + \left\{\beta_m\text{cosec } \pi \sqrt{\frac{N}{N_{cr,y}}} + \cot \pi \sqrt{\frac{N}{N_{cr,y}}}\right\}^2\right]^{0.5},$$

$$(7.3)$$

when $\beta_m < -\cos \pi \sqrt{(N/N_{cr,y})}$ and the point of maximum moment lies in the span, and is given by the end moment M, that is

$$M_{max} = M,$$

$$(7.4)$$

when $\beta_m \geq -\cos \pi \sqrt{(N/N_{cr,y})}$.

The variations of M_{max}/M with the axial load ratio $N/N_{cr,y}$ and the end moment ratio β_m are shown in Figure 7.4. In general, M_{max} remains equal to M for low values of $N/N_{cr,y}$ but increases later. The value of $N/N_{cr,y}$ at which M_{max} begins to increase above M is lowest for $\beta_m = -1$ (uniform bending), and increases with increasing β_m. Once M_{max} departs from M, it increases slowly at first, then more rapidly, and reaches very high values as N approaches the column buckling load $N_{cr,y}$.

For beam-columns which are bent in single curvature by equal and opposite end moments ($\beta_m = -1$), the maximum deflection δ_{max} is given by (see Section 7.5.1)

$$\frac{\delta_{max}}{\delta} = \frac{8/\pi^2}{N/N_{cr,y}}\left[\sec \frac{\pi}{2}\sqrt{\frac{N}{N_{cr,y}}} - 1\right]$$

$$(7.5)$$

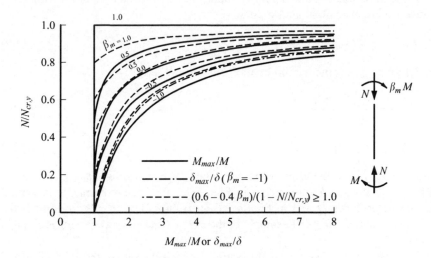

Figure 7.4 Maximum moments and deflections in elastic beam-columns.

in which $\delta = ML^2/8EI_y$ is the value of δ_{max} when $N = 0$, and the maximum moment by

$$M_{max} = M \sec \frac{\pi}{2} \sqrt{\frac{N}{N_{cr,y}}}. \tag{7.6}$$

These two equations are shown non-dimensionally in Figure 7.4. It can be seen that they may be approximated by using a factor $1/(1 - N/N_{cr,y})$ to amplify the deflection δ and the moment M due to the applied moments alone $(N = 0)$. This same approximation was used in Section 3.6.3 to estimate the reduced stiffness of an axially loaded restraining member.

For beam-columns with unequal end moments $(\beta_m > -1)$, the value of M_{max} may be approximated by using

$$\frac{M_{max}}{M} = \frac{C_m}{1 - N/N_{cr,y}} \geq 1.0 \tag{7.7}$$

in which

$$C_m = 0.6 - 0.4\beta_m. \tag{7.8}$$

These approximations are also shown in Figure 7.4. It can be seen that they are generally conservative for $\beta_m > 0$, and only a little unconservative for $\beta_m < 0$.

Approximations for the maximum moments M_{max} in beam-columns with transverse loads can be obtained by using

$$M_{max} = \delta M_{max,0} \tag{7.9}$$

in which $M_{max,0}$ is the maximum moment when $N = 0$,

$$\delta = \frac{\gamma_m(1 - \gamma_s N/N_{cr})}{(1 - \gamma_n N/N_{cr})}, \tag{7.10}$$

$$N_{cr} = \pi^2 EI/(k_{cr}^2 L^2), \tag{7.11}$$

and $k_{cr} = L_{cr}/L$ is the effective length ratio. Expressions for $M_{max,0}$ and values of γ_m, γ_n, γ_s, and k_{cr} are given in Figure 7.5 for beam-columns with central concentrated loads Q and in Figure 7.6 for beam-columns with uniformly distributed loads q. A worked example of the use of these approximations is given in Section 7.7.1.

The maximum stress σ_{max} in the beam-column is the sum of the axial stress and the maximum bending stress caused by the maximum moment M_{max}. It is therefore given by

$$\sigma_{max} = \sigma_{ac} + \sigma_{bcy}\frac{M_{max}}{M} \tag{7.12}$$

Beam-column	First-order moment ($N=0$)	k_{cr}	γ_m	γ_n	γ_s	$M_{max,0}$
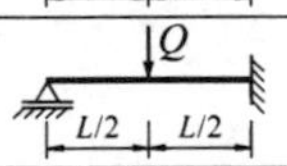 Q, $L/2$ $L/2$	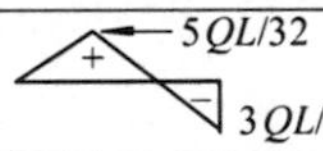 $QL/4$ (+)	1.0	1.0	1.0	0.18	$\dfrac{QL}{4}$
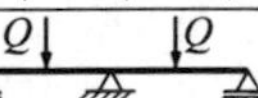 Q	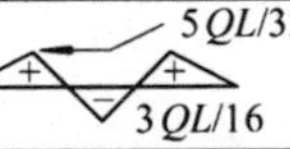 $-QL$	2.0	1.0	1.0	0.19	QL
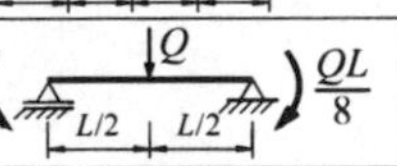 Q, $\dfrac{3QL}{16}$, $L/2$ $L/2$	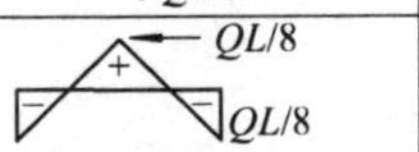 $5QL/32$ (+), $3QL/16$	1.0	$\dfrac{5}{6}$	1.0	0.45	$\dfrac{3QL}{16}$
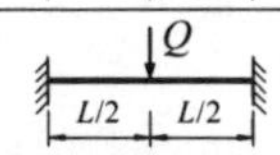 Q, $L/2$ $L/2$	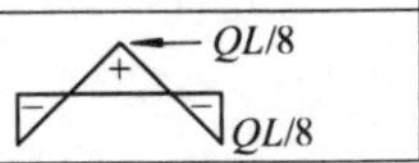 $5QL/32$ (+), $3QL/16$	0.7	1.0	1.0	0.28	$\dfrac{3QL}{16}$
 Q Q, $L/2$ $L/2$ $L/2$ $L/2$	 $5QL/32$ (+ +), $3QL/16$	1.0	1.0	0.49	0.14	$\dfrac{3QL}{16}$
$\dfrac{QL}{8}$ (Q) $\dfrac{QL}{8}$, $L/2$ $L/2$	$QL/8$ (+), $QL/8$	1.0	1.0	1.0	0.62	$\dfrac{QL}{8}$
Q, $L/2$ $L/2$	$QL/8$ (+), $QL/8$	0.5	1.0	1.0	0.18	$\dfrac{QL}{8}$

Figure 7.5 Approximations for beam-columns with central concentrated loads.

Beam-column	First-order moment ($N = 0$)	k_{cr}	γ_m	γ_n	γ_s	$M_{max,0}$
	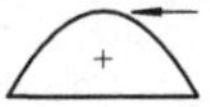 $qL^2/8$	1.0	1.0	1.0	−0.03	$\dfrac{qL^2}{8}$
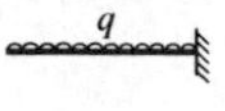	 $qL^2/2$	2.0	1.0	1.0	0.4	$\dfrac{qL^2}{2}$
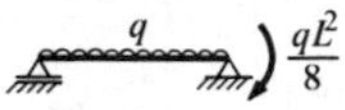 $\dfrac{qL^2}{8}$	 $9qL^2/128$ $qL^2/8$	1.0	$\dfrac{9}{16}$	1.0	0.29	$\dfrac{qL^2}{8}$
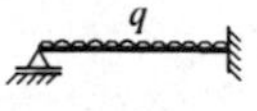	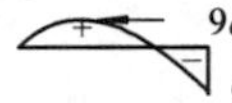 $9qL^2/128$ $qL^2/8$	0.7	1.0	1.0	0.36	$\dfrac{qL^2}{8}$
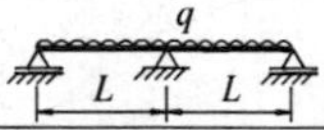 L L	 $9qL^2/128$ $qL^2/8$	1.0	1.0	0.49	0.18	$\dfrac{qL^2}{8}$
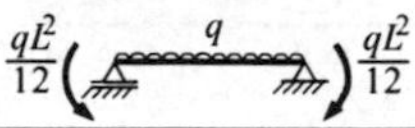 $\dfrac{qL^2}{12}$ $\dfrac{qL^2}{12}$	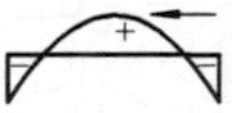 $qL^2/24$ $qL^2/12$	1.0	0.5	1.0	0.4	$\dfrac{qL^2}{12}$
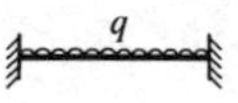	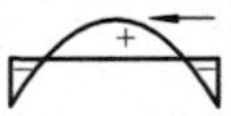 $qL^2/24$ $qL^2/12$	0.5	1.0	1.0	0.37	$\dfrac{qL^2}{12}$

Figure 7.6 Approximations for beam-columns with distributed loads.

in which $\sigma_{ac} = N/A$ and $\sigma_{bcy} = M/W_{el,y}$. If the member has no residual stresses, then it will remain elastic while σ_{max} is less than the yield stress f_y, and so the above results are valid while

$$\frac{N}{N_y} + \frac{M}{M_y}\frac{M_{max}}{M} \leq 1.0 \qquad (7.13)$$

in which $N_y = Af_y$ is the squash load and $M_y = f_y W_{el}$ the nominal first yield moment. A typical elastic limit of this type is shown by the first yield point marked on curve 3 of Figure 7.2. It can be seen that this limit provides a lower bound estimate of the resistance of a straight beam-column, while the elastic buckling load $N_{cr,y}$ provides an upper bound.

Variations of the first yield limits of N/N_y determined from equation 7.13 with M/M_y and β_m are shown in Figure 7.7a for the case where $N_{cr,y} = N_y/1.5$. For a beam-column in double curvature bending ($\beta_m = 1$), $M_{max} = M$, and so N varies linearly with M until the elastic buckling load $N_{cr,y}$ is reached. For a beam-column in uniform bending ($\beta_m = -1$), the relationship between N and M is non-linear and concave due to the amplification of the bending moment from M to M_{max} by the axial load. Also shown in Figure 7.7a are solutions of equation 7.13 based on the use of the approximations of equations 7.7 and 7.8. A comparison of these with the accurate solutions also shown in Figure 7.7a again demonstrates the comparative accuracy of the approximate equations 7.7 and 7.8.

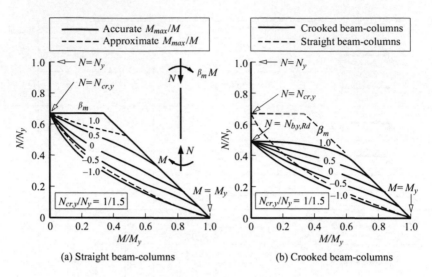

Figure 7.7 First yield of beam-columns with unequal end moments.

The elastic in-plane behaviour of members in which the axial forces cause tension instead of compression can also be analysed. The maximum moment in such a member never exceeds M, and so a conservative estimate of the elastic limit can be obtained from

$$\frac{N}{N_y} + \frac{M}{M_y} \le 1.0. \tag{7.14}$$

This forms the basis for the methods of designing tension members for in-plane bending, as discussed in Section 2.4.

7.2.2 Fully plastic beam-columns

An upper bound estimate of the resistance of an I-section beam-column bent about its major axis can be obtained from the combination of bending moment $M_{pl,r}$ and axial force $N_{y,r}$ which causes the cross-section to become fully plastic. A particular example is shown in Figure 7.8, for which the distance z_n from the centroid to the unstrained fibre is less than $(h - 2t_f)/2$. This combination of moment and force lies between the two extreme combinations for members with bending moment only ($N = 0$), which become fully plastic at

$$M_{pl} = f_y b_f t_f (h - t_f) + f_y t_w \left(\frac{h - 2t_f}{2}\right)^2, \tag{7.15}$$

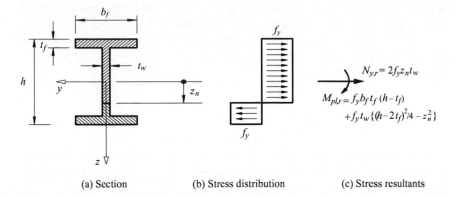

(a) Section   (b) Stress distribution   (c) Stress resultants

Figure 7.8 Fully plastic cross-section.

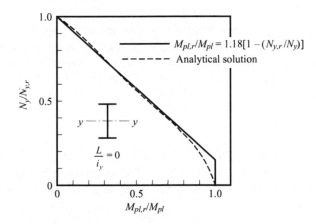

Figure 7.9 Interaction curves for 'zero length' members.

and members with axial force only ($M = 0$), which become fully plastic at

$$N_y = f_y[2b_f t_f + (h - 2t_f)t_w].$$ (7.16)

The variations of $M_{pl,r}$ and $N_{y,r}$ are shown by the dashed interaction curve in Figure 7.9. These combinations can be used directly for very short beam-columns for which the bending moment at the fully plastic cross-section is approximately equal to the applied end moment M.

Also shown in Figure 7.9 is the approximation

$$\frac{M_{pl,r,y}}{M_{pl,y}} = 1.18\left[1 - \frac{N_{y,r}}{N_y}\right] \leq 1.0,$$ (7.17)

which is generally close to the accurate solution, except when $N_{y,r} \approx 0.15N_y$, when the maximum error is of the order of 5%. This small error is quite acceptable since the accurate solutions ignore the strengthening effects of strain-hardening.

A similar analysis may be made of an I-section beam-column bent about its minor axis. In this case, a satisfactory approximation for the load and moment at full plasticity is given by

$$\frac{M_{pl,r,z}}{M_{pl,z}} = 1.19 \left[1 - \left(\frac{N_{y,r}}{N_y}\right)^2 \right] \leq 1.0. \tag{7.18}$$

Beam-columns of more general cross-section may also be analysed to determine the axial load and moment at full plasticity. In general, these may be safely approximated by the linear interaction equation

$$\frac{M_{pl,r}}{M_{pl}} = 1 - \frac{N_{y,r}}{N_y}. \tag{7.19}$$

In a longer beam-column, instability effects become important, and failure occurs before any section becomes fully plastic. A further complication arises from the fact that as the beam-column deflects by δ, its maximum moment increases from the nominal value M to $(M + N\delta)$. Thus, the value of M at which full plasticity occurs is given by

$$M = M_{pl,r} - N\delta, \tag{7.20}$$

and so the value of M for full plasticity decreases from $M_{pl,r}$ as the deflection δ increases, as shown by curve 7 in Figure 7.2. This curve represents an upper bound to the behaviour of beam-columns which is only approached after the maximum strength is reached.

7.2.3   Ultimate resistance

7.2.3.1   General

An isolated beam-column reaches its ultimate resistance at a load which is greater than that which causes first yield (see Section 7.2.1), but is less than that which causes a cross-section to become fully plastic (see Section 7.2.2), as indicated in Figure 7.2. These two bounds are often far apart, and when a more accurate estimate of the resistance is required, an elastic–plastic analysis of the imperfect beam-column must be made. Two different approximate analytical approaches to beam-column strength may be used, and these are related to the initial crookedness and residual stress methods discussed in Sections 3.2.2 and 3.3.4 of allowing for the effects of imperfections on the resistances of real compression members. These two approaches are discussed below.

7.2.3.2 Elastic–plastic resistances of straight beam-columns

The resistance of an initially straight beam-column with residual stresses may be found by analysing numerically its non-linear elastic–plastic behaviour and determining its maximum resistance [1]. Some of these resistances [2, 3] are shown as interaction plots in Figure 7.10. Figure 7.10a shows how the ultimate load N and the end moment M vary with the major axis slenderness ratio L/i_y for beam-columns with equal and opposite end moments ($\beta_m = -1$), while Figure 7.10b shows how these vary with the end moment ratio β_m for a slenderness ratio of $L/i_y = 60$.

It has been proposed that these ultimate resistances can be simply and closely approximated by using the values of M and N which satisfy the interaction equations of both equation 7.17 and

$$\frac{N}{N_{b,y,Rd}} + \frac{C_m}{(1 - N/N_{cr,y})}\frac{M}{M_{pl,y}} \leq 1 \tag{7.21}$$

in which $N_{b,y,Rd}$ is the ultimate resistance of a concentrically loaded column (the beam-column with $M = 0$) which fails by deflecting about the major axis, $N_{cr,y}$ is the major axis elastic buckling load of this concentrically loaded column, and C_m is given by

$$C_m = 0.6 - 0.4\beta_m \geq 0.4. \tag{7.22}$$

Equation 7.17 ensures that the reduced plastic moment $M_{pl,r}$ is not exceeded by the end moment M, and represents the resistances of very short members

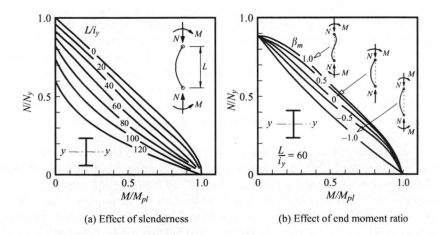

(a) Effect of slenderness

(b) Effect of end moment ratio

Figure 7.10 Ultimate resistance interaction curves.

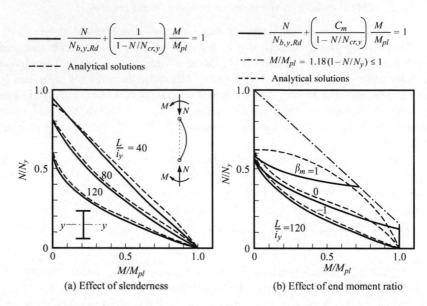

$$\rule{2cm}{0.4pt} \qquad \frac{N}{N_{b,y,Rd}} + \left(\frac{1}{1 - N/N_{cr,y}}\right)\frac{M}{M_{pl}} = 1$$

$- - - -$ Analytical solutions

$$\rule{2cm}{0.4pt} \qquad \frac{N}{N_{b,y,Rd}} + \left(\frac{C_m}{1 - N/N_{cr,y}}\right)\frac{M}{M_{pl}} = 1$$

$-\cdot - \cdot -$ $M/M_{pl} = 1.18(1 - N/N_y) \le 1$

$- - - -$ Analytical solutions

(a) Effect of slenderness

(b) Effect of end moment ratio

Figure 7.11 Interaction formulae.

$(L/i_y \to 0)$, and of some members which are not bent in single curvature (i.e. $\beta_m > 0$), as shown, for example, in Figure 7.11b.

Equation 7.21 represents the interaction between buckling and bending which determines the resistance of more slender members. If the case of equal and opposite end moments is considered first ($\beta_m = -1$ and $C_m = 1$), then it will be seen that equation 7.21 includes the same approximate amplification factor $1/(1 - N/N_{cr,y})$ as does equation 7.7 for the maximum moments in elastic beam-columns. It is of a similar form to the first yield condition of equation 7.13, except that a conversion to ultimate resistance has been made by using the ultimate resistance $N_{b,y,Rd}$, and moment $M_{pl,y}$ instead of the squash load N_y and the yield moment M_y. These substitutions ensure that this interaction formula gives the same limit predictions for concentrically loaded columns ($M = 0$) and for beams ($N = 0$) as do the treatments given in Chapters 3 and 5. The accuracy of equation 7.21 for beam-columns with equal and opposite end moments ($\beta_m = -1$) is demonstrated in Figure 7.11a.

The effects of unequal end moments ($\beta_m > -1$) on the ultimate resistances of beam-columns are allowed for approximately in equation 7.21 by the coefficient C_m which converts the unequal end moments M and $\beta_m M$ into equivalent equal and opposite end moments $C_m M$. Although this conversion is for in-plane behaviour, it is remarkably similar to the conversion used for beams with unequal end moments which buckle out of the plane of bending (see Section 6.2.1.2). The accuracy of equation 7.21 for beam-columns with unequal end moments is demonstrated in

Figure 7.11b, and it can be seen that it is good except for high values of β_m when it tends to be oversafe, and for high values of M when it tends to overestimate the analytical solutions for the ultimate resistance (although this overestimate is reduced if strain-hardening is accounted for in the analytical solutions). The over-conservatism for high values of β_m is caused by the use of the 0.4 cut-off in equation 7.22 for C_m, but more accurate solutions can be obtained by using equation 7.8 for which the minimum value of C_m is 0.2.

Thus, the interaction equations (equations 7.17 and 7.21) provide a reasonably simple method of estimating the ultimate resistances of I-section beam-columns bent about the major axis which fail in the plane of the applied moments. The interaction equations have been successfully applied to a wide range of sections, including solid and hollow circular sections as well as I-sections bent about the minor axis. In the latter case, $M_{pl,y}$, $N_{b,y,Rd}$, and $N_{cr,y}$ must be replaced by their minor axis equivalents, while equation 7.17 should be replaced by equation 7.18.

It should be noted that the interaction equations provide a convenient way of estimating the ultimate resistances of beam-columns. An alternative method has been suggested [4] using simple design charts to show the variation of the moment ratio $M/M_{pl,r}$ with the end moment ratio β_m and the length ratio L/L_c, in which L_c is the length of a column which just fails under the axial load N alone (the effects of strain-hardening must be included in the calculation of L_c).

7.2.3.3 First yield of crooked beam-columns

In the second approximate approach to the resistance of a real beam-column, the first yield of an initially crooked beam-column without residual stresses is used. For this, the magnitude of the initial crookedness is increased to allow approximately for the effects of residual stresses. One logical way of doing this is to use the same crookedness as is used in the design of the corresponding compression member, since this has already been increased so as to allow for residual stresses.

The first yield of a crooked beam-column is analysed in Section 7.5.2, and particular solutions are shown in Figure 7.7b for a beam-column with $N_{cr,y}/N_y = 1/1.5$ and whose initial crookedness is defined by $\eta = 0.290(L/i - 0.152)$. It can be seen that as M decreases to zero, the axial load at first yield increases to the column resistance $N_{b,y,Rd}$ which is reduced below the elastic buckling load $N_{cr,y}$ by the effects of imperfections. As N decreases, the first yield loads for crooked beam-columns approach the corresponding loads for straight beam-columns.

For beam-columns in uniform bending ($\beta_m = -1$), the interaction between M and N is non-linear and concave, as it was for straight beam-columns (Figure 7.7a). If this interaction is plotted using the maximum moment M_{max} instead of the end moment M, then the relationships between M_{max} and N becomes slightly convex, since the non-linear effects of the amplification of M are then included in M_{max}. Thus a linear relationship between M_{max}/M_y and $N/N_{b,y,Rd}$ will provide a simple and conservative approximation for first yield which is of good accuracy.

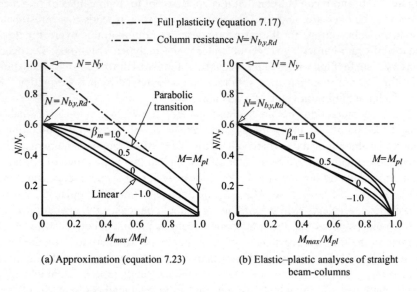

(a) Approximation (equation 7.23)

(b) Elastic–plastic analyses of straight beam-columns

Figure 7.12 Resistances of beam-columns.

However, if this approximation is used for the resistance, then it will become increasingly conservative as M_{max} increases, since it approaches the first yield moment M_y instead of the fully plastic moment M_{pl}. This difficulty may be overcome by modifying the approximation to a linear relationship between M_{max}/M_{pl} and $N/N_{b,y,Rd}$ which passes through $M_{max}/M_{pl} = 1$ when $N/N_{b,y,Rd} = 0$, as shown in Figure 7.12a. Such an approximation is also in good agreement with the uniform bending ($\beta_m = -1$) analytical strengths [2, 3] shown in Figure 7.12b for initially straight beam-columns with residual stresses.

For beam-columns in double curvature bending ($\beta_m = 1$), the first yield interaction shown in Figure 7.7b forms an approximately parabolic transition between the bounds of the column resistance $N_{b,y,Rd}$ and the first yield condition of a straight beam-column. In this case, the effects of initial curvature are greatest near mid-span, and have little effect on the maximum moment which is generally greatest at the ends. Thus the strength may be simply approximated by a parabolic transition from the column resistance $N_{b,y,Rd}$ to the full plasticity condition given by equation 7.17 for I-sections bent about the major axis, as shown in Figure 7.12a. This approximation is in good agreement with the double curvature bending ($\beta_m = 1$) analytical resistances [2, 3] shown in Figure 7.12b for straight beam-columns with residual stresses and zero strain-hardening.

The ultimate resistances of beam-columns with unequal end moments may be approximated by suitable interpolations between the linear and parabolic approximations for beam-columns in uniform bending ($\beta_m = -1$) and double curvature bending ($\beta_m = 1$). It has been suggested [5, 6] that this interpolation should be made according to

$$\frac{M}{M_{pl,y}} = \left\{ 1 - \left(\frac{1+\beta_m}{2} \right)^3 \right\} \left(1 - \frac{N}{N_{b,y,Rd}} \right)$$

$$+ 1.18 \left(\frac{1+\beta_m}{2} \right)^3 \sqrt{\left(1 - \frac{N}{N_{cr,y}} \right)} \leq 1. \tag{7.23}$$

This interpolation uses a conservative cubic weighting to shift the resulting curves downwards towards the linear curve for $\beta_m = -1$, as shown in Figure 7.12a. This is based on the corresponding shift shown in Figure 7.12b of the analytical results [2, 3] for initially straight beam-columns with residual stresses.

7.2.4 Design rules

7.2.4.1 Cross-section resistance

EC3 requires beam-columns to satisfy both cross-section resistance and overall member buckling resistance limitations. The cross-section resistance limitations are intended to prevent cross-section failure due to plasticity or local buckling.

The general cross-section resistance limitation of EC3 is given by a modification of the first yield condition of equation 7.14 to

$$\frac{N_{Ed}}{N_{c,Rd}} + \frac{M_{Ed}}{M_{c,Rd}} \leq 1 \tag{7.24}$$

in which N_{Ed} is the design axial force and M_{Ed} the design moment acting at the cross-section under consideration, $N_{c,Rd}$ is the cross-section axial resistance Af_y (obtained using the effective cross-sectional area A_{eff} for slender cross-sections in compression), and $M_{c,Rd}$ is the cross-section moment resistance (based on either the plastic, elastic or effective section modulus, depending on classification).

This linear interaction equation is often conservative, and so EC3 allows the use of the more economic alternative for Class 1 and 2 cross-sections of directly calculating the reduced plastic moment resistance $M_{N,Rd}$ in the presence of an axial load N_{Ed}. The following criterion should be satisfied:

$$M_{Ed} \leq M_{N,Rd}. \tag{7.25}$$

For doubly symmetrical I- and H-sections there is no reduction to the major axis plastic moment resistance (i.e. $M_{N,Rd} = M_{pl,Rd}$), provided both

$$N_{Ed} \leq 0.25 N_{pl,Rd} \tag{7.26a}$$

and

$$N_{Ed} \leq \frac{0.5 h_w t_w f_y}{\gamma_{M0}}. \tag{7.26b}$$

For larger values of N_{Ed}, the reduced major axis plastic moment resistance is given by

$$M_{N,y,Rd} = M_{pl,y,Rd} \frac{(1-n)}{(1-0.5a)} \leq M_{pl,y,Rd} \qquad (7.27)$$

in which

$$n = N_{Ed}/N_{pl,Rd} \qquad (7.28)$$

and

$$a = \frac{A - 2bt_f}{A} \leq 0.5 \qquad (7.29)$$

where b and t_f and the flange width and thickness respectively.

In minor axis bending, there is no reduction to the plastic moment resistance of doubly symmetrical I- and H-sections provided

$$N_{Ed} \leq \frac{h_w t_w f_y}{\gamma_{M0}}. \qquad (7.30)$$

For larger values of N_{Ed}, the reduced plastic minor axis moment resistance is given by

$$M_{N,z,Rd} = M_{pl,z,Rd} \qquad (7.31a)$$

for $n \leq a$, and by

$$M_{N,z,Rd} = M_{pl,z,Rd} \left[1 - \left(\frac{n-a}{1-a} \right)^2 \right] \qquad (7.31b)$$

for $n > a$. Similar formulae are also provided for box sections.

7.2.4.2   Member resistance

The general in-plane member resistance limitation of EC3 is given by a simplification of equation 7.21 to

$$\frac{N_{Ed}}{N_{b,y,Rd}} + k_{yy} \frac{M_{y,Ed}}{M_{c,y,Rd}} \leq 1 \qquad (7.32)$$

for major axis buckling, and

$$\frac{N_{Ed}}{N_{b,z,Rd}} + k_{zz} \frac{M_{z,Ed}}{M_{c,z,Rd}} \leq 1 \qquad (7.33)$$

for minor axis buckling, in which k_{yy} and k_{zz} are interaction factors whose values may be obtained from Annex A or Annex B of EC3 and $M_{c,y,Rd}$ and $M_{c,z,Rd}$ are the in-plane bending resistances about the major and minor axes respectively. The former is based on enhancing the elastically determined resistance to allow for partial plastification of the cross-section (i.e. following Section 7.2.3.3), whilst the latter reduced the plastically determined resistance to allow for instability effects (i.e. following Section 7.2.3.2). Lengthy formulae to calculate the interaction factors are provided in both cases.

7.3 Flexural–torsional buckling of isolated beam-columns

When an unrestrained beam-column is bent about its major axis, it may buckle by deflecting laterally and twisting at a load which is significantly less than the maximum load predicted by an in-plane analysis. This flexural–torsional buckling may occur while the member is still elastic, or after some yielding due to in-plane bending and compression has occurred, as indicated in Figure 7.13.

7.3.1 Elastic beam-columns

7.3.1.1 Beam-columns with equal end moments

Consider a perfectly straight elastic beam-column bent about its major axis (see Figure 7.1b) by equal and opposite end moments M (so that $\beta_m = -1$), and loaded by an axial force N. The ends of the beam-column are assumed to be simply

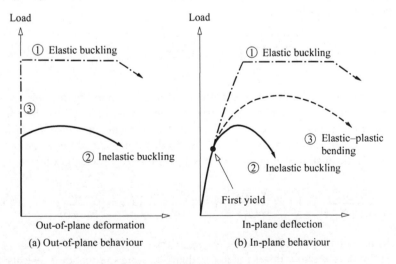

Figure 7.13 Flexural–torsional buckling of beam-columns.

supported and free to warp but end twist rotations are prevented. It is also assumed that the cross-section of the beam-column has two axes of symmetry so that the shear centre and centroid coincide.

When the applied load and moments reach the elastic buckling values $N_{cr,MN}$, $M_{cr,MN}$, a deflected and twisted equilibrium position is possible. It is shown in Section 7.6.1 that this position is given by

$$v = \frac{M_{cr,MN}}{N_{cr,z} - N_{cr,MN}} \phi = \delta \sin \frac{\pi x}{L} \tag{7.34}$$

in which δ is the undetermined magnitude of the central deflection, and that the elastic buckling combination $N_{cr,MN}$, $M_{cr,MN}$ is given by

$$\frac{M_{cr,MN}^2}{i_p^2 N_{cr,z} N_{cr,T}} = \left(1 - \frac{N_{cr,MN}}{N_{cr,z}}\right)\left(1 - \frac{N_{cr,MN}}{N_{cr,T}}\right) \tag{7.35}$$

in which $i_p = \sqrt{[(I_y + I_z)/A]}$ is the polar radius of gyration, and

$$N_{cr,z} = \pi^2 EI_z/L^2, \tag{7.36}$$

$$N_{cr,T} = \frac{GI_t}{i_p^2}\left(1 + \frac{\pi^2 EI_w}{GI_t L^2}\right) \tag{7.37}$$

are the minor axis and torsional buckling loads of an elastic axially loaded column (see Sections 3.2.1 and 3.7.5). It can be seen that for the limiting case when $M_{cr,MN} = 0$, the beam-column buckles as a compression member at the lower of $N_{cr,z}$ and $N_{cr,T}$, and when $N_{cr,MN} = 0$, it buckles as a beam at the elastic buckling moment (see equation 6.3)

$$M_{cr} = M_{zx} = \sqrt{\left(\frac{\pi^2 EI_z}{L^2}\right)\left(GI_t + \frac{\pi^2 EI_w}{L^2}\right)} = i_p\sqrt{N_{cr,z} N_{cr,T}}. \tag{7.38}$$

Some more general solutions of equation 7.35 are shown in Figure 7.14.

The derivation of equation 7.35 given in Section 7.6.1 neglects the approximate amplification by the axial load of the in-plane moments to $M_{cr,MN}/(1 - N_{cr,MN}/N_{cr,y})$ (see Section 7.2.1). When this is accounted for, equation 7.35 for the elastic buckling combination of $N_{cr,MN}$ and $M_{cr,MN}$ changes to

$$\frac{M_{cr,MN}^2}{i_p^2 N_{cr,z} N_{cr,T}} = \left(1 - \frac{N_{cr,MN}}{N_{cr,y}}\right)\left(1 - \frac{N_{cr,MN}}{N_{cr,z}}\right)\left(1 - \frac{N_{cr,MN}}{N_{cr,T}}\right) \tag{7.39}$$

The maximum possible value of $N_{cr,MN}$ is the lowest value of $N_{cr,y}$, $N_{cr,z}$, and $N_{cr,T}$. For most sections this is much less than $N_{cr,y}$ and so equation 7.39 is usually very close to equation 7.35. For most hot-rolled sections, $N_{cr,z}$ is less than $N_{cr,T}$,

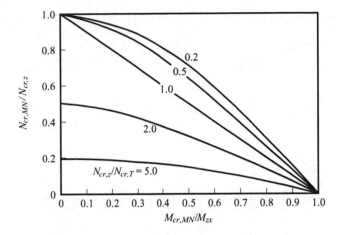

Figure 7.14 Elastic buckling load combinations for beam-columns with equal end moments.

so that $(1 - N_{cr,MN}/N_{cr,T}) > (1 - N_{cr,MN}/N_{cr,z})(1 - N_{cr,MN}/N_{cr,y})$. In this case, equation 7.39 can be safely approximated by the interaction equation

$$\frac{N_{cr,MN}}{N_{cr,z}} + \frac{1}{(1 - N_{cr,MN}/N_{cr,y})}\frac{M_{cr,MN}}{M_{zx}} = 1. \tag{7.40}$$

7.3.1.2   Beam-columns with unequal end moments

The elastic flexural–torsional buckling of simply supported beam-columns with unequal major axis end moments M and $\beta_m M$ has been investigated numerically, and many solutions are available [7–11]. The conservative interaction equation

$$\left(\frac{M/\sqrt{F}}{M_E}\right)^2 + \left(\frac{N}{N_{cr,z}}\right) = 1 \tag{7.41}$$

has also been proposed [9], in which

$$M_E^2 = \pi^2 EI_z GI_t/L^2. \tag{7.42}$$

The factor $1/\sqrt{F}$ in equation 7.41 varies with the end moment ratio β_m as shown in Figure 7.15, and allows the unequal end moments to be treated as equivalent equal end moments $M/\sqrt{F}$. Thus the elastic buckling of beam-columns with unequal end moments can also be approximated by modifying equation 7.35 to

$$\frac{(M_{cr,MN}/\sqrt{F})^2}{M_{zx}^2} = \left(1 - \frac{N_{cr,MN}}{N_{cr,z}}\right)\left(1 - \frac{N_{cr,MN}}{N_{cr,T}}\right), \tag{7.43}$$

or by a similar modification of equation 7.40.

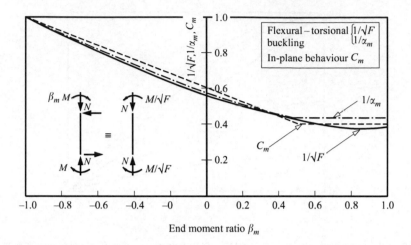

Figure 7.15 Equivalent end moments for beam-columns.

It is of interest to note that the variation of the factor $1/\sqrt{F}$ for the flexural–torsional buckling of beam-columns is very close to that of the coefficient $1/\alpha_m$ obtained from

$$\alpha_m = 1.75 + 1.05\beta_m + 0.3\beta_m^2 \le 2.56 \tag{7.44}$$

used for the flexural–torsional buckling of beams (see Section 6.2.1.2), and close to that of the coefficient C_m (see equation 7.22) used for the in-plane behaviour of beam-columns.

However, the results shown in Figure 7.16 of a more recent investigation [12] of the elastic flexural–torsional buckling of beam-columns have indicated that the factor for converting unequal end moments into equivalent equal end moments should vary with $N/N_{cr,z}$ as well as with β_m. More accurate predictions have been obtained using

$$\left(\frac{M_{cr,MN}}{\alpha_{bc}M_{zx}}\right)^2 = \left(1 - \frac{N_{cr,MN}}{N_{cr,z}}\right)\left(1 - \frac{N_{cr,MN}}{N_{cr,T}}\right), \tag{7.45}$$

with

$$\frac{1}{\alpha_{bc}} = \left(\frac{1-\beta_m}{2}\right) + \left(\frac{1+\beta_m}{2}\right)^3\left(0.40 - 0.23\frac{N_{cr,MN}}{N_{cr,z}}\right). \tag{7.46}$$

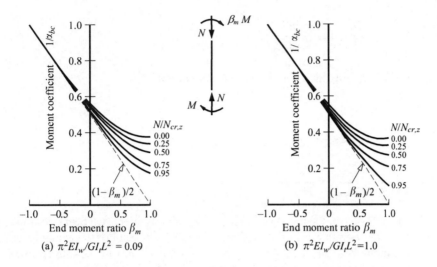

Figure 7.16 Flexural–torsional buckling coefficients $1/\alpha_{bc}$ in equation 7.45.

7.3.1.3 Beam-columns with elastic end restraints

The elastic flexural–torsional buckling of symmetrically restrained beam-columns with equal and opposite end moments ($\beta_m = -1$) is analysed in Section 7.6.2. The particular example considered is one for which the minor axis bending and end warping restraints are equal. This corresponds to the situation in which both ends of both flanges have equal elastic restraints whose stiffnesses are such that

$$\frac{\text{Flange end moment}}{\text{Flange end rotation}} = \frac{-EI_z}{L} \frac{R}{1 - R} \tag{7.47}$$

in which the restraint parameter R varies from 0 (flange unrestrained) to 1 (flange rigidly restrained).

It is shown in Section 7.6.2 that the elastic buckling values $M_{cr,MN}$ and $N_{cr,MN}$ of the end moments and axial load for which the restrained beam-column buckles elastically are given by

$$\frac{M_{cr,MN}^2}{i_p^2 N_{cr,z} N_{cr,T}} = \left(1 - \frac{N_{cr,MN}}{N_{cr,z}}\right)\left(1 - \frac{N_{cr,MN}}{N_{cr,T}}\right) \tag{7.48}$$

in which

$$N_{cr,z} = \pi^2 EI_z / L_{cr}^2, \tag{7.49}$$

and

$$N_{cr,T} = \frac{GI_t}{i_p^2} \left(1 + \frac{\pi^2 EI_w}{GI_t L_{cr}^2}\right),$$ (7.50)

where L_{cr} is the effective length of the beam-column

$$L_{cr} = k_{cr}L,$$ (7.51)

and the effective length ratio k_{cr} is the solution of

$$\frac{R}{1-R} = \frac{-\pi}{2k_{cr}} \cot \frac{\pi}{2k_{cr}}.$$ (7.52)

Equation 7.48 is the same as equation 7.35 for simply supported beam-columns, except for the familiar use of the effective length L_{cr} instead of the actual length L in equations 7.49 and 7.50 defining $N_{cr,z}$ and $N_{cr,T}$. Thus the solutions shown in Figure 7.14 can also be applied to end-restrained beam-columns.

The close relationships between the flexural–torsional buckling condition of equation 7.35 for unrestrained beam-columns with equal end moments ($\beta_m = -1$) and the buckling loads $N_{cr,z}$, $N_{cr,T}$ of columns and moments M_{zx} of beams have already been discussed. It should also be noted that equation 7.52 for the effective lengths of beam-columns with equal end restraints is exactly the same as equation 3.43 for columns when $(1-R)/R$ is substituted for γ_1 and γ_2, and equation 6.52 for beams. These suggest that the flexural–torsional buckling condition for beam-columns with unequal end restraints could well be approximated by equations 7.48–7.51 if Figure 3.21a is used to determine the effective length ratio k_{cr} in equation 7.51.

A further approximation may be suggested for restrained beam-columns with unequal end moments ($\beta_m \neq -1$), in which modifications of equations 7.45 and 7.46 (after substituting $i_p^2 N_{cr,z} N_{cr,T}$ for M_{zx}^2) are used (with equations 7.49–7.51) instead of equation 7.48.

7.3.2 Inelastic beam-columns

The solutions obtained in Section 7.3.1 for the flexural–torsional buckling of straight isolated beam-columns are only valid while they remain elastic. When the combination of the residual stresses with those induced by the in-plane loading causes yielding, the effective rigidities of some sections of the member are reduced, and buckling may occur at a load which is significantly less than the in-plane maximum load or the elastic buckling load, as indicated in Figure 7.13.

A method of analysing the inelastic buckling of I-section beam-columns with residual stresses has been developed [13], and used [14] to obtain the predictions shown in Figure 7.17 for the inelastic buckling loads of isolated beam-columns

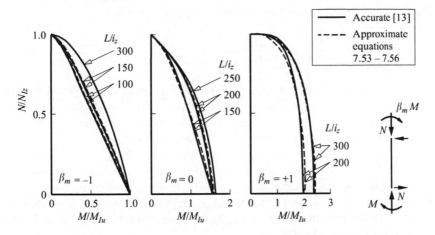

Figure 7.17 Inelastic flexural–torsional buckling of beam-columns.

with unequal end moments. It was found that these could be closely approximated by modifying the elastic equations 7.45 and 7.46 to

$$\left(\frac{M}{\alpha_{bcI}M_{Iu}}\right)^2 = \left(1 - \frac{N}{N_{Iz}}\right)\left(1 - \frac{N}{N_{cr,T}}\right), \tag{7.53}$$

$$\frac{1}{\alpha_{bcI}} = \left(\frac{1 - \beta_m}{2}\right) + \left(\frac{1 + \beta_m}{2}\right)^3 \left(0.40 - 0.23\frac{N}{N_{Iz}}\right) \tag{7.54}$$

in which

$$M_{Iu}/M_{pl} = 1.008 - 0.245M_{pl}/M_{zx} \leq 1.0 \tag{7.55}$$

is an approximation for the inelastic buckling of a beam in uniform bending (equation 7.55 is similar to equation 6.23 when $\beta_m = -1$), and

$$N_{Iz}/N_y = 1.035 - 0.181\sqrt{N_y/N_{cr,z}} - 0.128N_y/N_{cr,z} \leq 1.0 \tag{7.56}$$

provides an approximation for the inelastic buckling of a column.

7.3.3 Ultimate resistance

The ultimate resistances of real beam-columns which fail by flexural–torsional buckling are reduced below their elastic and inelastic buckling loads by the presence of geometrical imperfections such as initial crookedness, just as are those of columns (Section 3.4) and beams (Section 6.4).

One method of predicting these reduced resistances is to modify the approximate interaction equation for in-plane bending (equation 7.21) to

$$\frac{N}{N_{b,z,Rd}} + \frac{C_{my}}{(1 - N/N_{cr,y})} \frac{M_y}{M_{b0,y,Rd}} \leq 1, \tag{7.57}$$

where $N_{b,z,Rd}$ is the minor axis strength of a concentrically loaded column ($M_y = 0$), and $M_{b0,y,Rd}$ the resistance of a beam ($N = 0$) with equal and opposite end moments ($\beta_m = -1$) which fails either by in-plane plasticity at $M_{pl,y}$, or by flexural–torsional buckling. The basis for this modification is the remarkable similarity between equation 7.57 and the simple linear approximation of equation 7.40 for the elastic flexural–torsional buckling of beam-columns with equal and opposite end moments ($\beta_m = -1$). The application of equation 7.57 to beam-columns with unequal end moments is simplified by the similarity shown in Figure 7.15 between the values of C_m for in-plane bending and $1/\sqrt{F}$ and $1/\alpha_m$ for flexural–torsional buckling.

However, the value of this simple modification is lost if the in-plane strength of a beam-column is to be predicted more accurately than by equation 7.21. If, for example, equation 7.23 is to be used for the in-plane strength as suggested in Section 7.2.3.3, then it would be more appropriate to modify the flexural–torsional buckling equations 7.45 and 7.53 to obtain out-of-plane strength predictions. It has been suggested [15] that the modified equations should take the form

$$\left(\frac{M}{\alpha_{bcu}M_{b0,y,Rd}}\right)^2 = \left(1 - \frac{N}{N_{b,z,Rd}}\right)\left(1 - \frac{N}{N_{cr,T}}\right) \tag{7.58}$$

in which

$$\frac{1}{\alpha_{bcu}} = \left(\frac{1 - \beta_m}{2}\right) + \left(\frac{1 + \beta_m}{2}\right)^3 \left(0.40 - 0.23\frac{N}{N_{b,z,Rd}}\right). \tag{7.59}$$

7.3.4 Design rules

The design of beam-columns against out-of-plane buckling will generally be governed by

$$\frac{N_{Ed}}{N_{b,z,Rd}} + k_{zy}\frac{M_{y,Ed}}{M_{b,Rd}} \leq 1 \tag{7.60}$$

in which k_{zy} is an interaction factor that may be determined from Annex A or Annex B of EC3. Equation 7.60 is similar to equation 7.32 for in-plane major axis buckling resistance, except for the allowance for lateral torsional buckling in the second term and the use of the minor axis column buckling resistance in the first term.

For members with the intermediate lateral restraints against minor axis buckling, failure by major axis buckling with lateral torsional buckling between the restraints becomes a possibility. This leads to requirement

$$\frac{N_{Ed}}{N_{b,y,Rd}} + k_{yy}\frac{M_{y,Ed}}{M_{b,Rd}} \leq 1 \tag{7.61}$$

which incorporates the major axis flexural buckling resistance $N_{b,y,Rd}$ and the lateral torsional buckling resistance $M_{b,Rd}$. For members with no lateral restraints along the span, equation 7.60 will always govern over equation 7.61.

The design of members subjected to bending and tension is discussed in Section 2.4.

A worked example of checking the out-of-plane resistance of a beam-column is given in Section 7.7.4.

7.4 Biaxial bending of isolated beam-columns

7.4.1 Behaviour

The geometry and loading of most framed structures are three-dimensional, and the typical member of such a structure is compressed, bent about both principal axes and twisted by the other members connected to it, as shown in Figure 7.1c. The structure is usually arranged so as to produce significant bending about the major axis of the member, but the minor axis deflections and twists are often significant as well, because the minor axis bending and torsional stiffnesses are small. In addition, these deformations are amplified by the components of the axial load and the major axis moment which are induced by the deformations of the member.

The elastic biaxial bending of isolated beam-columns with equal and opposite end moments acting about each principal axis has been analysed [16–22], but the solutions obtained cannot be simplified without approximation. These analyses have shown that the elastic biaxial bending of a beam-column is similar to its in-plane behaviour (see curve 3 of Figure 7.2), in that the major and minor axis deflections and the twist all begin at the commencement of loading, and increase rapidly as the elastic buckling load (which is the load which causes elastic flexural–torsional buckling when there are no minor axis moments and torques acting) is approached. First yield predictions based on these analyses have given conservative estimates of the member resistances. The elastic biaxial bending of beam-columns with unequal end moments has also been investigated. In one of these investigations [16], approximate solutions were obtained by changing both sets of unequal end moments into equivalent equal end moments by multiplying them by the conversion factor $1/\sqrt{F}$ (see Figure 7.15) for flexural–torsional buckling.

Although the first yield predictions for the resistances of slender beam-columns are of reasonable accuracy, they are rather conservative for stocky members in which considerable yielding occurs before failure. Sophisticated numerical analyses have been made [18, 20–24] of the biaxial bending of inelastic beam-columns, and good agreement with test results has been obtained. However, such an analysis requires a specialised computer program, and is only useful as a research tool.

A number of attempts have been made to develop approximate methods of predicting the resistances of inelastic beam-columns. One of the simplest of these uses a linear extension

$$\frac{N}{N_{b,Rd}} + \frac{C_{my}}{(1 - N/N_{cr,y})}\frac{M_y}{M_{b0,y,Rd}} + \frac{C_{mz}}{(1 - N/N_{cr,z})}\frac{M_z}{M_{c0,z,Rd}} \leq 1 \qquad (7.62)$$

of the linear interaction equations for in-plane bending and flexural–torsional buckling. In this extension, $M_{b0,y,Rd}$ is the ultimate moment which the beam-column can support when $N = M_z = 0$ for the case of equal end moments ($\beta_m = -1$), while $M_{c0,z,Rd}$ is similarly defined. Thus equation 7.62 reduces to an equation similar to equation 7.21 for in-plane behaviour when $M_y = 0$, and to equation 7.57 for flexural–torsional buckling when $M_z = 0$. Equation 7.62 is used in conjunction with

$$\frac{M_y}{M_{pl,r,y}} + \frac{M_z}{M_{pl,r,z}} \leq 1, \qquad (7.63)$$

where $M_{pl,r,y}$ is given by equation 7.17 and $M_{pl,r,z}$ by equation 7.18. Equation 7.63 represents a linear approximation to the biaxial bending full plasticity limit for the cross-section of the beam-column, and reduces to equation 7.18 or 7.17 which provides the uniaxial bending full plasticity limit when $M_y = 0$ or $M_z = 0$.

A number of criticisms may be made of these extensions of the interaction equations. First there is no allowance in equation 7.62 for the amplification of the minor axis moment M_z by the major axis moment M_y (the term $(1 - N/N_{cr,z})$ only allows for the amplification caused by axial load N). A simple method of allowing for this would be to replace the term $(1 - N/N_{cr,z})$ by either $(1 - N/N_{cr,MN})$ or by $(1 - M_y/M_{cr,MN})$ where the flexural–torsional buckling load $N_{cr,MN}$ and moment $M_{cr,MN}$ are given by equation 7.45.

Second, studies [22–26] have shown that the linear additions of the moment terms in equations 7.62 and 7.63 generally lead to predictions which are too conservative, as indicated in Figure 7.18. It has been proposed that for column type hot-rolled I-sections, the section full plasticity limit of equation 7.63 should be replaced by

$$\left(\frac{M_y}{M_{pl,r,y}}\right)^{\alpha_0} + \left(\frac{M_z}{M_{pl,r,z}}\right)^{\alpha_0} \leq 1, \qquad (7.64)$$

where $M_{pl,r,y}$ and $M_{pl,r,z}$ are given in equations 7.17 and 7.18 as before, and the index α_0 by

$$\alpha_0 = 1.60 - \frac{N/N_y}{2\ln(N/N_y)}. \qquad (7.65)$$

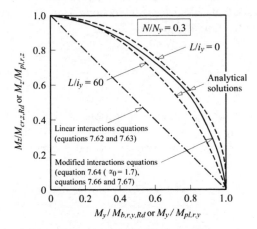

Figure 7.18 Interaction curves for biaxial bending.

At the same time, the linear member interaction equation (equation 7.62) should be replaced by

$$\left(\frac{M_y}{M_{b,r,y,Rd}}\right)^{\alpha_L} + \left(\frac{M_z}{M_{c,r,z,Rd}}\right)^{\alpha_L} \le 1,$$

(7.66)

where $M_{b,r,y,Rd}$, the maximum value of M_y when N acts but $M_z = 0$, is obtained from equation 7.57, $M_{c,r,z,Rd}$ is similarly obtained from an equation similar to equation 7.21, and the index α_L is given by

$$\alpha_L = 1.40 + \frac{N}{N_y}.$$

(7.67)

The approximations of equations 7.64–7.67 have been shown to be of reasonable accuracy when compared with test results [27]. This conclusion is reinforced by the comparison of some analytical solutions with the approximations of equations 7.64–7.67 (but with $\alpha_0 = 1.70$ instead of 1.725 approximately for $N/N_y = 0.3$) shown in Figure 7.18. Unfortunately, these approximations are of limited application, although it seems to be possible that their use may be extended.

7.4.2 Design rules

7.4.2.1 Cross-section resistance

The general biaxial bending cross-section resistance limitation of EC3 is given by

$$\frac{N_{Ed}}{N_{c,Rd}} + \frac{M_{y,Ed}}{M_{c,y,Rd}} + \frac{M_{z,Ed}}{M_{c,z,Rd}} \le 1$$

(7.68)

which is a simple linear extension of the uniaxial cross-section resistance limitation of equation 7.24. For Class 1 plastic and Class 2 compact sections, a more economic alternative is provided by a modification of equation 7.63 to

$$\left(\frac{M_{y,Ed}}{M_{N,y,Rd}}\right)^{\alpha} + \left(\frac{M_{z,Ed}}{M_{N,z,Rd}}\right)^{\beta} \leq 1 \tag{7.69}$$

in which $\alpha = 2.0$ and $\beta = 5n \geq 1$ for equal flanged I-sections, $\alpha = \beta = 2.0$ for circular hollow sections, and $\alpha = \beta = 1.66/(1 - 1.13n^2) \leq 6)$ for rectangular hollow sections, where $n = N_{Ed}/N_{pl,Rd}$. Conservatively, α and β may be taken as unity.

7.4.2.2   Member resistance

The general biaxial bending member resistance limitations are given by extensions of equations 7.61 and 7.60 for uniaxial bending to

$$\frac{N_{Ed}}{N_{b,y,Rd}} + k_{yy}\frac{M_{y,Ed}}{M_{b,Rd}} + k_{yz}\frac{M_{z,Ed}}{M_{z,Rd}} \leq 1 \tag{7.70}$$

and

$$\frac{N_{Ed}}{N_{b,z,Rd}} + k_{zy}\frac{M_{y,Ed}}{M_{b,Rd}} + k_{zz}\frac{M_{z,Ed}}{M_{z,Rd}} \leq 1, \tag{7.71}$$

both of which must be satisfied.

Alternative expressions for the interaction factors k_{yy}, k_{yz}, k_{zy}, and k_{zz} are provided in Annexes A and B of EC3. The Annex B formulations for the determination of the interaction factors k_{ij} are less complex than those set out in Annex A, resulting in quicker and simpler calculations, though generally at the expense of some structural efficiency [28]. Annex B may therefore be regarded as the simplified approach whereas Annex A represents a more exact approach. The background to the development of the Annex A interaction factors has been described in [29], and that of the Annex B interaction factors in [30].

For columns in simple construction, the bending moments arising as a result of the eccentric loading from the beams are relatively small, and much simpler interaction expressions can be developed with no significant loss of structural efficiency. Thus, it has been shown [31] that for simple construction with hot-rolled I- and H-sections and with no intermediate lateral restraints along the member length, equations 7.70 and 7.71 may be replaced by the single equation

$$\frac{N_{Ed}}{N_{b,z,Rd}} + \frac{M_{y,Ed}}{M_{b,Rd}} + 1.5\frac{M_{z,Ed}}{M_{c,z,Rd}} \leq 1 \tag{7.72}$$

provided the ratio of end moments is less than zero (i.e. a triangular, or less severe, bending moment diagram).

A worked example for checking the biaxial bending of a beam-column is given in Section 7.7.5. Further worked examples of the application of the EC3 beam-column formulae are given in [30] and [32].

7.5   Appendix – in-plane behaviour of elastic beam-columns

7.5.1   Straight beam-columns

The bending moment in the beam-column shown in Figure 7.3 is the sum of the moment $[M - M(1 + \beta_m)x/L]$ due to the end moments and reactions and the moment Nw due to the deflection w. Thus the differential equation of bending is obtained by equating this bending moment to the internal moment of resistance $-EI_y d^2w/dx^2$, whence

$$-EI_y \frac{d^2w}{dx^2} = M - M(1 + \beta_m)\frac{x}{L} + Nw. \tag{7.73}$$

The solution of this equation which satisfies the support conditions $(w)_0 = (w)_L = 0$ is

$$w = \frac{M}{N}\left[\cos \mu x - (\beta_m \mathrm{cosec}\, \mu L + \cot \mu L) \sin \mu x - 1 + (1 + \beta_m)\frac{x}{L}\right], \tag{7.1}$$

where

$$\mu^2 = \frac{N}{EI_y} = \frac{\pi^2}{L^2}\frac{N}{N_{cr,y}}. \tag{7.2}$$

The bending moment distribution can be obtained by substituting equation 7.1 into equation 7.73. The maximum bending moment can be determined by solving the condition

$$\frac{d(-EI_y d^2w/dx^2)}{dx} = 0$$

for the position x_{max} of the maximum moment, whence

$$\tan \mu x_{max} = -\beta_m \mathrm{cosec}\, \mu L - \cot \mu L. \tag{7.74}$$

The value of x_{max} is positive while

$$\beta_m < -\cos \pi \sqrt{\frac{N}{N_{cr,y}}},$$

and the maximum moment obtained by substituting equation 7.74 into equation 7.73 is

$$
M_{max} = M \left[1 + \left\{ \beta_m \text{cosec}\, \pi \sqrt{\frac{N}{N_{cr,y}}} + \cot \pi \sqrt{\frac{N}{N_{cr,y}}} \right\}^2 \right]^{0.5} .
\tag{7.3}
$$

When $\beta_m > -\cos \pi \sqrt{(N/N_{cr,y})}$, x_{max} is negative, in which case the maximum moment in the beam-column is the end moment M at $x = 0$, whence

$$
M_{max} = M,
\tag{7.4}
$$

For beam-columns which are bent in single curvature by equal and opposite end moments ($\beta_m = -1$), equation 7.1 for the deflected shape simplifies, and the central deflection δ_{max} can be expressed non-dimensionally as

$$
\frac{\delta_{max}}{\delta} = \frac{8/\pi^2}{N/N_{cr,y}} \left[\sec \frac{\pi}{2} \sqrt{\frac{N}{N_{cr,y}}} - 1 \right]
\tag{7.5}
$$

in which $\delta = ML^2/8EI_y$ is the value of δ_{max} when $N = 0$. The maximum moment occurs at the centre of the beam-column, and is given by

$$
M_{max} = M \sec \frac{\pi}{2} \sqrt{\frac{N}{N_{cr,y}}}.
\tag{7.6}
$$

Equations 7.5 and 7.6 are plotted in Figure 7.4.

7.5.2   Beam-columns with initial curvature

If the beam-column shown in Figure 7.3 is not straight, but has an initial crookedness given by

$$
w_0 = \delta_0 \sin \pi x/L,
$$

then the differential equation of bending becomes

$$
-EI_y \frac{\mathrm{d}^2 w}{\mathrm{d}x^2} = M - M(1 + \beta_m)\frac{x}{L} + N(w + w_0).
\tag{7.75}
$$

The solution of this equation which satisfies the support conditions of $(w)_0 = (w)_L = 0$ is

$$
w = (M/N)[\cos \mu x - (\beta_m \text{cosec}\, \mu L + \cot \mu L) \sin \mu x - 1
$$
$$
+ (1 + \beta_m)x/L] + [(\mu L/\pi)^2/\{1 - (\mu L/\pi)^2\}]\delta_0 \sin \pi x/L,
\tag{7.76}
$$

where

$$(\mu L/\pi)^2 = N/N_{cr,y}. \tag{7.77}$$

The bending moment distribution can be obtained by substituting equation 7.76 into equation 7.75. The position of the maximum moment can then be determined by solving the condition $d(-EI_y d^2 w/dx^2)dx = 0$, whence

$$\frac{N\delta_0}{M} = \frac{\mu L}{\pi}\left(1 - \frac{\mu^2 L^2}{\pi^2}\right)\frac{\{(\beta_m \cosec \mu L + \cot \mu L)\cos \mu x_{max} + \sin \mu x_{max}\}}{\cos \pi x_{max}/L}. \tag{7.78}$$

The value of the maximum moment M_{max} can be obtained from equations 7.75 and 7.76 (with $x = x_{max}$) and from equation 7.78, and is given by

$$\frac{M_{max}}{M} = \cos \mu x_{max} - (\beta_m \cosec \mu L + \cot \mu L)\sin \mu x_{max}$$

$$+ \frac{(N\delta_0/M)}{(1 - \mu^2 L^2/\pi^2)}\sin \frac{\pi x_{max}}{L}. \tag{7.79}$$

First yield occurs when the maximum stress is equal to the yield stress f_y, in which case

$$\frac{N}{N_y} + \frac{M_{max}}{M_y} = 1,$$

or

$$\frac{N}{N_y}\left\{1 + \frac{M_{max}}{M}\frac{N_y\delta_0}{M_y}\frac{M}{N\delta_0}\right\} = 1. \tag{7.80}$$

In the special case of $M = 0$, the first yield solution is the same as that given in Section 3.2.2 for a crooked column. Equation 7.78 is replaced by $x_{max} = L/2$, and equation 7.79 by

$$M_{max} = \frac{N\delta_0}{(1 - \mu^2 L^2/\pi^2)}.$$

The value N_L of N which then satisfies equation 7.80 is given by the solution of

$$\frac{N_L}{N_y} + \frac{N_L\delta_0/M_y}{\{1 - (N_L/N_y)/(N_{cr,y}/N_y)\}} = 0. \tag{7.81}$$

The value of M/M_y at first yield can be determined for any specified set of values of $N_{cr,y}/N_y$, β_m, $\delta_0 N_y/M_y$, and N/N_y. An initial guess for x_{max}/L allows $N\delta_0/M$ to be determined from equation 7.78, and M_{max}/M from equation 7.79, and these can then be used to evaluate the left-hand side of equation 7.80. This process can be repeated until the values are found which satisfy equation 7.80. Solutions obtained in this way are plotted in Figure 7.7b.

7.6 Appendix – flexural–torsional buckling of elastic beam-columns

7.6.1 Simply supported beam-columns

The elastic beam-column shown in Figure 7.1b is simply supported and prevented from twisting at its ends so that

$$(v)_{0,L} = (\phi)_{0,L} = 0, \tag{7.82}$$

and is free to warp (see Section 10.8.3) so that

$$\left(\frac{d^2\phi}{dx^2}\right)_{0,L} = 0. \tag{7.83}$$

The combination at elastic buckling of the axial load $N_{cr,MN}$ and equal and opposite end moments $M_{cr,MN}$ (i.e. $\beta_m = -1$) can be determined by finding a deflected and twisted position which is one of equilibrium. The differential equilibrium equations for such a position are

$$EI_z \frac{d^2v}{dx^2} + N_{cr,MN} v = -M_{cr,MN}\phi \tag{7.84}$$

for minor axis bending, and

$$\left(GJ - N_{cr,MN} i_p^2\right)\frac{d\phi}{dx} - EI_w \frac{d^3\phi}{dx^3} = M_{cr,MN}\frac{dv}{dx} \tag{7.85}$$

for torsion, where $i_p^2 = \sqrt{[(I_y+I_z)/A]}$ is the polar radius of gyration. Equation 7.84 reduces to equation 3.59 for compression member buckling when $M_{cr,MN} = 0$, and to equation 6.82 for beam buckling when $N_{cr,MN} = 0$. Equation 7.85 can be used to derive equation 3.73 for torsional buckling of a compression member when $M_{cr,MN} = 0$, and reduces to equation 6.82 for beam buckling when $N_{cr,MN} = 0$.
    The buckled shape of the beam-column is given by

$$v = \frac{M_{cr,MN}}{N_{cr,z} - N_{cr,MN}}\phi = \delta \sin\frac{\pi x}{L}, \tag{7.34}$$

where $N_{cr,z} = \pi^2 EI_z/L^2$ is the minor axis elastic buckling load of an axially loaded column, and δ the undetermined magnitude of the central deflection. The boundary conditions of equations 7.82 and 7.83 are satisfied by this buckled shape, as is equation 7.84, while equation 7.85 is satisfied when

$$\frac{M_{cr,MN}^2}{i_p^2 N_{cr,z} N_{cr,T}} = \left(1 - \frac{N_{cr,MN}}{N_{cr,z}}\right)\left(1 - \frac{N_{cr,MN}}{N_{cr,T}}\right) \tag{7.35}$$

where

$$N_{cr,T} = \frac{GI_t}{i_p^2}\left(1 + \frac{\pi^2 EI_w}{GI_t L^2}\right) \tag{7.37}$$

is the elastic torsional buckling load of an axially loaded column (see Section 3.7.5). Equation 7.35 defines the combinations of axial load $N_{cr,MN}$ and end moments $M_{cr,MN}$ which cause the beam-column to buckle elastically.

7.6.2 Elastically restrained beam-columns

If the elastic beam-column shown in Figure 7.1b is not simply supported but is elastically restrained at its ends against minor axis rotations dv/dx and against warping rotations $(d_f/2)d\phi/dx$ (where d_f is the distance between flange centroids), then the boundary conditions at the end $x = L/2$ of the beam-column can be expressed in the form

$$\left.\begin{array}{c} \dfrac{M_B + M_T}{(dv/dx)_{L/2}} = \dfrac{-EI_z}{L}\dfrac{2R_2}{1 - R_2} \\[2ex] \dfrac{M_T - M_B}{(d_f/2)(d\phi/dx)_{L/2}} = \dfrac{-EI_z}{L}\dfrac{2R_4}{1 - R_4} \end{array}\right\}, \tag{7.86}$$

where M_T and M_B are the flange minor axis end restraining moments

$$M_T = 1/2EI_z\left(d^2v/dx^2\right)_{L/2} + (d_f/4)\,EI_z\left(d^2\phi/dx^2\right)_{L/2},$$

$$M_B = 1/2EI_z\left(d^2v/dx^2\right)_{L/2} - (d_f/4)\,EI_z\left(d^2\phi/dx^2\right)_{L/2},$$

and R_2 and R_4 are dimensionless minor axis bending and warping end restraint parameters, which vary between zero (no restraint) and one (rigid restraint). If the beam-column is symmetrically restrained, similar conditions apply at the end $x = -L/2$. The other support conditions are

$$(v)_{\pm L/2} = (\phi)_{\pm L/2} = 0 \tag{7.87}$$

The particular case for which the minor axis and warping end restraints are equal, so that

$$R_2 = R_4 = R \tag{7.88}$$

can be analysed. When the restrained beam-column has equal and opposite end moments ($\beta_m = -1$), the differential equilibrium equations for a buckled position v, ϕ are

$$\left. \begin{array}{l} EI_z \dfrac{d^2 v}{dx^2} + N_{cr,MN} v = -M_{cr,MN}\phi + (M_B + M_T) \\[2mm] (GI_t - N_{cr,MN} i_p^2) \dfrac{d\phi}{dx} - EI_w \dfrac{d^3\phi}{dx^3} = M_{cr,MN} \dfrac{dv}{dx} \end{array} \right\} . \tag{7.89}$$

These differential equations and the boundary conditions of equations 7.86–7.88 are satisfied by the buckled shape

$$v = \frac{M_{cr,MN}\phi}{N_{cr,z} - N_{cr,MN}} = A \left(\cos \frac{\pi x}{k_{cr}L} - \cos \frac{\pi}{2k_{cr}} \right),$$

where

$$N_{cr,z} = \pi^2 EI_z / L_{cr}^2, \tag{7.49}$$

$$L_{cr} = k_{cr} L, \tag{7.51}$$

when the effective length ratio k_{cr} satisfies

$$\frac{R}{1-R} = \frac{-\pi}{2k_{cr}} \cot \frac{\pi}{2k_{cr}}. \tag{7.52}$$

This is the same as equation 6.52 for the buckling of restrained beams, whose solutions are shown in Figure 6.18c. The values $M_{cr,MN}$ and $N_{cr,MN}$ for flexural–torsional buckling of the restrained beam-column are obtained by substituting the buckled shape v, ϕ into equations 7.89, and are related by

$$\frac{M_{cr,MN}^2}{i_p^2 N_{cr,z} N_{cr,T}} = \left(1 - \frac{N_{cr,MN}}{N_{cr,z}}\right) \left(1 - \frac{N_{cr,MN}}{N_{cr,T}}\right) \tag{7.48}$$

where

$$N_{cr,T} = \frac{GI_t}{i_p^2} \left(1 + \frac{\pi^2 EI_w}{GI_t L_{cr}^2}\right). \tag{7.51}$$

7.7 Worked examples

7.7.1 Example 1 – approximating the maximum elastic moment

Problem. Use the approximations of Figures 7.5 and 7.6 to determine the maximum moments $M_{max,y}$, $M_{max,z}$ for the 254 × 146 UB 37 beam-columns shown in Figure 7.19d and c.

Solution for $M_{max,y}$ for the beam-column of Figure 7.19d.

Using Figure 7.5, $k_{cr} = 1.0$, $\gamma_m = 1.0$, $\gamma_n = 1.0$, $\gamma_s = 0.18$
$M_y = 20 \times 9/4 = 45$ kNm.

$$N_{cr,y} = \pi^2 \times 210\,000 \times 5537 \times 10^4/(1.0 \times 9000)^2 \text{ N} = 1417\,\text{kN}.$$

Using equation 7.10, $\delta_y = \dfrac{1.0 \times (1 - 0.18 \times 200/1417)}{(1 - 1.0 \times 200/1417)} = 1.135.$

Using equation 7.9, $M_{max,y} = 1.135 \times 45 = 51.1$ kNm.

Solution for $M_{max,z}$ for the beam-column of Figure 7.19c.

Using Figure 7.6, $k_{cr} = 1.0$, $\gamma_m = 1.0$, $\gamma_n = 0.49$, $\gamma_s = 0.18$
$M_z = 3.2 \times 4.5^2/8 = 8.1$ kNm.
$$N_{cr,z} = \pi^2 \times 210\,000 \times 571 \times 10^4/(1.0 \times 4500)^2 \text{ N} = 584.4 \text{ kN}.$$

Using equation 7.10, $\delta_z = \dfrac{1.0 \times (1 - 0.18 \times 200/584.4)}{(1 - 0.49 \times 200/584.4)} = 1.127.$

Using equation 7.9, $M_{max,z} = 1.127 \times 8.1 = 9.1$ kNm.

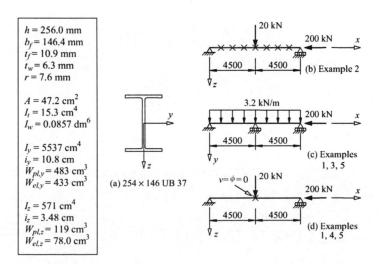

$h = 256.0$ mm
$b_f = 146.4$ mm
$t_f = 10.9$ mm
$t_w = 6.3$ mm
$r = 7.6$ mm

$A = 47.2$ cm^2
$I_t = 15.3$ cm^4
$I_w = 0.0857$ dm^6

$I_y = 5537$ cm^4
$i_y = 10.8$ cm
$W_{pl.y} = 483$ cm^3
$W_{el.y} = 433$ cm^3

$I_z = 571$ cm^4
$i_z = 3.48$ cm
$W_{pl.z} = 119$ cm^3
$W_{el.z} = 78.0$ cm^3

(a) 254 × 146 UB 37

20 kN
200 kN
4500 | 4500
(b) Example 2

3.2 kN/m
200 kN
4500 | 4500
(c) Examples 1, 3, 5

$v = \phi = 0$ 20 kN
200 kN
4500 | 4500
(d) Examples 1, 4, 5

Figure 7.19 Examples 1–5.

7.7.2   Example 2 – checking the major axis in-plane resistance

Problem. The 9 m long simply supported beam-column shown in Figure 7.19b has a factored design axial compression force of 200 kN and a design concentrated load of 20 kN (which includes an allowance for self-weight) acting in the major principal plane at mid-span. The beam-column is the 254 × 146 UB 37 of S275 steel shown in Figure 7.19a. The beam-column is continuously braced against lateral deflections v and twist rotations ϕ. Check the adequacy of the beam-column.

Simplified approach for cross-section resistance.

$$t_f = 10.9 \text{ mm}, \ f_y = 275 \text{ N/mm}^2 \qquad\qquad \text{EN 10025-2}$$

$$\varepsilon = (235/275)^{0.5} = 0.924 \qquad\qquad \text{T5.2}$$

$$c_f/(t_f\varepsilon) = ((146.4 - 6.3 - 2 \times 7.6)/2)/(10.9 \times 0.924) = 6.20 < 9 \quad \text{T5.2}$$

and the flange is Class 1.

$$c_w = 256.0 - (2 \times 10.9) - (2 \times 7.6) = 219.0 \text{ mm}.$$

The compression proportion of web is

$$
\alpha = \left(\frac{h}{2} - (t_f + r) + \frac{1}{2}\frac{N_{Ed}}{t_w f_y} \right) \Big/ c_w
$$

$$
= \left(\frac{256}{2} - (10.9 + 7.6) + \frac{1}{2} \times \frac{200 \times 10^3}{6.3 \times 275} \right) \Big/ 219.0
$$

$$
= 0.76 > 0.5 \qquad\qquad \text{T5.2}
$$

$$
c_w/t_w = 219.0/6.3 = 34.8 < 41.3 = \frac{396\varepsilon}{13\alpha - 1} \qquad\qquad \text{T5.2}
$$

and the web is Class 1.

$$M_{c,y,Rd} = 275 \times 483 \times 10^3/1.0 \text{ Nmm} = 132.8 \text{ kNm.} \qquad\qquad 6.2.5(2)$$

$$M_{y,Ed} = 20 \times 9/4 = 45.0 \text{ kNm.}$$

$$
\frac{200 \times 10^3}{47.2 \times 10^2 \times 275/1.0} + \frac{45.0}{132.8} = 0.493 \le 1 \qquad\qquad 6.2.1(7)
$$

and the cross-section resistance is adequate.

Alternative approach for cross-section resistance.

Because the section is Class 1, Clause 6.2.9.1 can be used.
No reduction in plastic moment resistance is required provided both

$$N_{Ed} = 200 \text{ kN} < 324.5 \text{ kN} = (0.25 \times 47.2 \times 10^2 \times 275/1.0)/10^3$$

$$= 0.25 N_{pl,Rd} \text{ and} \qquad \text{6.2.9.1(4)}$$

$$N_{Ed} = 200 \text{ kN} < 202.9 \text{ kN} = \frac{0.5 \times (256.0 - 2 \times 10.9) \times 6.3 \times 275}{1.0 \times 10^3}$$

$$= \frac{0.5 h_w t_w f_y}{\gamma_{M0}} \qquad \text{6.2.9.1(4)}$$

and so no reduction in the plastic moment resistance is required.
Thus $M_{N,y,Rd} = M_{pl,y,Rd} = 132.8 \text{ kNm} > 45.0 \text{ kNm} = M_{y,Ed}$
and the cross-section resistance is adequate.

Compression member buckling resistance.

Because the member is continuously braced, beam lateral buckling and column
minor axis buckling need not be considered.

$$\overline{\lambda}_y = \sqrt{\frac{A f_y}{N_{cr,y}}} = \frac{L_{cr,y}}{i_y} \frac{1}{\lambda_1} = \frac{9000}{(10.8 \times 10)} \frac{1}{93.9 \times 0.924} = 0.960 \quad \text{6.3.1.3(1)}$$

For a rolled UB section (with $h/b > 1.2$ and $t_f \leq 40$ mm), buckling about the
y-axis, use buckling curve (a) with $\alpha = 0.21$ \hfill T6.2, T6.1

$$\Phi_y = 0.5[1 + 0.21(0.960 - 0.2) + 0.960^2] = 1.041 \qquad \text{6.3.1.2(1)}$$

$$\chi_y = 1/(1.041 + \sqrt{1.041^2 - 0.960^2}) = 0.693 \qquad \text{6.3.1.2(1)}$$

$$N_{b,y,Rd} = \chi_y A f_y/\gamma_{M1} = 0.693 \times 47.2 \times 10^2 \times 275/1.0 \text{ N}$$

$$= 900 \text{ kN} > 200 \text{ kN} = N_{Ed} \qquad \text{6.3.1.1(3)}$$

Beam-column member resistance – simplified approach (Annex B).

$$\alpha_h = M_h/M_s = 0, \quad \psi = 0, \quad C_{my} = 0.90 + 0.1\alpha_h = 0.90 \qquad \text{TB.3}$$

$$C_{my}\left(1 + (\overline{\lambda}_y - 0.2)\frac{N_{Ed}}{\chi_y N_{Rk}/\gamma_{M1}}\right) = 0.90$$

$$\times \left(1 + (0.960 - 0.2) \times \frac{200 \times 10^3}{0.693 \times 47.2 \times 10^2 \times 275/1.0}\right) = 1.052$$

$$< C_{my} \left(1 + 0.8 \frac{N_{Ed}}{\chi_y N_{Rk}/\gamma_{M1}} \right) = 0.90$$

$$\times \left(1 + 0.8 \times \frac{200 \times 10^3}{0.693 \times 47.2 \times 10^2 \times 275/1.0} \right) = 1.060 \qquad \text{TB.1}$$

so that $k_{yy} = 1.052$

$$\frac{N_{Ed}}{N_{b,y,Rd}} + k_{yy} \frac{M_{y,Ed}}{M_{c,y,Rd}} = \frac{200}{900} + 1.052 \times \frac{45.0}{132.8}$$

$$= 0.222 + 0.356 = 0.579 < 1 \qquad \text{6.3.3(4)}$$

and the member resistance is adequate.

Beam-column member resistance – more exact approach (Annex A).

$$\bar{\lambda}_{max} = \bar{\lambda}_y = 0.960 \qquad \text{TA.1}$$

$$N_{cr,y} = \pi^2 EI_y/L_{cr}^2 = \pi^2 \times 21\,0000 \times 5537 \times 10^4/9000^2 = 1417\,\text{kN}.$$

Since there is no lateral buckling, $\bar{\lambda}_0 = 0, b_{LT} = 0, C_{mLT} = 1.0$ \quad TA.1

$$C_{my} = C_{my,0} = 1 - 0.18 N_{Ed}/N_{cr,y} = 1 - 0.18 \times 200/1417 = 0.975 \;\; \text{TA.2}$$

$$w_y = \frac{W_{pl,y}}{W_{el,y}} = \frac{483}{433} = 1.115,$$

$$n_{pl} = \frac{N_{Ed}}{N_{Rk}/\gamma_{M1}} = \frac{200 \times 10^3}{47.2 \times 10^2 \times 275/1.0} = 0.154 \qquad \text{TA.1}$$

$$C_{yy} = 1 + (w_y - 1) \left[\left(2 - \frac{1.6}{w_y} C_{my}^2 \bar{\lambda}_{max} - \frac{1.6}{w_y} C_{my}^2 \bar{\lambda}_{max}^2 \right) n_{pl} - b_{LT} \right]$$

$$\geq \frac{W_{el,y}}{W_{pl,y}}$$

$$= 1 + (1.115 - 1) \left[\left(2 - \frac{1.6}{1.115} \times 0.975^2 \times 0.960 - \frac{1.6}{1.115} \right. \right.$$

$$\left. \left. \times 0.975^2 \times 0.960^2 \right) \times 0.154 - 0 \right]$$

$$= 0.990 > 0.896 = 1/1.115 \qquad \text{TA.1}$$

$$\mu_y = \frac{1 - N_{Ed}/N_{cr,y}}{1 - \chi_y N_{Ed}/N_{cr,y}} = \frac{1 - 200/1417}{1 - 0.693 \times 200/1417} = 0.952 \qquad \text{TA.1}$$

$$k_{yy} = C_{my} C_{mLT} \frac{\mu_y}{(1 - N_{Ed}/N_{cr,y})} \frac{1}{C_{yy}}$$

$$= 0.975 \times 1.0 \times \frac{0.952}{1 - 200/1417} \times \frac{1}{0.990} = 1.091 \qquad \text{TA.1}$$

$$\frac{N_{Ed}}{N_{b,y,Rd}} + k_{yy} \frac{M_{y,Ed}}{M_{c,y,Rd}} = \frac{200}{900} + 1.091 \times \frac{45.0}{132.8}$$

$$= 0.222 + 0.370 = 0.592 < 1 \qquad 6.3.3(4)$$

and the member resistance is adequate.

7.7.3 Example 3 – checking the minor axis in-plane resistance

Problem. The 9 m long two span beam-column shown in Figure 7.19c has a factored design axial compression force of 200 kN and a factored design uniformly distributed load of 3.2 kN/m acting in the minor principal plane. The beam-column is the 254 × 146 UB 37 of S275 steel shown in Figure 7.19a. Check the adequacy of the beam-column.

Simplified approach for cross-section resistance.

$$M_{z,Ed} = 3.2 \times 4.5^2/8 = 8.1 \text{ kNm.}$$

The flange is Class 1 under uniform compression, and so remains Class 1 under combined loading.

$$c_w/(t_w \varepsilon) = (256.0 - 2 \times 10.9 - 2 \times 7.6)/(6.3 \times 0.924) = 37.6 < 38 \qquad \text{T5.2}$$

and the web is Class 2. Thus the overall cross-section is Class 2.

$$M_{c,z,Rd} = 275 \times 119 \times 10^3/1.0 \text{ Nmm} = 32.7 \text{ kNm.} \qquad 6.2.5(2)$$

$$\frac{200 \times 10^3}{47.2 \times 10^2 \times 275/1.0} + \frac{8.1}{32.7} = 0.402 < 1 \qquad 6.2.1(7)$$

and the cross-section resistance is adequate.

Alternative approach for cross-section resistance.

Because the section is Class 2, Clause 6.2.9.1 can be used.

$$N_{Ed} = 200 \text{ kN} < 406 \text{ kN} = \frac{(256 - 2 \times 10.9) \times 6.3 \times 275}{1.0 \times 10^3} = \frac{h_w t_w f_y}{\gamma_{M0}}.$$

6.2.9.1(4)

and so no reduction in plastic moment resistance is required.

$$\text{Thus } M_{N,z,Rd} = M_{pl,z,Rd} = 32.7 \text{ kNm} > 8.1 \text{ kNm} = M_{z,Ed} \qquad 6.2.5(2)$$

and the cross-section resistance is adequate.

Compression member buckling resistance.

$$\bar{\lambda}_z = \sqrt{\frac{Af_y}{N_{cr,z}}} = \frac{L_{cr,z}}{i_z} \frac{1}{\lambda_1} = \frac{4500}{(3.48 \times 10)} \frac{1}{93.9 \times 0.924} = 1.490 \quad 6.3.1.3(1)$$

For a rolled UB section (with $h/b > 1.2$ and $t_f \le 40$ mm), buckling about the z-axis, use buckling curve (b) with $\alpha = 0.34$ 　　　　　　　T6.2, T6.1

$$\Phi_z = 0.5[1 + 0.34\ (1.490 - 0.2) + 1.490^2] = 1.829 \qquad 6.3.1.2(1)$$

$$\chi_z = 1/(1.829 + \sqrt{1.829^2 - 1.490^2}) = 0.346 \qquad 6.3.1.2(1)$$

$$N_{b,z,Rd} = \chi_z Af_y/\gamma_{M1} = 0.346 \times 47.2 \times 10^2 \times 275/1.0 \text{ N}$$

$$= 449 \text{ kN} > 200 \text{ kN} = N_{Ed} \qquad 6.3.1.1(3)$$

Beam-column member resistance – simplified approach (Annex B).

Because the member is bent about the minor axis, beam lateral buckling need not be considered, and $e_{LT} = 0$.

$$M_h = -8.1 \text{ kNm}, \quad M_s = 9 \times 3.2 \times 4.5^2/128 = 4.56 \text{ kNm}, \quad \psi = 0 \quad \text{TB.3}$$

$$\alpha_s = M_s/M_h = 4.56/(-8.1) = -0.563, \qquad \text{TB.3}$$

$$C_{mz} = 0.1 - 0.8\alpha_s = 0.1 - 0.8 \times (-0.563) = 0.550 > 0.40 \qquad \text{TB.3}$$

$$C_{mz}\left(1 + (2\bar{\lambda}_z - 0.6)\frac{N_{Ed}}{\chi_z N_{Rk}/\gamma_{M1}}\right) = 0.550 \times \left(1 + (2 \times 1.490 - 0.6)\right.$$

$$\left. \times \frac{200 \times 10^3}{0.346 \times 47.2 \times 10^2 \times 275/1.0}\right) = 1.133$$

$$> C_{mz}\left(1 + 1.4\frac{N_{Ed}}{\chi_z N_{Rk}/\gamma_{M1}}\right)$$

$$= 0.550 \times \left(1 + 1.4 \times \frac{200 \times 10^3}{0.346 \times 47.2 \times 10^2 \times 275/1.0}\right) = 0.893 \quad \text{TB.1}$$

$$k_{zz} = 0.893 \hspace{8cm} \text{TB.1}$$

$$\frac{N_{Ed}}{N_{b,z,Rd}} + k_{zz}\frac{M_{z,Ed}}{M_{c,y,Rd}} = \frac{200}{449} + 0.893\frac{8.1}{32.7} = 0.445 + 0.221 = 0.666 \leq 1$$

$$6.3.3(4)$$

and the member resistance is adequate.

Beam-column member resistance – more exact approach (Annex A).
From Section 7.7.2, $\bar{\lambda}_y = 0.960$

$$\bar{\lambda}_{max} = \bar{\lambda}_z = 1.490 > 0.960 = \bar{\lambda}_y \hspace{4cm} \text{TA.1}$$

$$N_{cr,z} = \pi^2 EI_z/L_{cr,z}^2 = \pi^2 \times 210\,000 \times 571 \times 10^4/4500^2 \text{ N} = 584.4\,\text{kN}.$$

$$|\delta_x| = \frac{qL^4}{185EI_z} = \frac{3.2 \times 4500^4}{185 \times 210\,000 \times 571 \times 10^4} = 5.92\,\text{mm},$$

$$|M_{z,Ed}(x)| = 8.1\,\text{kNm}$$

$$C_{mz} = C_{mz,0} = 1 + \left(\frac{\pi^2 EI_z\,|\delta_x|}{L^2\,|M_{z,Ed}(x)|} - 1\right)\frac{N_{Ed}}{N_{cr,z}}$$

$$= 1 + \left(\frac{\pi^2 \times 210\,000 \times 571 \times 10^4 \times 5.9}{4500^2 \times 8.1 \times 10^6} - 1\right)\frac{200}{584.4} = 0.804 \quad \text{TA.2}$$

$$w_z = \frac{W_{pl,z}}{W_{el,z}} = \frac{119}{78} = 1.526 > 1.5,$$

$$n_{pl} = \frac{N_{Ed}}{N_{Rk}/\gamma_{M1}} = \frac{200 \times 10^3}{47.2 \times 10^2 \times 275/1.0} = 0.154 \hspace{3cm} \text{TA.1}$$

$$C_{zz} = 1 + (w_z - 1)\left[\left(2 - \frac{1.6}{w_z}C_{mz}^2\bar{\lambda}_{max} - \frac{1.6}{w_z}C_{mz}^2\bar{\lambda}_{max}^2\right)n_{pl} - e_{LT}\right] \geq \frac{W_{el,z}}{W_{pl,z}}$$

$$= 1 + (1.5 - 1)\left[\left(2 - \frac{1.6}{1.5} \times 0.804^2 \times 1.490 - \frac{1.6}{1.5} \times 0.804^2 \times 1.490^2\right)\right.$$

$$\left. \times 0.154 - 0\right] = 0.958 > 0.667 = 1/1.5 \hspace{3cm} \text{TA.1}$$

$$\mu_z = \frac{1 - N_{Ed}/N_{cr,z}}{1 - \chi_z N_{Ed}/N_{cr,z}} = \frac{1 - 200/584.4}{1 - 0.346 \times 200/584.4} = 0.746 \qquad \text{TA.1}$$

$$k_{zz} = C_{mz} \frac{\mu_z}{(1 - N_{Ed}/N_{cr,z})} \frac{1}{C_{zz}} = 0.804 \times \frac{0.746}{1 - 200/584} \times \frac{1}{0.958} = 0.952$$

$$\text{TA.1}$$

$$\frac{N_{Ed}}{N_{b,z,Rd}} + k_{zz} \frac{M_{y,Ed}}{M_{c,y,Rd}} = \frac{200}{449} + 0.952 \times \frac{8.1}{32.7} = 0.445 + 0.236 = 0.681 < 1$$

$$6.3.3(4)$$

and the member resistance is adequate.

7.7.4   Example 4 – checking the out-of-plane resistance

Problem. The 9 m long simply supported beam-column shown in Figure 7.19d has a factored design axial compression force of 200 kN and a factored design concentrated load of 20 kN (which includes an allowance for self-weight) acting in the major principal plane at mid-span. Lateral deflections v and twist rotations ϕ are prevented at the ends and at mid-span. The beam-column is the 254 × 146 UB 37 of S275 steel shown in Figure 7.19a. Check the adequacy of the beam-column.

Design bending moment.

   $M_{y,Ed} = 45.0$ kNm as in Section 7.7.2.

Simplified approach for cross-section resistance.

The cross-section resistance was checked in Section 7.7.2.

Beam-column member buckling resistance – simplified approach (Annex B).

While the member major axis in-plane resistance without lateral buckling was checked in Section 7.7.2, the lateral buckling resistance between supports should be used in place of the in-plane bending resistance to check against equation 6.61 of EC3.

   From Section 6.15.2, $M_{b,Rd} = 121.4$ kNm.
   From Section 7.7.2, $N_{b,y,Rd} = 900$ kN, $k_{yy} = 1.052$, and $M_{y,Ed} = 45.0$ kNm.

$$\frac{N_{Ed}}{N_{b,y,Rd}} + k_{yy} \frac{M_{y,Ed}}{M_{b,Rd}} = \frac{200}{900} + 1.052 \times \frac{45.0}{121.4} = 0.222 + 0.390$$

$$= 0.612 \le 1 \qquad 6.3.3(4)$$

and the member in-plane resistance is adequate.
   For the member out-of-plane resistance (equation 6.62 of EC3),

From Section 7.7.3, $N_{b,z,Rd} = 449$ kN, $\chi_z = 0.346$, $\bar{\lambda}_z = 1.490$, $\bar{\lambda}_{max} = 1.490$, $N_{cr,z} = 584.4$ kN, and $w_z = 1.5$.

$$\psi = 0, \quad C_{mLT} = 0.6 + 0.4\psi = 0.6 > 0.4 \qquad\qquad \text{TB.3}$$

$$\left(1 - \frac{0.1\bar{\lambda}_z}{(C_{mLT} - 0.25)} \frac{N_{Ed}}{\chi_z N_{Rk}/\gamma_{M1}}\right)$$

$$= \left(1 - \frac{0.1 \times 1.490}{(0.6 - 0.25)} \times \frac{200 \times 10^3}{0.346 \times 47.2 \times 10^2 \times 275/1.0}\right) = 0.810$$

$$\left(1 - \frac{0.1}{(C_{mLT} - 0.25)} \frac{N_{Ed}}{\chi_z N_{Rk}/\gamma_{M1}}\right)$$

$$= \left(1 - \frac{0.1}{(0.6 - 0.25)} \times \frac{200 \times 10^3}{0.346 \times 47.2 \times 10^2 \times 275/1.0}\right) = 0.873 \quad \text{TB.1}$$

$$k_{zy} = 0.873 \qquad\qquad \text{TB.1}$$

$$\frac{N_{Ed}}{N_{b,z,Rd}} + k_{zy}\frac{M_{y,Ed}}{M_{b,Rd}} = \frac{200}{449} + 0.873 \times \frac{45.0}{121.4} = 0.445 + 0.324 = 0.769 < 1$$

$$6.3.3(4)$$

and the member out-of-plane resistance is adequate.

Beam-column member buckling resistance – more exact approach (Annex A).
The member in-plane resistance was checked in Section 7.7.2, but as for the simplified approach, equation 6.61 should be checked by taking the lateral buckling resistance in place of the in-plane major axis bending resistance. The interaction factor k_{yy} also requires re-calculating due to the possibility of lateral buckling.

$$\bar{\lambda}_y = 0.960, \quad \bar{\lambda}_{max} = \bar{\lambda}_z = 1.490 \qquad\qquad \text{TA.1}$$

Using $M_{zx} = 111.2$ kNm from Section 6.15.2,

$$\bar{\lambda}_0 = \sqrt{132.8/111.2} = 1.093 \qquad\qquad \text{TA.1}$$

Since cross-section is doubly-symmetrical, using equation 7.37 with

$$i_p = \sqrt{(I_y + I_z)/A} = \sqrt{(5537 \times 10^4 + 571 \times 10^4)/47.2 \times 10^2}$$

$$= 113.8 \text{ mm,}$$

$$N_{cr,T} = \frac{1}{i_p^2}\left(G I_t + \frac{\pi^2 E I_w}{L_{cr,T}^2}\right)$$

$$= \frac{1}{113.8^2}\left(81000 \times 15.3 \times 10^4 + \frac{\pi^2 \times 210\,000 \times 0.0857 \times 10^{12}}{4500^2}\right)$$

$$= 1636 \text{ kN.}$$

Using α_m of equation 6.5 for C_1 with $\beta_m = 0$ leads to $C_1 = 1.75$

$$0.2\sqrt{C_1}\sqrt[4]{\left(1 - \frac{N_{Ed}}{N_{cr,z}}\right)\left(1 - \frac{N_{Ed}}{N_{cr,T}}\right)}$$

$$= 0.2 \times \sqrt{1.75} \times \sqrt[4]{\left(1 - \frac{200}{584.4}\right)\left(1 - \frac{200}{1636}\right)}$$

$$= 0.231 < 1.093 = \overline{\lambda}_0 \qquad \text{TA.1}$$

$$a_{LT} = 1 - I_t/I_y = 1 - 15.4/5537 = 0.997 > 0 \qquad \text{TA.1}$$

$$\varepsilon_y = \frac{M_{y,Ed}}{N_{Ed}} \frac{A}{W_{el,y}} = \frac{45 \times 10^6}{200 \times 10^3} \times \frac{47.2 \times 10^2}{433 \times 10^3} = 2.453 \qquad \text{TA.1}$$

Using $N_{cr,y} = 1417$ kN from Section 7.7.2,

$$C_{my,0} = 1 - 0.18N_{Ed}/N_{cr,y} = 1 - 0.18 \times 200/1417 = 0.975 \qquad \text{TA.2}$$

$$C_{my} = C_{my,0} + (1 - C_{my,0})\frac{\sqrt{\varepsilon_y}a_{LT}}{1 + \sqrt{\varepsilon_y}a_{LT}}$$

$$= 0.975 + (1 - 0.975) \times \frac{\sqrt{2.453} \times 0.997}{1 + \sqrt{2.453} \times 0.997} = 0.990 \qquad \text{TA.1}$$

$$C_{mLT} = C_{my}^2 \frac{a_{LT}}{\sqrt{\left(1 - \frac{N_{Ed}}{N_{cr,z}}\right)\left(1 - \frac{N_{Ed}}{N_{cr,T}}\right)}}$$

$$= 0.990^2 \times \frac{0.997}{\sqrt{\left(1 - \frac{200}{584.4}\right)\left(1 - \frac{200}{1636}\right)}} = 1.287 > 1 \qquad \text{TA.1}$$

$$w_y = \frac{W_{pl,y}}{W_{el,y}} = \frac{483}{433} = 1.115,$$

$$n_{pl} = \frac{N_{Ed}}{N_{Rk}/\gamma_{M1}} = \frac{200 \times 10^3}{47.2 \times 10^2 \times 275/1.0} = 0.154 \qquad \text{TA.1}$$

$$C_{yy} = 1 + (w_y - 1)\left[\left(2 - \frac{1.6}{w_y}C_{my}^2\overline{\lambda}_{max} - \frac{1.6}{w_y}C_{my}^2\overline{\lambda}_{max}^2\right)n_{pl} - b_{LT}\right]$$

$$\geq \frac{W_{el,y}}{W_{pl,y}}$$

$$= 1 + (1.115 - 1) \left[\left(2 - \frac{1.6}{1.115} \times 0.990^2 \times 1.490 - \frac{1.6}{1.115} \right. \right.$$

$$\left. \left. \times 0.990^2 \times 1.490^2 \right) \times 0.154 - 0 \right]$$

$$= 0.942 > 0.896 \qquad\qquad \text{TA.1}$$

Using $\chi_y = 0.693$ from Section 7.7.2,

$$\mu_y = \frac{1 - N_{Ed}/N_{cr,y}}{1 - \chi_y N_{Ed}/N_{cr,y}} = \frac{1 - 200/1417}{1 - 0.693 \times 200/1417} = 0.952 \qquad\qquad \text{TA.1}$$

$$k_{yy} = C_{my} C_{mLT} \frac{\mu_y}{1 - N_{Ed}/N_{cr,y}} \frac{1}{C_{yy}} = 0.9900 \times 1.287$$

$$\times \frac{0.952}{1 - 200/1417} \frac{1}{0.942} = 1.499 \qquad\qquad \text{TA.1}$$

$$\frac{N_{Ed}}{N_{b,y,Rd}} + k_{yy} \frac{M_{y,Ed}}{M_{b,Rd}} = \frac{200}{900} + 1.499 \times \frac{45.0}{121.4}$$

$$= 0.222 + 0.556 = 0.778 < 1 \qquad\qquad 6.3.3(4)$$

and the member in-plane resistance (with lateral buckling between the lateral supports) is adequate.

For the member out-of-plane resistance (equation 6.62 of EC3),

$$C_{zy} = 1 + (w_y - 1) \left[\left(2 - 14 \frac{C_{my}^2 \bar{\lambda}_{max}^2}{w_y^5} \right) n_{pl} - d_{LT} \right] \geq 0.6 \sqrt{\frac{w_y}{w_z}} \frac{W_{el,y}}{W_{pl,y}}$$

$$= 1 + (1.115 - 1) \left[\left(2 - 14 \times \frac{0.990^2 \times 1.490^2}{1.115^5} \right) \times 0.154 - 0 \right]$$

$$= 0.722 > 0.46 \qquad\qquad \text{TA.1}$$

Using $\chi_z = 0.346$ from Section 7.7.3,

$$\mu_z = \frac{1 - N_{Ed}/N_{cr,z}}{1 - \chi_z N_{Ed}/N_{cr,z}} = \frac{1 - 200/584.4}{1 - 0.346 \times 200/584.4} = 0.746 \qquad\qquad \text{TA.1}$$

$$k_{zy} = C_{my} C_{mLT} \frac{\mu_z}{1 - \dfrac{N_{Ed}}{N_{cr,y}}} \frac{1}{C_{zy}} 0.6 \sqrt{\frac{w_y}{w_z}} = 0.990 \times 1.287$$

$$\times \frac{0.746}{1 - \dfrac{200}{1417}} \times \frac{1}{0.722} \times 0.6 \times \sqrt{\frac{1.115}{1.5}} = 0.793 \qquad\qquad \text{TA.1}$$

$$\frac{N_{Ed}}{N_{b,z,Rd}} + k_{zy} \frac{M_{y,Ed}}{M_{b,Rd}} = \frac{200}{449} + 0.793 \times \frac{45.0}{121.4} = 0.445 + 0.294 = 0.739 < 1$$

6.3.3(4)

and the member out-of-plane resistance is adequate.

7.7.5 Example 5 – checking the biaxial bending capacity

Problem. The 9 m long simply supported beam-column shown in Figure 7.19c and d has a factored design axial compression force of 200 kN, a factored design concentrated load of 20 kN (which includes an allowance for self-weight) acting in the major principal plane, and a factored design uniformly distributed load of 3.2 kN/m acting in the minor principal plane. Lateral deflections v and twist rotations ϕ are prevented at the ends and at mid-span. The beam-column is the 254 × 146 UB 37 of S275 steel shown in Figure 7.19a. Check the adequacy of the beam-column.

Design bending moments.

$$M_{y,Ed} = 45.0 \, \text{kNm(Section 7.7.2)}, M_{z,Ed} = 8.1 \, \text{kNm(Section 7.7.3)}$$

Simplified approach for cross-section resistance.

Combining the calculations of Sections 7.7.2 and 7.7.3,

$$\frac{200 \times 10^3}{47.2 \times 10^2 \times 275/1.0} + \frac{45.0}{132.8} + \frac{8.1}{32.7} = 0.740 < 1$$

6.2.1(7)

and the cross-section resistance is adequate.

Alternative approach for cross-section resistance.

As shown in Sections 7.7.2 and 7.7.3, the axial load is sufficiently low for there is to be no reduction in the moment resistance about either the major or the minor axis.

$$n = N_{Ed}/N_{pl,Rd} = 200 \times 10^3/(47.2 \times 10^2 \times 275/1.0) = 0.154,$$
$$\beta = 5n = 0.770$$

6.2.9.1(6)

$$\left[\frac{M_{y,Ed}}{M_{N,y,Rd}}\right]^{\alpha} + \left[\frac{M_{z,Ed}}{M_{N,z,Rd}}\right]^{\beta} = \left[\frac{45.0}{132.8}\right]^2 + \left[\frac{8.1}{32.7}\right]^{0.770} = 0.456 \le 1$$

6.2.9.1(6)

and the cross-section resistance is adequate.

Beam-column member resistance – simplified approach (Annex B).

Checks are required against equations 6.61 and 6.62 of EC3. From the appropriate parts of Sections 7.7.2–7.7.4,

$k_{yy} = 1.052$, $k_{zz} = 0.893$, $k_{zy} = 0.873$, $N_{b,y,Rd} = 900$ kN,

$M_{b,Rd} = 121.4$ kNm, $M_{z,Rd} = 32.7$ kNm, and $N_{b,z,Rd} = 449$ kN.

$k_{yz} = 0.6 k_{zz} = 0.6 \times 0.893 = 0.536$ TB.1

$$\frac{N_{Ed}}{N_{b,y,Rd}} + k_{yy}\frac{M_{y,Ed}}{M_{b,Rd}} + k_{yz}\frac{M_{z,Ed}}{M_{z,Rd}} = \frac{200}{900} + 1.052 \times \frac{45.0}{121.4} + 0.536$$

$$\times \frac{8.1}{32.7} = 0.222 + 0.390 + 0.133 = 0.745 < 1$$

$$\frac{N_{Ed}}{N_{b,z,Rd}} + k_{zy}\frac{M_{y,Ed}}{M_{b,Rd}} + k_{zz}\frac{M_{z,Ed}}{M_{z,Rd}} = \frac{200}{449} + 0.873 \times \frac{45.0}{121.4}$$

$$+ 0.893 \times \frac{8.1}{32.7} = 0.445 + 0.324 + 0.221 = 0.990 < 1$$

and the member resistance is adequate.

Beam-column member resistance – more exact approach (Annex A).

Checks are required against equations 6.61 and 6.62 of EC3. Relevant details are taken from the appropriate parts of Sections 7.7.2, 7.7.3, and 7.7.4.

$$b_{LT} = 0.5 a_{LT} \bar{\lambda}_0^2 \frac{M_{y,Ed}}{\chi_{LT} M_{pl,y,Rd}} \frac{M_{z,Ed}}{M_{pl,z,Rd}} = 0.5 \times 0.997 \times 1.093^2$$

$$\times \frac{45}{121.4} \times \frac{8.1}{32.7} = 0.0546 \qquad \text{TA.1}$$

$$c_{LT} = 10 a_{LT} \frac{\bar{\lambda}_0^2}{5 + \bar{\lambda}_z^4} \frac{M_{y,Ed}}{C_{my}\chi_{LT} M_{pl,y,Rd}} = 10 \times 0.997$$

$$\times \frac{1.093^2}{5 + 1.490^4} \times \frac{45}{0.990 \times 121.4} = 0.449 \qquad \text{TA.1}$$

$$d_{LT} = 2 a_{LT} \frac{\bar{\lambda}_0}{0.1 + \bar{\lambda}_z^4} \frac{M_{y,Ed}}{C_{my}\chi_{LT} M_{pl,y,Rd}} \frac{M_{z,Ed}}{C_{mz}M_{pl,z,Rd}}$$

$$= 2 \times 0.997 \times \frac{1.093}{0.1 + 1.490^4} \times \frac{45}{0.990 \times 121.4} \times \frac{8.1}{0.804 \times 32.7}$$

$$= 0.0500 \qquad \text{TA.1}$$

$$e_{LT} = 1.7 a_{LT} \frac{\bar{\lambda}_0}{0.1 + \bar{\lambda}_z^4} \frac{M_{y,Ed}}{C_{my}\chi_{LT} M_{pl,y,Rd}} = 1.7 \times 0.997 \times \frac{1.093}{0.1 + 1.490^4}$$

$$\times \frac{45}{0.990 \times 121.4} = 0.138 \qquad \text{TA.1}$$

$$C_{yy} = 1 + (w_y - 1)\left[\left(2 - \frac{1.6}{w_y}C_{my}^2\bar{\lambda}_{max} - \frac{1.6}{1.12}C_{my}^2\bar{\lambda}_{max}^2\right)n_{pl} - b_{LT}\right]$$

$$\geq \frac{W_{el,y}}{W_{pl,y}}$$

$$= 1 + (1.115 - 1)\left[\left(2 - \frac{1.6}{1.115} \times 0.990^2 \times 1.490 - \frac{1.6}{1.115}\right.\right.$$

$$\left.\times 0.990^2 \times 1.490^2\right) \times 0.154 - 0.0546] = 0.936 > 0.896 \qquad \text{TA.1}$$

$$C_{yz} = 1 + (w_z - 1)\left[\left(2 - 14\frac{C_{mz}^2\bar{\lambda}_{max}^2}{w_z^5}\right)n_{pl} - c_{LT}\right] \geq 0.6\sqrt{\frac{w_z}{w_y}}\frac{W_{el,z}}{W_{pl,z}}$$

$$= 1 + (1.5 - 1)\left[\left(2 - 14 \times \frac{0.804^2 \times 1.490^2}{1.5^5}\right) \times 0.154 - 0.449\right]$$

$$= 0.726 > 0.456 \qquad \text{TA.1}$$

$$C_{zy} = 1 + (w_y - 1)\left[\left(2 - 14\frac{C_{my}^2\bar{\lambda}_{max}^2}{w_y^5}\right)n_{pl} - d_{LT}\right] \geq 0.6\sqrt{\frac{w_y}{w_z}}\frac{W_{el,y}}{W_{pl,y}}$$

$$= 1 + (1.115 - 1)\left[\left(2 - 14 \times \frac{0.990^2 \times 1.490^2}{1.115^5}\right) \times 0.154 - 0.0500\right]$$

$$= 0.716 > 0.464 \qquad \text{TA.1}$$

$$C_{zz} = 1 + (w_z - 1)\left[\left(2 - \frac{1.6}{w_z}C_{mz}^2\bar{\lambda}_{max} - \frac{1.6}{w_z}C_{mz}^2\bar{\lambda}_{max}^2\right)n_{pl} - e_{LT}\right] \geq \frac{W_{el,z}}{W_{pl,z}}$$

$$= 1 + (1.5 - 1)\left[\left(2 - \frac{1.6}{1.5} \times 0.804^2 \times 1.490 - \frac{1.6}{1.5} \times 0.804^2\right.\right.$$

$$\left.\times 1.490^2\right) \times 0.154 - 0.138\right] = 0.888 > 0.655 \qquad \text{TA.1}$$

$$k_{yy} = C_{my}C_{mLT}\frac{\mu_y}{1 - N_{Ed}/N_{cr,y}}\frac{1}{C_{yy}}$$

$$= 0.990 \times 1.287 \times \frac{0.952}{1 - 200/1417} \times \frac{1}{0.936} = 1.508 \qquad \text{TA.1}$$

$$k_{yz} = C_{mz}\frac{\mu_y}{1 - N_{Ed}/N_{cr,z}}\frac{1}{C_{yz}}0.6\sqrt{\frac{w_z}{w_y}} = 0.804 \times \frac{0.952}{1 - 200/584.4}$$

$$\times \frac{1}{0.726} \times 0.6 \times \sqrt{\frac{1.5}{1.115}} = 1.115 \qquad \text{TA.1}$$

$$k_{zy} = C_{my} C_{mLT} \frac{\mu_z}{1 - N_{Ed}/N_{cr,y}} \frac{1}{C_{zy}} 0.6 \sqrt{\frac{w_y}{w_z}} = 0.990$$

$$\times 1.287 \times \frac{0.746}{1 - 200/1417} \times \frac{1}{0.716} \times 0.6 \times \sqrt{\frac{1.115}{1.5}} = 0.800 \quad \text{TA.1}$$

$$k_{zz} = C_{mz} \frac{\mu_z}{1 - N_{Ed}/N_{cr,z}} \frac{1}{C_{zz}} = 0.804 \times \frac{0.746}{1 - 200/584.4} \times \frac{1}{0.888} = 1.027$$

$$\text{TA.1}$$

$$\frac{N_{Ed}}{N_{b,y,Rd}} + k_{yy} \frac{M_{y,Ed}}{M_{b,Rd}} + k_{yz} \frac{M_{z,Ed}}{M_{z,Rd}} = \frac{200}{900} + 1.508 \times \frac{45.0}{121.4}$$

$$+ 1.115 \times \frac{8.1}{32.7} = 0.222 + 0.559 + 0.276 = 1.057 > 1$$

$$\frac{N_{Ed}}{N_{b,z,Rd}} + k_{zy} \frac{M_{y,Ed}}{M_{b,Rd}} + k_{zz} \frac{M_{z,Ed}}{M_{z,Rd}} = \frac{200}{449} + 0.800 \times \frac{45.0}{121.4} + 1.027$$

$$\times \frac{8.1}{32.7} = 0.445 + 0.296 + 0.254 = 0.995 < 1$$

and the member resistance appears to be inadequate.

7.8   Unworked examples

7.8.1   *Example 6 – non-linear analysis*

Analyse the non-linear elastic in-plane bending of the simply supported beam-column shown in Figure 7.20a, and show that the maximum moment M_{max} may be closely approximated by using the greater of $M_{max} = qL^2/8$ and

$$M_{max} = \frac{9qL^2}{128} \frac{(1 - 0.29N/N_{cr,y})}{(1 - N/N_{cr,y})}$$

in which $N_{cr,y} = \pi^2 EI_y/L^2$.

7.8.2   *Example 7 – non-linear analysis*

Analyse the non-linear elastic in-plane bending of the propped cantilever shown in Figure 7.20b, and show that the maximum moment M_{max} may be closely approximated by using

$$\frac{M_{max}}{qL^2/8} \approx \frac{(1 - 0.18N/N_{cr,y})}{(1 - 0.49N/N_{cr,y})}$$

in which $N_{cr,y} = \pi^2 EI_y/L^2$.

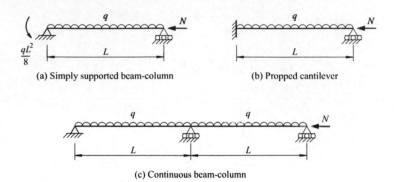

(a) Simply supported beam-column   (b) Propped cantilever

(c) Continuous beam-column

Figure 7.20 Examples 6–8.

7.8.3   Example 8 – non-linear analysis

Analyse the non-linear elastic in-plane bending of the two-span beam-column shown in Figure 7.20c, and show that the maximum moment M_{max} may be closely approximated by using

$$\frac{M_{max}}{qL^2/8} \approx \frac{(1 - 0.18N/N_{cr,y})}{(1 - 0.49N/N_{cr,y})}$$

while $N \leq N_{cr,y}$, in which $N_{cr,y} = \pi^2 EI_y/L^2$.

7.8.4   Example 9 – in-plane design

A 7.0 m long simply supported beam-column, which is prevented from swaying and from deflecting out of the plane of bending, is required to support a factored design axial load of 1200 kN and factored design end moments of 300 and 150 kNm which bend the beam-column in single curvature about the major axis. Design a suitable UB or UC beam-column of S275 steel.

7.8.5   Example 10 – out-of-plane design

The intermediate restraints of the beam-column of example 9 which prevent deflection out of the plane of bending are removed. Design a suitable UB or UC beam-column of S275 steel.

7.8.6   Example 11 – biaxial bending design

The beam-column of example 10 has to carry additional factored design end moments of 75 and 75 kNm which cause double curvature bending about the minor axis. Design a suitable UB or UC beam-column of S275 steel.

References

1. Chen, W-F. and Atsuta, T. (1976) *Theory of Beam-Columns, Vol. 1 In-Plane Behavior and Design*, McGraw-Hill, New York.
2. Galambos, T.V. and Ketter, R.L. (1959) Columns under combined bending and thrust, *Journal of the Engineering Mechanics Division, ASCE*, **85**, No. EM2, April, pp. 1–30.
3. Ketter, R.L. (1961) Further studies of the strength of beam-columns, *Journal of the Structural Division, ASCE*, **87**, No. ST6, August, pp. 135–52.
4. Young, B.W. (1973) The in-plane failure of steel beam-columns, *The Structural Engineer*, **51**, pp. 27–35.
5. Trahair, N.S. (1986) Design strengths of steel beam-columns, *Canadian Journal of Civil Engineering*, **13**, No. 6, December, pp. 639–46.
6. Bridge, R.Q. and Trahair, N.S. (1987) Limit state design rules for steel beam-columns, *Steel Construction*, Australian Institute of Steel Construction, **21**, No. 2, September, pp. 2–11.
7. Salvadori, M.G. (1955) Lateral buckling of I-beams, *Transactions, ASCE*, **120**, pp. 1165–77.
8. Salvadori, M.G. (1955) Lateral buckling of eccentrically loaded I-columns, *Transactions, ASCE*, **121**, pp. 1163–78.
9. Horne, M.R. (1954) The flexural–torsional buckling of members of symmetrical I-section under combined thrust and unequal terminal moments, *Quarterly Journal of Mechanics and Applied Mathematics*, **7**, pp. 410–26.
10. Column Research Committee of Japan (1971) *Handbook of Structural Stability*, Corona, Tokyo.
11. Trahair, N.S. (1993) *Flexural–Torsional Buckling of Structures*, E. & F.N. Spon, London.
12. Cuk, P.E. and Trahair, N.S. (1981) Elastic buckling of beam-columns with unequal end moments, *Civil Engineering Transactions*, Institution of Engineers, Australia, **CE 23**, No. 3, August, pp. 166–71.
13. Bradford, M.A., Cuk, P.E., Gizejowski, M.A., and Trahair, N.S. (1987) Inelastic lateral buckling of beam-columns, *Journal of Structural Engineering, ASCE*, **113**, No. 11, November, pp. 2259–77.
14. Bradford, M.A. and Trahair, N.S. (1985) Inelastic buckling of beam-columns with unequal end moments, *Journal of Constructional Steel Research*, **5**, No. 3, pp. 195–212.
15. Cuk, P.E., Bradford, M.A., and Trahair, N.S. (1986) Inelastic lateral buckling of steel beam-columns, *Canadian Journal of Civil Engineering*, **13**, No. 6, December, pp. 693–9.
16. Horne, M.R. (1956) The stanchion problem in frame structures designed according to ultimate load carrying capacity, *Proceedings of the Institution of Civil Engineers*, Part III, **5**, pp. 105–46.
17. Culver, C.G. (1966) Exact solution of the biaxial bending equations, *Journal of the Structural Division, ASCE*, **92**, No. ST2, pp. 63–83.
18. Harstead, G.A., Birnsteil, C., and Leu, K-C. (1968) Inelastic H-columns under biaxial bending, *Journal of the Structural Division, ASCE*, **94**, No. ST10, pp. 2371–98.
19. Trahair, N.S. (1969) Restrained elastic beam-columns, *Journal of the Structural Division, ASCE*, **95**, No. ST12, pp. 2641–64.

20. Vinnakota, S. and Aoshima, Y. (1974) Inelastic behaviour of rotationally restrained columns under biaxial bending, *The Structural Engineer*, **52**, pp. 245–55.
21. Vinnakota, S. and Aysto, P. (1974) Inelastic spatial stability of restrained beam-columns, *Journal of the Structural Division, ASCE*, **100**, No. ST11, pp. 2235–54.
22. Chen, W-F. and Atsuta, T. (1977) *Theory of Beam-Columns, Vol. 2 Space Behavior and Design*, McGraw-Hill, New York.
23. Pi, Y-L. and Trahair, N.S. (1994) Nonlinear inelastic analysis of steel beam-columns. I: Theory, *Journal of Structural Engineering, ASCE*, **120**, No. 7, pp. 2041–61.
24. Pi, Y-L. and Trahair, N.S. (1994) Nonlinear inelastic analysis of steel beam-columns. II: Applications, *Journal of Structural Engineering, ASCE*, **120**, No. 7, pp. 2062–85.
25. Tebedge, N. and Chen, W-F. (1974) Design criteria for H-columns under biaxial loading, *Journal of the Structural Division, ASCE*, **100**, No. ST3, pp. 579–98.
26. Young, B.W. (1973) Steel column design, *The Structural Engineer*, **51**, pp. 323–36.
27. Bradford, M.A. (1995) Evaluation of design rules of the biaxial bending of beam-columns, *Civil Engineering Transactions*, Institution of Engineers, Australia, **CE 37**, No. 3, pp. 241–45.
28. Nethercot, D. A. and Gardner, L. (2005) The EC3 approach to the design of columns, beams and beam-columns, *Journal of Steel and Composite Structures*, **5**, pp. 127–40.
29. Boissonnade, N., Jaspart, J.-P., Muzeau, J.-P., and Villette, M. (2004) New interaction formulae for beam-columns in Eurocode 3: The French-Belgian approach, *Journal of Constructional Steel Research*, **60**, pp. 421–31.
30. Greiner, R. and Lindner, J. (2006) Interaction formulae for members subjected to bending and axial compression in EUROCODE 3 – the Method 2 approach, *Journal of Constructional Steel Research*, **62**, pp. 757–70.
31. SCI (2007) Verification of columns in simple construction – a simplified interaction criterion, *UK localised NCCI*, The Steel Construction Institute, www.access-steel.com.
32. Gardner, L. and Nethercot, D. A. (2005) *Designers' Guide to EN 1993-1-1: Eurocode 3: Design of Steel Structures*, Thomas Telford Publishing, London.

Chapter 8

Frames

8.1 Introduction

Structural frames are composed of one-dimensional members connected together in skeletal arrangements which transfer the applied loads to the supports. While most frames are three-dimensional, they may often be considered as a series of parallel two-dimensional frames, or as two perpendicular series of two-dimensional frames. The behaviour of a structural frame depends on its arrangement and loading, and on the type of connections used.

Triangulated frames with joint loading only have no primary bending actions, and the members act in simple axial tension or compression. The behaviour and design of these members have already been discussed in Chapters 2 and 3.

Frames which are not triangulated include rectangular frames, which may be multi-storey, or multi-bay, or both, and pitched-roof portal frames. The members usually have substantial bending actions (Chapters 5 and 6), and if they also have significant axial forces, then they must be designed as beam-ties (Chapter 2) or beam-columns (Chapter 7). In frames with simple connections (see Figure 9.3a), the moments transmitted by the connections are small, and often can be neglected, and the members can be treated as isolated beams, or as eccentrically loaded beam-ties or beam-columns. However, when the connections are semi-rigid (referred to as semi-continuous in EC3) or rigid (referred to as continuous in EC3), then there are important moment interactions between the members.

When a frame is or can be considered as two-dimensional then its behaviour is similar to that of the beam-columns of which it is composed. With in-plane loading only, it will fail either by in-plane bending, or by flexural – torsional buckling out of its plane. If, however, the frame or its loading is three-dimensional, then it will fail in a mode in which the individual members are subjected to primary biaxial bending and torsion actions.

In this chapter, the in-plane, out-of-plane, and biaxial behaviour, analysis, and design of two- and three-dimensional frames are treated, and related to the behaviour and design of isolated tension members, compression members, beams, and beam-columns discussed in Chapters 2, 3, 5, 6, and 7.

8.2 Triangulated frames

8.2.1 Statically determinate frames

The primary actions in triangulated frames whose members are concentrically connected and whose loads act concentrically through the joints are those of axial compression or tension in the members, and any bending actions are secondary only. These bending actions are usually ignored, in which case the member forces may be determined by a simple analysis for which the member connections are assumed to be made through frictionless pin-joints.

If the assumed pin-jointed frame is statically determinate, then each member force can be determined by statics alone, and is independent of the behaviour of the remaining members. Because of this, the frame may be assumed to fail when its weakest member fails, as indicated in Figure 8.1a and b. Thus, each member may be designed independently of the others by using the procedures discussed in Chapters 2 (for tension members) or 3 (for compression members).

If all the members of a frame are designed to fail simultaneously (either by tension yielding or by compression buckling), there will be no significant moment interactions between the members at failure, even if the connections are not pin-jointed. Because of this, the effective lengths of the compression members should be taken as equal to the lengths between their ends. If, however, the members of a rigid-jointed frame are not designed to fail simultaneously, there will be moment interactions between them at failure. While it is a common and conservative practice to ignore these interactions and to design each member as if pin-ended, some account may be taken of these by using one of the procedures discussed in Section 8.3.5.3 to estimate the effective lengths of the compression members, and exemplified in Section 8.5.1.

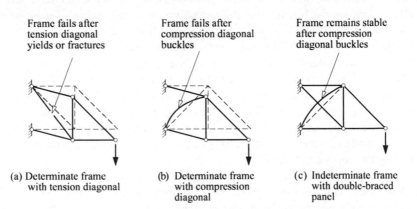

Frame fails after tension diagonal yields or fractures

Frame fails after compression diagonal buckles

Frame remains stable after compression diagonal buckles

(a) Determinate frame with tension diagonal

(b) Determinate frame with compression diagonal

(c) Indeterminate frame with double-braced panel

Figure 8.1 Determinate and indeterminate triangulated frames.

8.2.2 Statically indeterminate frames

When an assumed pin-jointed triangulated frame without primary bending actions is statically indeterminate, then the forces transferred by the members will depend on their axial stiffnesses. Since these decrease with tension yielding or compression buckling, there is usually some force redistribution as failure is approached. To design such a frame, it is necessary to estimate the member forces at failure, and then to design the individual tension and compression members for these forces, as in Chapters 2 and 3. Four different methods may be used to estimate the member forces.

In the first method, a sufficient number of members are ignored so that the frame becomes statically determinate. For example, it is common to ignore completely the compression diagonal in the double-braced panel of the indeterminate frame of Figure 8.1c, in which case the strength of the now determinate frame is controlled by the strength of one of the other members. This method is simple and conservative.

A less conservative method can be used when each indeterminate member can be assumed to be ductile, so that it can maintain its maximum strength over a considerable range of axial deformations. In this method, the frame is converted to a statically determinate frame by replacing a sufficient number of early failing members by sets of external forces equal to their maximum load capacities. For the frame shown in Figure 8.1c, this might bc done for the diagonal compression member, by replacing it by forces equal to its buckling strength. This simple method is very similar in principle to the plastic method of analysing the collapse of flexural members (Section 5.5) and frames (Section 8.3.5.7). However, the method can only be used when the ductilities of the members being replaced are assured.

In the third method, each member is assumed to behave linearly and elastically, which allows the use of a linear elastic method to analyse the frame and determine its member force distribution. The frame strength is then controlled by the strength of the most severely loaded member. For the frame shown in Figure 8.1c, this would be the compression diagonal if this reaches its buckling strength before any of the other members fail. This method ignores redistribution, and is conservative when the members are ductile, or when there are no lacks-of-fit at member connections.

In the fourth method, account is taken of the effects of changes in the member stiffnesses on the force distribution in the frame. Because of its complexity, this method is rarely used.

When the members of a triangulated frame are eccentrically connected, or when the loads act eccentrically or are applied to the members between the joints, the flexural effects are no longer secondary, and the frame should be treated as a flexural frame. Flexural frames are discussed in the following sections.

8.3   Two-dimensional flexural frames

8.3.1   General

The structural behaviour of a frame may be classified as two-dimensional when there are a number of independent two-dimensional frames with in-plane loading, or when this is approximately so. The primary actions in a two-dimensional frame in which the members support transverse loads are usually flexural, and are often accompanied by significant axial actions. The structural behaviour of a flexural frame is influenced by the behaviour of the member joints, which are usually considered to be either simple, semi-rigid (or semi-continuous), or rigid (or continuous), according to their ability to transmit moment. The EC3 classifications of joints are given in [1].

In the following sub-sections, the behaviour, analysis, and design of two-dimensional flexural frames with simple, semi-rigid, or rigid joints are discussed, together with the corresponding EC3 requirements. An introduction to the EC3 requirements is given in [2].

8.3.2   Frames with simple joints

Simple joints may be defined as being those that will not develop restraining moments which adversely affect the members and the structure as a whole, in which case the structure may be designed as if pin-jointed. It is usually assumed, therefore, that no moment is transmitted through a simple joint, and that the members connected to the joint may rotate. The joint should therefore have sufficient rotation capacity to permit member end rotations to occur without causing failure of the joint or its elements. An example of a typical simple joint between a beam and column is shown in Figure 9.3a.

If there are a sufficient number of pin-joints to make the structure statically determinate, then each member will act independently of the others, and may be designed as an isolated tension member (Chapter 2), compression member (Chapter 3), beam (Chapters 5 and 6), or beam-column (Chapter 7). However, if the pin-jointed structure is indeterminate, then some part of it may act as a rigid-jointed frame. The behaviour, analysis, and design of rigid-jointed frames are discussed in Sections 8.3.4, 8.3.5, and 8.3.6.

One of the most common methods of designing frames with simple joints is often used for rectangular frames with vertical (column) and horizontal (beam) members under the action of vertical loads only. The columns in such a frame are assumed to act as if eccentrically loaded. A minimum eccentricity of 100 mm from the face of the column was specified in [3], but there is no specific guidance in EC3. Approximate procedures for distributing the moment caused by the eccentric beam reaction between the column lengths above and below the connection are given in [1]. It should be noted that such a pin-jointed frame is usually incapable of resisting transverse forces, and must therefore be provided with an independent bracing or shear wall system.

8.3.3   Frames with semi-rigid joints

Semi-rigid joints are those which have dependable moment capacities and which partially restrain the relative rotations of the members at the joints. The action of these joints in rectangular frames is to reduce the maximum moments in the beams, and so the semi-rigid design method offers potential economies over the simple design method [4–10]. An example of a typical semi-rigid joint between a beam and column is shown in Figure 9.3b.

A method of semi-rigid design is permitted by EC3. In this method, the stiffness, strength, and rotation capacities of the joints based on experimental evidence are suggested in [1], and may be used to assess the moments and forces in the members. However, this method has not found great favour with designers, and therefore will not be discussed further.

8.3.4   In-plane behaviour of frames with rigid joints

A rigid joint may be defined as a joint which has sufficient rigidity to virtually prevent relative rotation between the members connected. Properly arranged welded and high-strength friction grip bolted joints are usually assumed to be rigid (tensioned high strength friction grip bolts are referred to in [1] as preloaded bolts). An example of a typical rigid joint between a beam and column is shown in Figure 9.3c. There are important interactions between the members of frames with rigid joints, which are generally stiffer and stronger than frames with simple or semi-rigid joints. Because of this, rigid frames offer significant economies, while many difficulties associated with their analysis have been greatly reduced by the widespread availability of standard computer programs.

The in-plane behaviour of rigid frames is discussed in general terms in Section 1.4.2, where it is pointed out that, although a rigid frame may behave in an approximately linear fashion while its service loads are not exceeded, especially when the axial forces are small, it becomes non-linear near its in-plane ultimate load because of yielding and buckling effects. When the axial compression forces are small, then failure occurs when a sufficient number of plastic hinges has developed to cause the frame to form a collapse mechanism, in which case the load resistance of the frame can be determined by plastic analysis of the collapse mechanism. More generally, however, the in-plane buckling effects associated with the axial compression forces significantly modify the behaviour of the frame near its ultimate loads. The in-plane analysis of rigid-jointed frames is discussed in Section 8.3.5, and their design in 8.3.6.

8.3.5   In-plane analysis of frames with rigid joints

8.3.5.1   General

The most common reason for analysing a rigid-jointed frame is to determine the moments, shears, and axial forces acting on its members and joints. These may then be used with the design rules of EC3 to determine if the members and joints are adequate.

Such an analysis must allow for any significant second-order moments arising from the finite deflections of the frame. Two methods may be used to allow for these second-order moments. In the first of these, the moments determined by a *first-order elastic analysis* are amplified by using the results of an *elastic buckling analysis*. In the second and more accurate method, a full *second-order elastic analysis* is made of the frame.

A less common reason for analysing a rigid-jointed frame is to determine whether the frame can reach an equilibrium position under the factored loads, in which case the frame is adequate. A *first-order plastic analysis* may be used for a frame with negligible second-order effects. When there are significant second-order effects, then an *advanced analysis* may be made in which account is taken of second-order effects, inelastic behaviour, and residual stresses and geometrical imperfections, although this is rarely done in practice.

These various methods of analysis are discussed in more detail in the following sub-sections.

8.3.5.2   First-order elastic analysis

A first-order (linear) elastic analysis of a rigid-jointed frame is based on the assumptions that:

(a) the material behaves linearly, so that all yielding effects are ignored,
(b) the members behave linearly, with no member instability effects such as those caused by axial compressions which reduce the members' flexural stiffnesses (these are often called the P-δ effects), and
(c) the frame behaves linearly, with no frame instability effects such as those caused by the moments of the vertical forces and the horizontal frame deflections (these are often called the P-Δ effects).

For example, for the portal frame of Figure 8.2, a first-order elastic analysis ignores all second-order moments such as $R_R(\delta + \Delta z/h)$ in the right-hand column, so that the bending moment distribution is linear in this case. First-order analyses predict linear behaviour in elastic frames, as shown in Figure 1.15.

Rigid-jointed frames are invariably statically indeterminate, and while there are many manual methods of first-order elastic analysis available [11–13], these are labour-intensive and error-prone for all but the simplest frames. In the past, designers were often forced to rely on approximate methods or available solutions for specific frames [14, 15]. However, computer methods of first-order elastic analysis [16, 17] have formed the basis of computer programs such as [18–20] which are now used extensively. These first-order elastic analysis programs require the geometry of the frame and its members to have been established (usually by

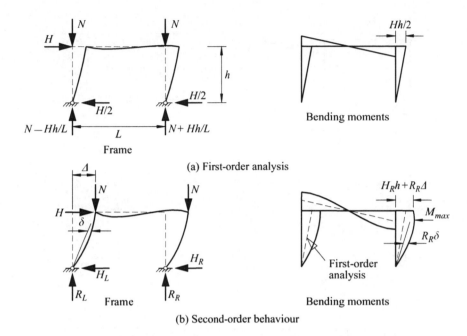

Figure 8.2 First-order analysis and second-order behaviour.

a preliminary design), and then compute the first-order member moments and forces, and the joint deflections for each specified set of loads. Because these are proportional to the loads, the results of individual analyses may be combined by linear superposition. The results of computer first-order elastic analyses [18] of a braced and an unbraced frame are given in Figures 8.3 and 8.4.

8.3.5.3 Elastic buckling of braced frames

The results of an elastic buckling analysis may be used to approximate any second-order effects (Figure 8.2). The elastic buckling analysis of a rigid-jointed frame is carried out by replacing the initial set of frame loads by a set which produces the same set of member axial forces without any bending, as indicated in Figure 8.5b. The set of member forces N_{cr} which causes buckling depends on the distribution of the axial forces in the frame, and is often expressed in terms of a load factor α_{cr} by which the initial set of axial forces N_i must be multiplied to obtain the member forces N_{cr} at frame buckling, so that

$$N_{cr} = \alpha_{cr} N_i \tag{8.1}$$

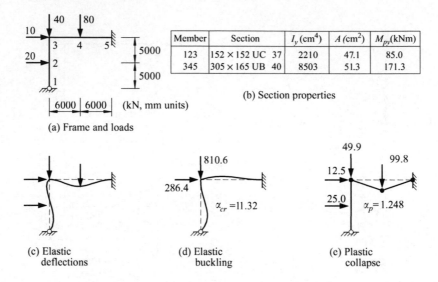

(a) Frame and loads

(b) Section properties

(c) Elastic deflections

(d) Elastic buckling

(e) Plastic collapse

Quantity	Units	First-order elastic	Elastic buckling	Second-order elastic	First-order plastic
M_2	kNm	23.5	0	24.6	19.9
M_3	kNm	−53.0	0	−53.6	−85.0
M_4	kNm	136.6	0	137.3	171.3
M_5	kNm	−153.8	0	−154.6	−171.3
N_{123}	kN	−71.6	−810.6	−71.6	−92.6
N_{345}	kN	−25.3	−286.4	−25.2	−33.5
v_4	mm	57.4	(0.154)	57.8	–
u_2	mm	18.5	(1.000)	20.6	–

(f) Analysis results

Figure 8.3 Analysis of a braced frame.

Alternatively, it may be expressed by a set of effective length factors $k_{cr} = L_{cr}/L$ which define the member forces at frame buckling by

$$N_{cr} = \pi^2 EI / (k_{cr}L)^2. \qquad (8.2)$$

For all but isolated members and very simple frames, the determination of the frame buckling factor α_{cr} is best carried out numerically, using a suitable computer program. The bases of frame elastic buckling programs are discussed in [16, 17]. The computer program PRFSA [18] first finds the initial member axial forces $\{N_i\}$

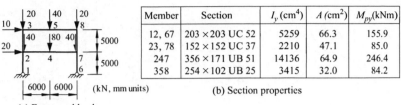

(a) Frame and loads

Member	Section	I_y (cm^4)	A (cm^2)	M_{py}(kNm)
12, 67	203 × 203 UC 52	5259	66.3	155.9
23, 78	152 × 152 UC 37	2210	47.1	85.0
247	356 × 171 UB 51	14136	64.9	246.4
358	254 × 102 UB 25	3415	32.0	84.2

(b) Section properties

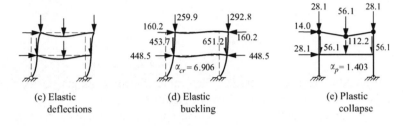

(c) Elastic deflections

(d) Elastic buckling

(e) Plastic collapse

Quantity	Units	First-order elastic	Elastic buckling	Second-order elastic	First-order plastic
M_4	kNm	153.1	0	153.9	236.0
M_5	kNm	71.7	0	72.9	84.2
M_{76}	kNm	−119.3	0	−130.5	−155.9
M_{74}	kNm	−172.8	0	−184.7	−240.9
M_{78}	kNm	53.5	0	54.2	85.0
M_8	kNm	−62.5	0	−64.4	−84.2
N_{12}	kN	−103.3	−713.6	−100.9	−144.9
N_{23}	kN	−37.6	−259.9	−37.6	−56.1
N_{67}	kN	−136.7	−943.8	−138.3	−191.7
N_{78}	kN	−42.4	−292.6	−42.3	−56.1
v_4	mm	44.7	(0.003)	46.1	–
v_5	mm	80.1	(0.001)	83.2	–
u_2	mm	85.5	(0.826)	100.3	–
u_3	mm	121.4	(1.000)	140.4	–

(f) Analysis results

Figure 8.4 Analysis of an unbraced frame.

by carrying out a first-order elastic analysis of the frame under the initial loads. It then uses a finite element method to determine the elastic buckling load factors α_{cr} for which the total frame stiffness vanishes [21], so that

$$[K]\{D\} - \alpha_{cr}[G]\{D\} = \{0\} \qquad (8.3)$$

in which $[K]$ is the elastic stiffness matrix, $[G]$ is the stability matrix associated with the initial axial forces $\{N_i\}$, and $\{D\}$ is the vector of nodal deformations which

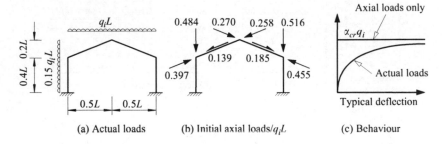

Figure 8.5 In-plane behaviour of a rigid frame.

define the buckled shape of the frame. The results of a computer elastic buckling analysis [18] of a braced frame are given in Figure 8.3.

For isolated braced members or very simple braced frames, a buckling analysis may also be made from first principles, as demonstrated in Section 3.10. Alternatively, the effective length factor k_{cr} of each compression member may be obtained by using estimates of the member relative end stiffness factors k_1, k_2 in a braced member chart such as that of Figure 3.21a. The direct application of this chart is limited to the vertical columns of regular rectangular frames with regular loading patterns in which each horizontal beam has zero axial force and all the columns buckle simultaneously in the same mode. In this case, the values of the column end stiffness factors can be obtained from

$$k = \frac{\Sigma(I_c/L_c)}{\Sigma(I_c/L_c) + \Sigma(\beta_e I_b/L_b)} \qquad (8.4)$$

where β_e is a factor which varies with the restraint conditions at the far end of the beam ($\beta_e = 0.5$ if the restraint condition at the far end is the same as that at the column, $\beta_e = 0.75$ if the far end is pinned, and $\beta_e = 1.0$ if the far end is fixed – note that the proportions 0.5:0.75:1.0 of these are the same as 2:3:4 of the appropriate stiffness multipliers of Figure 3.19), and where the summations are carried out for the columns (c) and beams (b) at the column end.

The effective length chart of Figure 3.21a may also be used to approximate the buckling forces in other braced frames. In the simplest application, an effective length factor k_{cr} is determined for each compression member of the frame by approximating its member end stiffness factors (see equation 3.46) by

$$k = \frac{I/L}{I/L + \Sigma(\beta_e \alpha_r I_r/L_r)} \qquad (8.5)$$

where α_r is a factor which allows for the effect of axial force on the flexural stiffness of each restraining member (Figure 3.19), and the summation is carried out for all of the members connected to that end of the compression member. Each effective length factor so determined is used with equations 8.1 and 8.2 to obtain an estimate of the frame buckling load factor as

$$\alpha_{cr,i} = \frac{\pi^2 EI_i / (k_{cr,i} L_i)^2}{N_i}.$$ (8.6)

The lowest of these provides a conservative approximation of the actual frame buckling load factor α_{cr}.

The accuracy of this method of calculating effective lengths using the stiffness approximations of Figure 3.19 is indicated in Figure 8.6c for the rectangular frame shown in Figure 8.6a and b. For the buckling mode shown in Figure 8.6a, the buckling of the vertical members is restrained by the horizontal members. These horizontal members are braced restraining members which bend in symmetrical single curvature, so that their stiffnesses are $(2EI_1/L_1)(1 - N_1/N_{cr1})$ in which $N_{cr1} = \pi^2 EI_1/L_1^2$, as indicated in Figure 3.19. It can be seen from Figure 8.6c that the approximate buckling loads are in very close agreement with the accurate values. Worked examples of the application of this method are given in Sections 8.5.1 and 8.5.2.

A more accurate iterative procedure [22] may also be used, in which the accuracies of the approximations for the member end stiffness factors k increase with each iteration.

8.3.5.4   Elastic buckling of unbraced frames

The determination of the frame buckling load factor α_{cr} of a rigid-jointed unbraced frame may also be carried out using a suitable computer program such as that described in [16, 17].

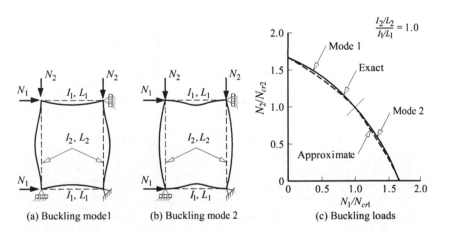

(a) Buckling mode1   (b) Buckling mode 2   (c) Buckling loads

Figure 8.6 Buckling of a braced frame.

For isolated unbraced members or very simple unbraced frames, a buckling analysis may also be made from first principles, as demonstrated in Section 3.10. Alternatively the effective length factor k_{cr} of each compression member may be obtained by using estimates of the end stiffness factors k_1, k_2 in a chart such as that of Figure 3.21b. The application of this chart is limited to the vertical columns of regular rectangular frames with regular loading patterns in which each horizontal beam has zero axial force and all the columns buckle simultaneously in the same mode and with the same effective length factor. The member end stiffness factors are then given by equation 8.4.

The effects of axial forces in the beams of an unbraced regular rectangular frame can be approximated by using equation 8.5 to approximate the relative end stiffness factors k_1, k_2 (but with $\beta_e = 1.5$ if the restraint conditions at the far end are the same as at the column, $\beta_e = 0.75$ if the far end is pinned, and $\beta_e = 1.0$ if the far end is fixed – note that the proportions 1.5:0.75:1.0 of these are the same as 6:3:4 of the appropriate stiffness multipliers of Figure 3.19) and Figure 3.21b.

The accuracy of this method of using the stiffness approximations of Figure 3.19 is indicated in Figure 8.7c for the rectangular frame shown in Figure 8.7a and b. For the sway buckling mode shown in Figure 8.7a, the sway buckling of the vertical members is restrained by the horizontal members. These horizontal members are braced restraining members which bend in antisymmetrical double curvature, so that their stiffnesses are $(6EI_1/L_1)(1 - N_1/4N_{cr1})$ as indicated in Figure 3.19. It can be seen from Figure 8.7c that the approximate buckling loads are in close agreement with the accurate values. A worked example of the application of this method is given in Section 8.5.4.

The effective length factor chart of Figure 3.21b may also be used to approximate the storey buckling load factor $\alpha_{cr,s}$ for each storey of an unbraced rectangular

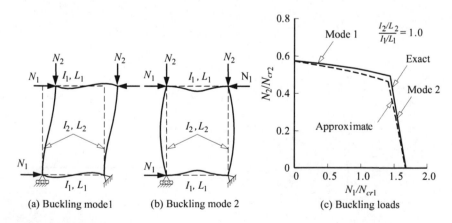

(a) Buckling mode1   (b) Buckling mode 2   (c) Buckling loads

Figure 8.7 Buckling of an unbraced frame.

frame with negligible axial forces in the beams as

$$\alpha_{cr,s} = \frac{\sum(N_{cr}/L)}{\sum(N_i/L)} \tag{8.7}$$

in which N_{cr} is the buckling load of a column in a storey obtained by using equation 8.4, and the summations are made for each column in the storey. For a storey in which all the columns are of the same length, this equation implies that the approximate storey buckling load factor depends on the total stiffness of the columns and the total load on the storey, and is independent of the distributions of stiffness and load. The frame buckling load factor α_{cr} may be approximated by the lowest of the values of $\alpha_{cr,s}$ calculated for all the stories of the frame. This method gives close approximations when the buckling pattern is the same in each storey, and is conservative when the horizontal members have zero axial forces and the buckling pattern varies from storey to storey.

A worked example of the application of this method is given in Section 8.5.3.

Alternatively, the storey buckling load factor $\alpha_{cr,s}$ may be approximated by using

$$\alpha_{cr,s} = \frac{H_s}{V_s} \frac{h_s}{\delta_{Hs}} \tag{8.8}$$

in which H_s is the storey shear, h_s is the storey height, V_s is the vertical load transmitted by the storey, and δ_{Hs} is the first-order shear displacement of the storey caused by H_s. This method is an adaptation of the sway buckling load solution given in Section 3.10.6 for a column with an elastic brace (of stiffness H_s/δ_{Hs}).

The approximate methods described above of analysing the elastic buckling of unbraced frames are limited to rectangular frames. While the buckling of non-rectangular unbraced frames may be analysed by using a suitable computer program such as that described in [18], there are many published solutions for specific frames [23].

Approximate solutions for the elastic buckling load factors of symmetrical portal frames have been presented in [24–27]. The approximations of [27] for portal frames with elastically restrained bases take the general form of

$$\alpha_{cr} = \left\{ \left(\frac{N_c}{\rho_c N_{cr,c}} \right)^C + \left(\frac{N_r}{\rho_r N_{cr,r}} \right)^C \right\}^{-1/C} \tag{8.9}$$

in which N_c, N_r are the column and rafter forces, $N_{cr,c}, N_{cr,r}$ are the reference buckling loads obtained from

$$N_{cr,c,r} = \pi^2 EI_{c,r}/L_{c,r}^2 \tag{8.10}$$

and the values of C, ρ_c, and ρ_r depend on the base restraint stiffness and the buckling mode.

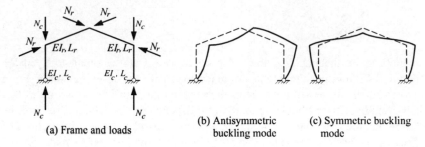

(a) Frame and loads

(b) Antisymmetric
buckling mode

(c) Symmetric buckling
mode

Figure 8.8 Pinned base portal frame.

For portal frames with pinned bases

$$C = 1.2 \tag{8.11}$$

and for antisymmetric buckling as shown in Figure 8.8b,

$$\rho_c = 3/(10R + 12) \tag{8.12a}$$

and

$$\rho_r = 1 \tag{8.12b}$$

in which

$$R = (I_c/L_c)/(I_r/L_r) \tag{8.13}$$

For symmetric buckling as shown in Figure 8.8c,

$$\rho_c = \frac{4.8 + 12R(1 + R_H)}{2.4 + 12R(1 + R_H) + 7R^2 R_H^2} \tag{8.14a}$$

and

$$\rho_r = \frac{12R + 8.4RR_H}{12R + 4} \tag{8.14b}$$

in which

$$R_H = (L_r/L_c) \sin \theta_R \tag{8.15}$$

in which θ_R is the inclination of the rafter. A worked example of the application of this method is given in Section 8.5.5.

For portal frames with fixed bases

$$C = 1.6 \tag{8.16}$$

and for antisymmetric buckling,

$$\rho_c = \frac{R+3}{4R+3} \tag{8.17a}$$

and

$$\rho_r = \frac{2R+4}{R+4} \tag{8.17b}$$

For symmetric buckling,

$$\rho_c = \frac{1 + 4.8R + 4.2RR_H}{2.4R + 2R^2 R_H^2} \tag{8.18a}$$

and

$$\rho_r = \frac{4R + 4.2RR_H + R^2 R_H^2}{4R + 2} \tag{8.18b}$$

Approximate methods for multi-bay frames are given in [25].

8.3.5.5 Second-order elastic analysis

Second-order effects in elastic frames include additional moments such as $R_R(\delta + \Delta z/h)$ in the right-hand column of Figure 8.2 which result from the finite deflections δ and Δ of the frame. The second-order moments arising from the member deflections from the straight line joining the member ends are often called the P-δ effects, while the second-order moments arising from the joint displacements Δ are often called the P-Δ effects. In braced frames, the joint displacements Δ are small, and only the P-δ effects are important. In unbraced frames, the P-Δ effects are important, and often much more so than the P-δ effects. Other second-order effects include those arising from the end-to-end shortening of the members due to stress ($= NL/EA$), due to bowing ($= \frac{1}{2} \int_0^L (d\delta/dx)^2 dx$), and from finite deflections. Second-order effects cause non-linear behaviour in elastic frames, as shown in Figure 1.14.

The second-order P-δ effects in a number of elastic beam-columns have been analysed, and approximations for the maximum moments are given in Section 7.2.1. A worked example of the use of these approximations is given in Section 7.7.1.

The P-δ and P-Δ second-order effects in elastic frames are most easily and accurately accounted for by using a computer second-order elastic analysis program such as those of [18–20]. The results of second-order elastic analyses [18] of braced and unbraced frames are given in Figures 8.3 and 8.4. For the braced frame of Figure 8.3, the second-order results are only slightly higher than the first-order results. This is usually true for well-designed braced frames that have substantial bending effects and small axial compressions. For the unbraced frame of Figure 8.4, the second-order sway deflections are about 18% larger than those of the first-order analysis, while the moment at the top of the right-hand lower-storey column is about 10% larger than that of the first-order analysis.

Second-order effects are often significant in unbraced frames.

8.3.5.6　Approximate second-order elastic analysis

There are a number of methods of approximating second-order effects which allow a general second-order analysis to be avoided. In many of these, the results of a first-order elastic analysis are amplified by using the results of an elastic buckling analysis (Sections 8.3.5.3 and 8.3.5.4).

For members without transverse forces in braced frames, the maximum member moment M_{max} determined by first-order analysis may be used to approximate the maximum second-order design moment M as

$$M = \delta M_{max} \tag{8.19}$$

in which δ is an amplification factor given by

$$\delta = \delta_b = \frac{c_m}{1 - N/N_{cr,b}} \geq 1, \tag{8.20}$$

in which $N_{cr,b}$ is the elastic buckling load calculated for the braced member, and

$$c_m = 0.6 - 0.4\beta_m \leq 1.0 \tag{8.21}$$

in which β_m is the ratio of the smaller to the larger end moment (equations 8.20 and 8.21 are related to equations 7.7 and 7.8 used for isolated beam-columns). A worked example of the application of this method is given in Section 8.5.6. A procedure for approximating the value of β_m to be used for members with transverse forces is available [28], while more accurate solutions are obtained by using equations 7.9–7.11 and Figures 7.5 and 7.6.

For unbraced frames, the amplification factor δ may be approximated by

$$\delta = \delta_s = \frac{1}{1 - 1/\alpha_{cr}} \tag{8.22}$$

in which α_{cr} is the elastic sway buckling load factor calculated for the unbraced frame (Section 8.3.5.4). A worked example of the application of this method is given in Section 8.5.7.

For unbraced rectangular frames with negligible axial forces in the beams, a more accurate approximation can be obtained by amplifying the column moments of each storey by using the sway buckling load factor $\alpha_{cr,s}$ for that storey obtained using equation 8.7 in

$$\delta = \delta_s = \frac{1}{1 - 1/\alpha_{cr,s}} \tag{8.23}$$

in which $\alpha_{cr,s}$ is the elastic sway buckling load factor calculated for the storey (Section 8.3.5.4). A worked example of the application of this method is given in Section 8.5.8.

Alternatively, the first-order end moment

$$M_f = M_{fb} + M_{fs} \tag{8.24}$$

may be separated into a braced frame component M_{fb} and a complementary sway component M_{fs} [29]. The second-order maximum moment M may be approximated by using these components in

$$M = M_{fb} + \frac{M_{fs}}{(1 - 1/\alpha_{cr})} \tag{8.25}$$

in which α_{cr} is the load factor at frame elastic sway buckling (Section 8.3.5.4).

A more accurate second-order analysis can be made by including the P-Δ effects of sway displacement directly in the analysis. An approximate method of doing this is to include fictitious horizontal forces equivalent to the P-Δ effects in an iterative series of first-order analyses [30, 31]. A series of analyses must be carried out because the fictitious horizontal forces increase with the deflections Δ. If the analysis series converges, then the frame is stable. An approximate method of anticipating convergence is to examine the value of $(\Delta_{n+1} - \Delta_n)/(\Delta_n - \Delta_{n-1})$ computed from the values of Δ at the steps $(n-1)$, n, and $(n+1)$. If this is less than $n/(n+1)$, then the analysis series probably converges. This method of analysing frame buckling is slow and clumsy, but it does allow the use of the widely available first-order elastic computer programs. The method ignores the decreases in the member stiffnesses caused by axial compressions (the P-δ effects), and therefore tends to underestimate the second-order moments.

8.3.5.7   First-order plastic analysis

In the first-order method of plastic analysis, all instability effects are ignored, and the collapse strength of the frame is determined by using the rigid-plastic assumption (Section 5.5.2) and finding the plastic hinge locations which first convert the frame to a collapse mechanism.

The methods used for the plastic analysis of frames are extensions of those discussed in Section 5.5.5 which incorporate reductions in the plastic moment capacity to account for the presence of axial force (Section 7.2.2). These methods are discussed in many texts, such as those [18–24: Chapter 5] referred to in

Section 5.14. The manual application of these methods is not as simple as it is for beams, but computer programs such as [18] are available. The results of computer first-order plastic analyses [18] of a braced and an unbraced frame are given in Figures 8.3 and 8.4.

8.3.5.8 Advanced analysis

Present methods of designing statically indeterminate steel structures rely on predictions of the stress resultants acting on the individual members, and on models of the strengths of these members. These member strength models include allowances for second-order effects and inelastic behaviour, as well as for residual stresses and geometrical imperfections. It is therefore possible that the common methods of second-order elastic analysis used to determine the member stress resultants, which do not account for inelastic behaviour, residual stresses, and geometrical imperfections, may lead to inaccurate predictions.

Second-order effects and inelastic behaviour cause redistributions of the member stress resultants in indeterminate structures as some of the more critical members lose stiffness and transfer moments to their neighbours. Non-proportional loading also causes redistributions. For example, frame shown in Figure 8.9, the column end moments induced initially by the beam loads decrease as the column load increases and the column becomes less stiff, due to second-order effects at first, and then due to inelastic behaviour. The end moments may reduce to zero, or even reverse in direction. In this case the end moments change from initially disturbing the column to stabilising it by resisting further end rotations. This change in the sense of the column end moments, which contributes significantly to the column strength, cannot be predicted unless an analysis is made which more closely models the behaviour of the real structure than does a simple plastic analysis (which ignores second-order effects) or a second-order elastic analysis (which ignores inelastic behaviour).

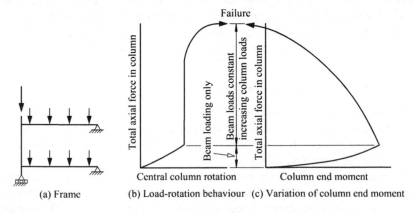

Figure 8.9 Reversal of column end moments in frames.

Ideally, the member stress resultants should be determined by a method of frame analysis which accounts for both second-order effects (P-δ and P-Δ), inelastic behaviour, residual stresses, and geometrical imperfections, and any local or out-of-plane buckling effects. Such a method has been described as an advanced analysis. Not only can an advanced analysis be expected to lead to predictions that are of high accuracy, it has the further advantage that the design process can be greatly simplified, since the structure can be considered to be satisfactory if the advanced analysis can show that it can reach an equilibrium position under the design loads.

Past research on advanced analysis and the prediction of the strengths of real structures has concentrated on frames for which local and lateral buckling are prevented (by using Class 1 sections and full lateral bracing). One of the earliest, and perhaps the simplest, methods of approximating the strengths of rigid-jointed frames was suggested in [32] where it was proposed that the ultimate load factor α_{ult} of a frame should be calculated from the load factor α_p at which plastic collapse occurs (all stability effects being ignored) and the load factor α_{cr} at which elastic in-plane buckling takes place (all plasticity effects being ignored) by using

$$\frac{1}{\alpha_{ult}} = \frac{1}{\alpha_p} + \frac{1}{\alpha_{cr}} \qquad (8.26)$$

or

$$\alpha_{ult}/\alpha_p = 1 - \alpha_{ult}/\alpha_{cr} \qquad (8.27)$$

However, it can be said that this method represents a considerable simplification of a complex relationship between the ultimate strength and the plastic collapse and elastic buckling strengths of a frame.

More accurate predictions of the strengths of two-dimensional rigid frames can be made using a sub-assemblage method of analysis. In this method, the braced or unbraced frame is considered as a number of subassemblies [33–35] such as those shown in Figure 8.10. The conditions at the ends of the sub-assemblage are approximated according to the structural experience and intuition of the designer (it is believed that these approximations are not critical), and the sub-assemblage is analysed by adding together the load-deformation characteristics ([36] and curve 5 in Figure 7.2) of the individual members. Although the application of this manual technique to the analysis of braced frames was developed to an advanced stage [37, 38], it does not appear to have achieved widespread popularity.

Further improvements in the prediction of frame strength are provided by computer methods of second-order plastic analysis, such as the deteriorated stiffness method of analysis, in which progressive modifications are made to a second-order elastic analysis to account for the formation of plastic hinges [16, 39, 40]. The use of such an analysis is limited to structures with appropriate material properties, and for which local and out-of-plane buckling effects are prevented. However, the use of such an analysis has in the past been largely limited to research studies.

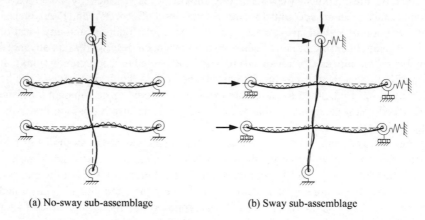

(a) No-sway sub-assemblage   (b) Sway sub-assemblage

Figure 8.10 Sub-assemblages for multi-storey frames.

More recent research [41] has concentrated on extending computer methods of analysis so that they allow for residual stresses and geometrical imperfections as well as for second-order effects and inelastic behaviour. Such methods can duplicate design code predictions of the strengths of individual members (through the incorporation of allowances for residual stresses and geometrical imperfections) as well as the behaviour of complete frames. Two general types of analysis have developed, called plastic zone analysis [42, 43] and concentrated plasticity analysis [44]. While plastic zone analysis is the more accurate, concentrated plasticity is the more economical of computer memory and time. It seems likely that advanced methods of analysis will play important roles in the future design of two-dimensional rigid-jointed frames in which local and lateral buckling are prevented.

8.3.6   In-plane design of frames with rigid joints

8.3.6.1   Residual stresses and geometrical imperfections

The effects of residual stresses are allowed for approximately in EC3 by using enhanced equivalent geometrical imperfections. These include both global (frame) imperfections (which are associated with initial sway or lack of verticality) and local (member) imperfections (which are associated with initial bow or crookedness). EC3 also allows the equivalent geometrical imperfections to be replaced by closed systems of equivalent fictitious forces. The magnitudes of the equivalent geometrical imperfections and equivalent fictitious forces are specified in Clause 5.3.2 of EC3.

In some low slenderness frames, the equivalent imperfections or fictitious forces have only small effects on the distributions of moments and axial forces in the frames, and may be ignored in the analysis (their effects on member resistance are included in Clause 6.3 of EC3). More generally, the global (frame) imperfections or fictitious forces are included in the analysis, but not the local (member) imperfections or fictitious forces. It is rare to include both the global and the local imperfections or fictitious forces in the analysis.

The EC3 methods used for the analysis and design of frames with rigid joints are discussed in the following sub-sections.

8.3.6.2   Strength design using first-order elastic analysis

When a braced or unbraced frame has low slenderness so that

$$\alpha_{cr} = F_{cr}/F_{Ed} \geq 10 \qquad (8.28)$$

in which F_{Ed} is the design loading on the frame and F_{cr} is the elastic buckling load, then EC3 allows the moments and axial forces in the frame to be determined using a first-order elastic analysis of the frame with all the equivalent imperfections or fictitious forces ignored. The frame members are adequate when their axial forces and moments satisfy the section and member buckling resistance requirements of Clauses 6.2 and 6.3.3 of EC3.

When a braced frame does not satisfy equation 8.28, Clause 5.2.2(3)b of EC3 appears to allow a first-order elastic analysis to be used which neglects all second-order effects and the imperfections associated with member crookedness, provided the member axial forces and moments satisfy the buckling resistance requirements of Clause 6.3.3. It is suggested, however, that amplified first-order analysis should be used for frames of moderate slenderness which satisfy equation 8.29 (Section 8.3.6.3), and that second-order analysis should be used for frames of high slenderness which satisfy equation 8.30 (Section 8.3.6.4).

8.3.6.3   Strength design using amplified first-order elastic analysis

When an unbraced frame has moderate slenderness so that

$$10 > \alpha_{cr} = F_{cr}/F_{Ed} \geq 3 \qquad (8.29)$$

then EC3 allows the moments in the frame to be determined by amplifying (Section 8.3.5.6) the moments determined by a first-order elastic analysis, provided the equivalent global imperfections or fictitious forces are included in the analysis. The frame members are adequate when their axial forces and moments satisfy the section resistance requirements of Clause 6.2 of EC3 and the member buckling resistance requirements of Clause 6.3.3. EC3 allows the member length to be used in Clause 6.3.3, but this may overestimate the resistances of members for which compression effects dominate. It is suggested that in this case, the member should also satisfy the member buckling requirements of Clause 6.3.1 in which the effective length is used.

If the member imperfections or fictitious forces are also included in the analysis, then the frame members are satisfactory when their axial forces and moments satisfy the section resistance requirements of Clause 6.2 of EC3.

8.3.6.4   Strength design using second-order elastic analysis

When an unbraced frame has high slenderness so that

$$3 > \alpha_{cr} = F_{cr}/F_{Ed} \tag{8.30}$$

then EC3 requires a more accurate analysis to be made of the second-order effects (Section 8.3.5.5) than by amplifying the moments determined by a first-order elastic analysis. EC3 requires the equivalent global imperfections or fictitious forces to be included in the analysis. The frame members are adequate when their axial forces and moments satisfy the section resistance requirements of Clause 6.2 of EC3 and the member buckling resistance requirements of Clause 6.3.3. EC3 allows the member length to be used in Clause 6.3.3, but this may overestimate the resistances of members for which compression effects dominate. It is suggested that in this case, the member should also satisfy the member buckling requirements of Clause 6.3.1 in which the effective length is used.

If the member imperfections or fictitious forces are also included in the analysis, then the frame members are satisfactory when the axial forces and moments satisfy the section resistance requirements of Clause 6.2 of EC3.

8.3.6.5   Strength design using first-order plastic collapse analysis

When a braced or unbraced frame has low slenderness so that

$$\alpha_{cr} = F_{cr}/F_{Ed} \geq 15 \tag{8.31}$$

then EC3 allows all the equivalent imperfections or fictitious forces to be ignored and a first-order rigid plastic collapse analysis (Section 8.3.5.7) to be used. This limit is reduced to $\alpha_{cr} \geq 10$ for clad structures, and to $\alpha_{cr} \geq 5$ for some portal frames subjected to gravity loads only.

All members forming plastic hinges must be ductile so that the plastic moment capacity can be maintained at each hinge over a range of hinge rotations sufficient to allow the plastic collapse mechanism to develop. Ductility is usually ensured by restricting the steel type to one which has a substantial yield plateau and significant strain-hardening (Section 1.3.1) and by preventing reductions in rotation capacity by local buckling effects (by satisfying the requirements of Clause 5.6 of EC3, including the use of Class 1 sections), by out-of-plane buckling effects (by satisfying the requirements of Clause 6.3.5 of EC3, including limiting the unbraced lengths), and by in-plane buckling effects (by using equation 8.31 to limiting the member in-plane slendernesses and axial compression forces).

The frame members are adequate when

$$\alpha_p = F_p/F_{Ed} \geq 1 \qquad (8.32)$$

in which F_p is the plastic collapse load. It should be noted that the use of reduced full plastic moments (to allow for the axial forces) will automatically ensure that Clause 6.2.9 of EC3 for the bending and axial force resistance is satisfied.

8.3.6.6   Strength design using advanced analysis

EC3 permits the use of a second-order plastic analysis. When the effects of residual stresses and geometrical imperfections are allowed for through the use of global and local equivalent imperfections or fictitious forces, and when local and lateral buckling are prevented (Clauses 5.6 and 6.3.5 of EC3), then this provides a method of design by advanced analysis.

The members of the structure are satisfactory when the section resistance requirements of Clause 6.2 of EC3 are met. Because these requirements are automatically satisfied by the prevention of local buckling and the accounting for inelastic behaviour, the members can be regarded as satisfactory if the analysis shows that the structure can reach an equilibrium position under the design loads.

8.3.6.7   Serviceability design

Because the service loads are usually substantially less than the factored loads used for strength design, the behaviour of a frame under its service loads is usually closely approximated by the predictions of a first-order elastic analysis (Section 8.3.5.2). For this reason, the serviceability design of a frame is usually carried out using the results of a first-order elastic analysis. The serviceability design of a frame is often based on the requirement that the service load deflections must not exceed specified values. Thus the member sizes are systematically changed until this requirement is met.

8.3.7   Out-of-plane behaviour of frames with rigid joints

Most two-dimensional rigid frames which have in-plane loading are arranged so that the stiffer planes of their members coincide with that of the frame. Such a frame deforms only in its plane until the out-of-plane (flexural–torsional) buckling loads are reached, and if these are less than the in-plane ultimate loads, then the members and the frame will buckle by deflecting out of the plane and twisting.

In some cases, the action of one loaded member dominates, and its elastic buckling load can be determined by evaluating the restraining effects of the remainder of the frame. For example, when the frame shown in Figure 8.11 has a zero beam

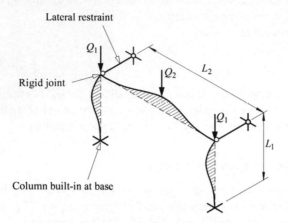

Lateral restraint

Rigid joint

Column built-in at base

Figure 8.11 Flexural–torsional buckling of a portal frame.

load Q_2, then the beam remains straight and does not induce moments in the columns, which buckle elastically out of the plane of the frame at a load given by $Q_1 \approx \pi^2 EI_z/(0.7L_1)^2$ (see Figure 3.15c). On the other hand, if the effects of the compressive forces in the columns are small enough to be neglected, then the elastic stiffnesses and restraining effects of the columns on the beam can be estimated, and the elastic flexural–torsional buckling load Q_2 of the beam can be evaluated. The solution of a related problem is discussed in Section 6.9.

In general, however, both the columns and the beams of a rigid frame buckle, and there is an interaction between them during out-of-plane buckling. This interaction is related to that which occurs during the in-plane buckling of rigid frames (see Sections 8.3.5.3 and 8.3.5.4 and Figures 8.6 and 8.7), and also to that which occurs during the elastic flexural–torsional buckling of continuous beams (see Section 6.8.2 and Figure 6.21). Because of the similarities between the elastic flexural–torsional buckling of restrained beam-columns (Section 7.3.1.3) and restrained columns (Section 3.6.4), it may prove possible to extend further the approximate method given in Section 8.3.5.3 for estimating the in-plane elastic buckling loads of rigid frames (which was applied to the elastic flexural–torsional buckling of some beams in Section 6.8.2). Thus, the elastic flexural–torsional buckling loads of some rigid frames might be calculated approximately by using the column effective length chart of Figure 3.21a. However, this development has not yet been investigated.

On the other hand, there have been developments [45–47] in the analytical techniques used to determine the elastic flexural–torsional buckling loads of general rigid plane frames with general in-plane loading systems, and a computer program [48] has been prepared which requires only simple data specifying the geometry of the frame, its supports and restraints, and the arrangement of the loads. The use

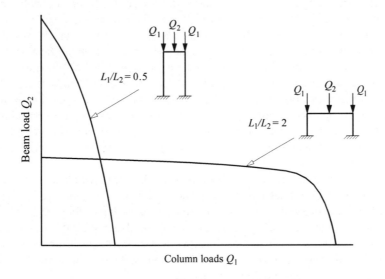

Figure 8.12 Elastic flexural–torsional buckling loads of portal frames.

of this program to calculate the elastic flexural–torsional buckling load of a frame will simplify the design problem considerably.

Interaction diagrams for the elastic flexural–torsional buckling loads of a number of frames have been determined [45], and two of these (for the portal frames shown in Figure 8.11) are given in Figure 8.12. These diagrams show that the region of stability is convex, as it is for the in-plane buckling of rigid frames (Figures 8.6 and 8.7) and for the flexural–torsional buckling of continuous beams (Figure 6.23b). Because of this convexity, linear interpolations (as in Figure 6.24) made between known buckling load sets are conservative.

Although the elastic flexural–torsional buckling of rigid frames has been investigated, the related problems of inelastic buckling and its influence on the strength of rigid frames have not yet been systematically studied. However, advanced computer programs for the inelastic out-of-plane behaviour of beam-columns [49–54] have been developed, and it seems likely that these will be extended in the near future to rigid-jointed frames.

8.3.8 Out-of-plane design of frames

While there is no general method yet available of designing for flexural–torsional buckling in rigid frames, design codes imply that the strength formulations for isolated beam-columns (see Section 7.3.4) can be used in conjunction with the moments and forces determined from an appropriate elastic in-plane analysis of the frame.

A rational design method has been developed [55] for the columns of two-dimensional building frames in which laterally braced horizontal beams form plastic hinges. Thus, when plastic analysis is used, columns which do not participate in the collapse mechanism may be designed as isolated beam-columns against flexural–torsional buckling. In later publications [56, 57] the method was extended so that the occurrence of plastic hinges at one or both ends of the column (instead of in the beams) could be allowed for. Because the assumption of plastic hinges at all the beam-column joints is only valid for frames with vertical loads, the analysis for lateral loads must be carried out independently. A frame analysed by this method must therefore have an independent bracing or shear wall system to resist the lateral loads.

A number of current research efforts are being directed towards the extension of advanced methods of analysing the in-plane behaviour of frames (Section 8.3.5.8) to include the out-of-plane flexural–torsional buckling effects [52–54]. The availability of a suitable computer program will greatly simplify the design of two-dimensional frames in which local buckling is prevented, since such a frame can be considered to be satisfactory if it can be shown that it can reach an equilibrium position under the factored design loads.

8.4   Three-dimensional flexural frames

The design methods commonly used when either the frame or its loading is three-dimensional are very similar to the methods used for two-dimensional frames. The actions caused by the design loads are usually calculated by allowing for second-order effects in the elastic analysis of the frame, and then compared with the design resistances determined for the individual members acting as isolated beam-columns. In the case of three-dimensional frames or loading, the elastic frame analysis is three-dimensional, while the design resistances are based on the biaxial bending resistances of isolated beam-columns (see Section 7.4).

A more rational method has been developed [55, 56] of designing three-dimensional braced rigid frames for which it could be assumed that plastic hinges form at all major and minor axis beam ends. This method is an extension of the method of designing braced two-dimensional frames ([55–57] and Section 8.3.8). For such frames, all the beams can be designed plastically, while the columns can again be designed as if independent of the rest of the frame. The method of designing these columns is based on a second-order elastic analysis of the biaxial bending of an isolated beam-column. Once again, the assumption of plastic hinges at all the beam ends is valid only for frames with vertical loads, and so an independent design must be made for the effects of lateral loads.

In many three-dimensional rigid frames, the vertical loads are carried principally by the major axis beams, while the minor axis beams are lightly loaded and do not develop plastic hinges at collapse, but restrain and strengthen the columns. In this case, an appropriate design method for vertical loading is one in which the major axis beams are designed plastically and the minor axis beams elastically. The chief difficulty in using such a method is in determining the minor axis column moments to be used in the design. In the method presented in [58], this is done by first making a first-order elastic analysis of the minor axis frame system to determine the column end moments. The larger of these is then increased to allow for the second-order effects of the axial forces, the amount of the increase depending on the column load ratio $N/N_{cr,z}$ (in which $N_{cr,z}$ is the minor axis buckling load of the elastically restrained column) and the end moment ratio β_m. The increased minor axis end moment is then used in a nominal first yield analysis of the column which includes the effects of initial crookedness.

However, it appears that this design method is erratically conservative [59], and that the small savings achieved do not justify the difficulty of using it. Another design method was proposed [60] which avoids the intractable problem of inelastic biaxial bending in frame structures by omitting altogether the calculation of the minor axis column moments. This omission is based on the observation that the minor axis moments induced in the columns by the working loads reduce to zero as the ultimate loads are approached, and even reverse in sign so that they finally restrain the column, as shown in Figure 8.9. With this simplification, the moments for which the column is to be designed are the same as those in a two-dimensional frame. However, the beneficial restraining effects of the minor axis beams must be allowed for, and a simple way of doing this has been proposed [60]. For this, the elastic minor axis stiffness of the column is reduced to allow for plasticity caused by the axial load and the major axis moments, and this reduced stiffness is used to calculate the effects of the minor axis restraining beams on the effective length L_{cr} of the column. The reduced stiffness is also used to calculate the effective minor axis flexural rigidity EI_{eff} of the column, and the ultimate strength N_{ult} of the column is taken as

$$N_{ult} = \pi^2 EI_{eff} / L_{cr}^2. \tag{8.33}$$

The results obtained from two series of full-scale three-storey tests indicate that the predicted ultimate loads obtained by this method vary between 84% and 104% of the actual ultimate loads.

8.5 Worked examples

8.5.1 Example 1 – buckling of a truss

Problem. All the members of the rigid-jointed triangulated frame shown in Figure 8.13a have the same in-plane flexural rigidity which is such that $\pi^2 EI / L^2 = 1$. Determine the effective length factor of member 12 and the elastic in-plane buckling loads of the frame.

Solution. The member 12 is one of the longest compression members and one of the most heavily loaded, and so at buckling it will be restrained by its adjacent members. An initial estimate is required of the buckling load, and this can be

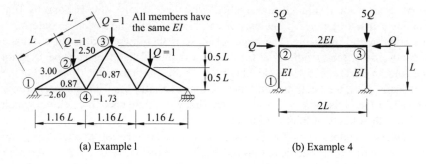

(a) Example 1   (b) Example 4

Figure 8.13 Worked examples 1 and 4.

obtained after observing that the end 1 will be heavily restrained by the tension member 14 and that the end 2 will be moderately restrained by the short lightly loaded compression member 24. Thus the initial estimate of the effective length factor of member 12 in this braced frame should be closer to the rigidly restrained value of 0.5 than to the unrestrained value of 1.0.

Accordingly, assume $k_{cr12} = 0.70$.

Then, by using equation 8.2, $N_{cr12} = \pi^2 EI/(0.70L)^2 = 2.04$, and $Q_{cr} = N_{cr12}/3.00 = 0.68$

Thus $N_{23} = 1.70, N_{24} = 0.59, N_{14} = -1.77$.

Using the form of equation 3.45, $\alpha_{12} = 2EI/L$. Using Figure 3.19,

$$\alpha_{23} = \frac{3EI}{L}\left[1 - \frac{1.70}{\pi^2 EI/L^2}\right] = -2.10\frac{EI}{L},$$

$$\alpha_{24} = \frac{4EI}{0.58L}\left[1 - \frac{0.59}{\pi^2 EI/(0.58L)^2}\right] = 6.22\frac{EI}{L},$$

$$\alpha_{14} = \frac{4EI}{1.16L}\left[1 - \frac{(-1.77)}{2\pi^2 EI/(1.16L)^2}\right] = 7.55\frac{EI}{L}.$$

Using equation 3.45,

$$k_1 = 2/(0.5 \times 7.55 + 2) = 0.346$$
$$k_2 = 2/\{0.5 \times (-2.10 + 6.22) + 2\} = 0.493$$

Using Figure 3.21a, $k_{cr12} = 0.65$

The calculation can be repeated using $k_{cr12} = 0.65$ instead of the initial estimate of 0.70, in which case the solution $k_{cr12} = 0.66$ will be obtained. The corresponding frame buckling loads are $Q_{cr} = \{\pi^2 EI/(0.66L)^2\}/3 = 0.77$.

8.5.2 Example 2 – buckling of a braced frame

Problem. Determine the effective length factors of the members of the braced frame shown in Figure 8.3a, and estimate the frame buckling load factor α_{cr}.

Solution.
For the vertical member 13, using equation 8.5,

$k_1 = 1.0$ (theoretical value for a frictionless hinge).

Using the form of equation 3.45 and Figure 3.19,

$$k_3 = \frac{2 \times 2210 \times 10^4 / 10\,000}{0.5 \times 4 \times 8503 \times 10^4 / 12\,000 + 2.0 \times 2210 \times 10^4 / 10\,000} = 0.238$$

$k_{cr13} = 0.74$ (using Figure 3.21a).

$(k_{cr13} = 0.73$ if a base stiffness ratio of 0.1 is used so that

$k_1 = (2 \times 2210 \times 10^4 / 10\,000)/(2 \times 1.1 \times 2210 \times 10^4 / 10\,000) = 0.909)$

$N_{cr13} = \pi^2 \times 210\,000 \times 2210 \times 10^4 / (0.74 \times 10\,000)^2 \text{N} = 836.5 \text{ kN}$

$\alpha_{cr13} = 836.5/71.6 = 11.7$

For the horizontal member 35, using equation 8.5,

$$k_3 = \frac{2 \times 8503 \times 10^4 / 12\,000}{0.5 \times 3 \times 2210 \times 10^4 / 10\,000 + 2 \times 8503 \times 10^4 / 12\,000} = 0.810.$$

$k_5 = 0$ (theoretical value for a fixed end).

$k_{cr35} = 0.66$ (using Figure 3.21a).

$(k_{cr35} = 0.77$ if a base stiffness ratio of 1.0 is used so that $k_5 = 0.5)$

$$N_{cr35} = \frac{\pi^2 \times 210\,000 \times 8503 \times 10^4}{(0.66 \times 12\,000)^2} \text{N} = 2809.6 \text{ kN}$$

$\alpha_{cr35} = 2809.6/25.3 = 111 > 11.7 = \alpha_{cr13}$

$\therefore \alpha_{cr} = 11.7$ (compare with the computer analysis value of $\alpha_{cr} = 11.32$ given in Figure 8.3d).

A slightly more accurate estimate might be obtained by using equation 8.5 with α_r obtained from Figure 3.19 as

$$\alpha_r = 1 - \frac{25.3 \times 10^3}{(2 \times \pi^2 \times 210\,000 \times 8503 \times 10^4 / 12\,000^2)} = 0.990$$

so that

$$k_3 = \frac{2 \times 2210 \times 10^4/10000}{0.5 \times 4 \times 0.990 \times 8503 \times 10^4/12\,000 + 2 \times 2210 \times 10^4/10\,000}$$
$$= 0.240$$

but k_{cr13} is virtually unchanged, and so therefore is α_{cr}.

8.5.3   Example 3 – buckling of an unbraced two-storey frame

Problem. Determine the storey buckling load factors of the unbraced two-storey frame shown in Figure 8.4a, and estimate the frame buckling load factor, by using

a.   the storey deflection method of equation 8.8, and
b.   the member buckling load method of equation 8.7.

(a) Solution using equation 8.8
    For the upper storey, using the first-order analysis results shown in Figure 8.4,

$$\alpha_{cr,su} = \frac{10}{(20 + 40 + 20)} \times \frac{5000}{(121.4 - 85.5)} = 17.41$$

For the lower storey, using the first-order analysis results shown in Figure 8.4,

$$\alpha_{cr,sl} = \frac{(10 + 20)}{(20 + 40 + 20 + 40 + 80 + 40)} \times \frac{5000}{85.5} = 7.31 < 17.41 = \alpha_{cr,su}$$

$\therefore \alpha_{cr} = 7.31$ (compare with the computer analysis value of $\alpha_{cr} = 6.905$ given in Figure 8.4d).
(b) Solution using equation 8.7
    For the upper-storey columns, using equation 8.4,

$$k_T = \frac{2210 \times 10^4/5000}{2210 \times 10^4/5000 + 1.5 \times 3415 \times 10^4/12\,000} = 0.508$$

$$k_B = \frac{2210 \times 10^4/5000 + 5259 \times 10^4/5000}{2210 \times 10^4/5000 + 5259 \times 10^4/5000 + 1.5 \times 14136 \times 10^4/12\,000}$$

$$= 0.458$$

$k_{cr,u} = 1.44$ (using Figure 3.21b)

$N_{cr,u} = \pi^2 \times 210\,000 \times 2210 \times 10^4/(1.44 \times 5000)^2 \text{N} = 883.6 \text{ kN}$

For the lower-storey columns, using equation 8.4,

$$k_T = \frac{2210 \times 10^4/5000 + 5259 \times 10^4/5000}{2210 \times 10^4/5000 + 5259 \times 10^4/5000 + 1.5 \times 14136 \times 10^4/12\,000}$$

$$= 0.458$$

$k_B = \infty$ (theoretical value for a frictionless hinge).

$k_{cr,l} = 2.4$ (using Figure 3.21b).

$N_{cr,l} = \pi^2 \times 210\,000 \times 5259 \times 10^4/(2.4 \times 5000)^2 \text{N} = 756.9 \text{ kN}$

Using equation 8.7 and the axial forces of Figure 8.4f,

$$\alpha_{cr,su} = \frac{2 \times 883.6/5.0}{(37.6 + 42.4)/5.0} = 22.1$$

$$\alpha_{cr,sl} = \frac{2 \times 756.9/5.0}{(103.3 + 136.7)/5.0} = 6.31 < 22.1 = \alpha_{cr,su}$$

$\therefore \alpha_{cr} = 6.31$ (compare with the storey deflection method value of $\alpha_{cr} = 7.31$ calculated above and the computer analysis value of $\alpha_{cr} = 6.905$ given in Figure 8.4d).

8.5.4   Example 4 – buckling of unbraced single storey frame

Problem. Determine the effective length factor of the member 12 of the unbraced frame shown in Figure 8.13b (and for which $\pi^2 EI/L^2 = 1$), and the elastic in-plane buckling loads.

Solution. The sway member 12 is unrestrained at end 1 and moderately restrained at end 2 by the lightly loaded member 23. Its effective length factor is therefore greater than 2, the value for a rigidly restrained sway member (see Figure 3.15e). Accordingly, assume $k_{cr,12} = 2.50$.
   Then, by using equation 8.2, $N_{cr,12} = \pi^2 EI/(2.50L)^2 = 0.16$,
   and $Q_{cr} = N_{cr,12}/5 = 0.032$,
   and $N_{23} = Q_{cr} = 0.032$.
   Using Figure 3.19,

$$\alpha_{b23} = \{6 \times (2EI)/(2L)\} \left(1 - \frac{0.032}{4\pi^2 (2EI)/(2L)^2}\right) = 5.904 EI/L$$

and using equation 3.49,

$$k_2 = \frac{6EI/L}{1.5 \times 5.904EI/L + 6EI/L} = 0.404$$

$k_1 = 1.0$ (theoretical value for a frictionless hinge).

Using Figure 3.21b, $k_{cr,12} = 2.3$

The calculation can be repeated using $k_{cr,12} = 2.3$ instead of the initial estimate of 2.50, in which case the solution $k_{cr,12} = 2.3$ will be obtained. The corresponding frame buckling loads are given by

$$Q_{cr} = \frac{\pi^2 EI/(2.3L)^2}{5} = 0.038.$$

8.5.5   Example 5 – buckling of a portal frame

Problem. A uniform section pinned-base portal frame is shown in Figure 8.14 with its member axial compression forces [42]. Determine the elastic buckling load factor of the frame.
Solution.

$$L_r = \sqrt{(12\,000^2 + 3000^2)} = 12\,369 \text{ mm}$$

The average axial forces are

$$N_c = (51.4 + 69.1)/2 = 60.3 \text{ kN},$$
$$N_r = (58.1 + 48.0 + 43.0 + 53.0)/4 = 50.5 \text{ kN}.$$

For a pinned base portal frame, $C = 1.2$ \hfill (8.11)

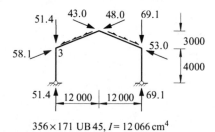

356×171 UB 45, $I = 12\,066 \text{ cm}^4$

Figure 8.14 Worked example 5.

For antisymmetric buckling,

$$R = (12\,080 \times 10^4/4000)/(12\,080 \times 10^4/12\,369) = 3.09 \tag{8.13}$$

$$\rho_c = 3/(10 \times 3.09 + 12) = 0.0699 \tag{8.12a}$$

$$\rho_r = 1 \tag{8.12b}$$

$$N_{cr,c} = \pi^2 \times 210\,000 \times 12\,080 \times 10^4/4000^2 \text{N} = 15\,648 \text{ kN} \tag{8.10}$$

$$N_{cr,r} = \pi^2 \times 210\,000 \times 12\,080 \times 10^4/12\,369^2 \text{N} = 1637 \text{ kN} \tag{8.10}$$

$$\alpha_{cr,as} = \left\{ \left(\frac{60.3}{0.0699 \times 15\,648} \right)^{1.2} + \left(\frac{50.5}{1 \times 1637} \right)^{1.2} \right\}^{-1/1.2} = 13.0 \tag{8.9}$$

For symmetric buckling,

$$R_H = (12\,369/4000) \times (3000/12\,369) = 0.75 \tag{8.15}$$

$$\rho_c = \frac{4.8 + 12 \times 3.09 \times (1 + 0.75)}{2.4 + 12 \times 3.09 \times (1 + 0.75) + 7 \times 3.09^2 \times 0.75^2}$$

$$= 0.664 \tag{8.14a}$$

$$\rho_r = \frac{12 \times 3.09 + 8.4 \times 3.09 \times 0.75}{12 \times 3.09 + 4} = 1.377 \tag{8.14b}$$

$$\alpha_{cr,s} = \left\{ \left(\frac{60.3}{0.664 \times 15\,648} \right)^{1.2} + \left(\frac{50.5}{1.377 \times 1637} \right)^{1.2} \right\}^{-1/1.2} = 38.4$$

$$> 13.0 = \alpha_{cr,as} \tag{8.9}$$

and so the frame buckling factor is $\alpha_{cr} = 13.0$. This is reasonably close to the value of 13.9 predicted by a computer elastic frame buckling analysis program [18].

8.5.6 Example 6 – moment amplification in a braced frame

Problem. Determine the amplified design moments for the braced frame shown in Figure 8.3. (Second-order effects are rarely important in braced members which have significant moment gradients. It can be inferred from Clause 5.2.2(3)b of EC3 that second-order effects need not be considered in braced frames. Nevertheless, an approximate method of allowing for second-order effects by estimating the amplified design moments is illustrated below.)

Solution. For the vertical member 123, the first-order moments are $M_1 = 0$, $M_2 = 23.5$ kNm, $M_3 = -53.0$ kNm, and for these, the method of [28] leads to $\beta_m = 0.585$, and so using equation 8.21,

$$c_m = 0.6 - 0.4 \times 0.585 = 0.366$$

Using $N_{cr,13} = 836.5$ kN from Section 8.5.2, and using equations 8.19 and 8.20,

$$M_{Ed,3} = \frac{0.366 \times 53.0}{1 - 71.6/836.5} = 21.2 \text{ kNm} \ < 53.0 \text{ kNm} = M_3$$

and so $M_{Ed,3} = 53.0$ kNm. This is close to the second-order moment of 53.6 kNm determined using the computer program of [18].

For the horizontal member 345, the first-order moments are $M_3 = -53.0$ kNm, $M_4 = 136.6$ kNm, $M_5 = -153.8$ kNm, and for these, the method of [28] leads to $\beta_m = 0.292$, and so

$$c_m = 0.6 - 0.4 \times 0.292 = 0.483$$

Using $N_{cr35} = 2809.6$ kN from Section 8.5.2, and using equations 8.19, and 8.20,

$$M_{Ed,5} = \frac{0.483 \times 153.8}{1 - 25.3/2809.6} = 75.0 \text{ kNm} < 153.8 \text{ kNm} = M_5$$

and so $M_{Ed,5} = 153.8$ kNm. This is close to the second-order moment of 154.6 kNm determined using the computer program of [18].

8.5.7   Example 7 – moment amplification in a portal frame

Problem. The maximum first-order elastic moment in the portal frame of Figure 8.14 whose buckling load factor was determined in Section 8.5.5 is 151.8 kNm [26]. Determine the amplified design moment.

Solution. Using the elastic frame buckling load factor determined in Section 8.5.5 of $\alpha_{cr,s} = 13.0$ and using equations 8.19 and 8.22,

$$M_{Ed} = \frac{151.8}{1 - 1/13.0} = 164.5 \text{ kNm}$$

which is about 6% higher than the value of 155.4 kNm obtained using the computer program of [18].

8.5.8   Example 8 – moment amplification
in an unbraced frame

Problem. Determine the amplified design moments for the unbraced frame shown in Figure 8.4.

Solution. For the upper-storey columns, using the value of $\alpha_{cr,su} = 22.1$ determined in Section 8.5.3(b) and using equation 8.23,

$$\delta_{su} = 1/(1 - 1/22.1) = 1.047, \text{ and so}$$
$$M_{Ed,8} = 1.047 \times 62.5 = 65.5 \text{ kNm,}$$

which is close to the second-order moment of 64.4 kNm determined using the computer program of [18].

For the lower-storey columns, using the value of $\alpha_{cr,sl} = 6.31$ determined in Section 8.5.3(b) and using equation 8.23,

$$\delta_{sl} = 1/(1 - 1/6.31) = 1.188, \text{ and so}$$
$$M_{Ed,76} = 1.188 \times 119.3 = 141.8 \text{ kNm,}$$

which is 9% higher than the value of 130.5 kNm determined using the computer program of [18].

For the lower beam, the second-order end moment $M_{Ed,74}$ may be approximated by adding the second-order end moments $M_{Ed,76}$ and $M_{Ed,78}$, so that

$$M_{Ed,74} = 141.8 + 53.5/(1 - 1/22.1) = 197.8 \text{ kNm,}$$

which is 7% higher than the value of 184.7 kNm determined using the computer program of [18].

Slightly different values of the second-order moments are obtained if the values of $\alpha_{cr,slu}$ and $\alpha_{cr,sl}$ determined by the storey deflection method in Section 8.5.3(a) are used.

8.5.9   Example 9 – plastic analysis of a braced frame

Problem. Determine the plastic collapse load factor for the braced frame shown in Figure 8.3, if the members are all of S275 steel.

Solution. For the horizontal member plastic collapse mechanism shown in Figure 8.3e,

$$\delta W = 80\alpha_{ph} \times (\delta\theta_h \times 6.0) = 480\alpha_{ph}\delta\theta_h \text{ for a virtual rotation } \delta\theta_h \text{ of } 34,$$

and

$$\delta U = (85.0 \times \delta\theta_h) + (171.3 \times 2\delta\theta_h) + (171.3 \times \delta\theta_h) = 598.9 \, \delta\theta_h$$

so that

$$\alpha_{ph} = 598.9/480 = 1.248.$$

Checking for reductions in M_p using equation 7.17,

$$M_{pr13} = 1.18 \times 85.0 \times \{1 - 89.5/(275 \times 47.1 \times 10^2/10^3)\}$$
$$= 93.4 \text{ kNm} > 85.0 \text{ kNm} = M_{p13}$$

and so there is no reduction in M_p for member 13.

$$M_{pr35} = 1.18 \times 171.3 \times \{1 - 32.1/(275 \times 51.3 \times 10^2/10^3)\}$$
$$= 197.5 \text{ kNm} > 171.3 \text{ kNm} = M_{p35}$$

and so there is no reduction in M_p for member 35.

Therefore $\alpha_{ph} = 1.248$.

For a plastic collapse mechanism in the vertical member with a frictionless hinge at 1 and plastic hinges at 2 and 3,

$$\alpha_{pv} = \frac{85.0 \times 2 \times \delta\theta_v + 85.0 \times \delta\theta_v}{20 \times 5.0 \times \delta\theta_v}$$
$$= 2.55 > 1.248 = \alpha_{ph}$$

and so

$$\alpha_p = 1.248.$$

8.5.10 Example 10 – plastic analysis of an unbraced frame

Problem. Determine the plastic collapse load factor for the unbraced frame shown in Figure 8.4, if the members are all of S275 steel.

Solution. For the beam plastic collapse mechanism shown in Figure 8.4e,

$$\delta W = 40\alpha_p \times 6\delta\theta = 240\alpha_p\delta\theta$$

for virtual rotations $\delta\theta$ of the half beams 35 and 58, and

$$\delta U = (84.2 \times \delta\theta) + (84.2 \times 2\delta\theta) + (84.2 \times \delta\theta) = 336.8\delta\theta$$

so that

$$\alpha_p = 36.8/240 = 1.403.$$

The frame is only partially determinate when this mechanism forms, and so the member axial forces cannot be determined by statics alone. The axial force N_{358} determined by a computer first-order plastic analysis (18) is $N_{358} = 33.8$ kN and the axial force ratio is $(N/N_y)_{358} = 33.8/(275 \times 32.0 \times 10^2/10^3) = 0.0384$, which is less than 0.15, the approximate value at which N/N_y begins to reduce M_p, and so α_p is unchanged at $\alpha_p = 1.403$.

The collapse load factors calculated for the other possible mechanisms of

- lower beam ($\alpha_p = 2 \times (155.9 + 85.0 + 246.4)/(80 \times 6) = 2.030$)
- lower-storey sway ($\alpha_p = 2 \times 155.9/(30 \times 5) = 2.079$)
- upper-storey sway ($\alpha_p = 2 \times (84.2 + 85.0)/(10 \times 5) = 6.768$)
- combined beam sway

$$\left(\alpha_p = \frac{2 \times (84.2 + 84.2 + 246.4 + 85.0 + 155.9)}{\{(40 + 80) \times 6 + (20 \times 5 + 10 \times 10)\}} = 1.425 \right)$$

are all greater than 1.403, and so $\alpha_p = 1.403$.

8.5.11   Example 11 – elastic member design

Problem. Check the adequacy of the lower-storey column 67 (of S275 steel) of the unbraced frame shown in Figure 8.4. The column is fully braced against deflection out of the plane of the frame and twisting.

Design actions.
The first-order actions are $N_{67} = 136.7$ kN and $M_{76} = 119.3$ kNm (Figure 8.4) and the amplified moment is 141.8 kNm (Section 8.5.8).

$$H_{Ed} = 10 + 20 = 30 \text{ kN and}$$
$$V_{Ed} = (20 + 40 + 20) + (40 + 80 + 40) = 240 \text{ kN.}$$
$$H_{Ed}/V_{Ed} = 30/240 = 0.125 < 0.15, \qquad\qquad 5.3.2(4)\text{B}$$

and so the effects of sway imperfections should be included.

$\phi_0 = 1/200$ 5.3.2(3)

$\alpha_h = 2/\sqrt{10} = 0.632 < 2/3$ and so $\alpha_h = 2/3$ 5.3.2(3)

$m = 1$ (one of the columns will have less than the average force),   5.3.2(3)

$\alpha_m = \sqrt{\{0.5(1 + 1/1)\}} = 1.0$, and   5.3.2(3)

$\phi = (1/200) \times (2/3) \times 1.0 = 0.00333$ 5.3.2(3)

The increased horizontal forces are   5.3.2(7)

$10 + 0.00333 \times 80 = 10.27$ kN for the upper storey, and
$20 + 0.00333 \times 240 = 20.80$ kN for the lower storey.

If the frame is reanalysed for these increased forces, then the axial force will increase from 136.7 to $N_{Ed} = 137.2$ kN, and the moment from $119.3 \times 1.188 = 141.8$ kNm to $M_{y,Ed} = 121.9 \times 1.188 = 144.8$ kNm.

Section resistance.

$t_f = 12.5$ mm, $f_y = 275$ N/mm^2 EN10025-2

$\varepsilon = \sqrt{(235/275)} = 0.924$ T5.2

$c_f/(t_f \varepsilon) = (204.3/2 - 7.9/2 - 10.2)/(12.5 \times 0.924) = 7.62 < 9$ T5.2

and the flange is Class 1.   T5.2
  The worst case for the web classification is when the web is fully plastic in compression, for which

$$c_w/(t_w \varepsilon) = (206.2 - 2 \times 12.5 - 2 \times 10.2)/(7.9 \times 0.924) = 22.0 < 33$$
                                                                                    T5.2

and the web is Class 1.   T5.2
  Thus the section is Class 1.

$\gamma_{M0} = 1.0$ 6.1

$N_{c,Rd} = 66.3 \times 10^2 \times 275/1.0$ N $= 1823$ kN $= N_{pl,Rd}$ 6.2.4(2)

$0.25 N_{pl,Rd} = 0.25 \times 1826 = 455.8$ kN > 137.2 kN $= N_{Ed}$,   6.2.9.1(4)

$0.5 h_w t_w f_y = 0.5(206.2 - 2 \times 12.5) \times 7.9 \times 275$ N $= 196.8$ kN   6.2.9.1(4)

$$> 137.2 \text{ kN} = N_{Ed},$$

and so no allowance need be made for the effect of axial force on the plastic resistance moment.   6.2.9.1(4)

$M_{N,Rd} = M_{pl,Rd} = 567 \times 10^3 \times 275/1.0$ Nmm

$$= 155.9 \text{ kNm} > 144.8 \text{ kNm} = M_{Ed} \qquad 6.2.9.1(5)$$

and the section resistance is adequate.   6.2.9.1(2)

Member resistance.
  Because the member is continuously braced out-of-plane, beam lateral buckling and column minor axis and torsional buckling need not be considered.
  Thus $M_{z,Ed} = 0$, $\chi_{LT} = 1.0$, and

$\bar{\lambda}_0 = 0$, $C_{mLT} = 1.0$, $a_{LT} = 0$, and $b_{LT} = 0$ TA.1

$L_{cr} = 5000$ 5.2.2.(7b)

$N_{cr} = \pi^2 \times 210\,000 \times 5259 \times 10^4/5000^2 \text{ N} = 4360 \text{ kN}$

$\dfrac{h}{b} = 206.2/204.3 = 1.009 < 1.2, t_f = 12.5 < 100,$ and so for a S275 steel

UC buckling about the y axis, use buckling curve b.   T6.2

$\alpha = 0.34$ T6.1

$\bar{\lambda}_y = \sqrt{(66.3 \times 10^2 \times 275)/(4360 \times 10^3)} = 0.647$ 6.3.1.2(1)

$\Phi = 0.5 \times [1 + 0.34 \times (0.647 - 0.2) + 0.647^2] = 0.785$ 6.3.1.2(1)

$\chi_y = \dfrac{1}{0.785 + \sqrt{0.785^2 - 0.647^2}} = 0.813$ 6.3.1.2(1)

Using Annex A,

$\mu_y = \dfrac{1 - 137.2/4360}{1 - 0.813 \times 137.2/4360} = 0.994$ TA.1

$\psi_y = 0/144.9 = 0$ TA.2

$C_{my} = C_{my,0} = 0.79 + 0.21 \times 0 + 0.36 \times (0 - 0.33) \times 137.2/4360$ TA.2

$\qquad = 0.786$

$w_y = 567 \times 10^3/510 \times 10^3 = 1.112 < 1.5$ TA.1

$\bar{\lambda}_{max} = \bar{\lambda}_y = 0.647$ TA.1

$N_{Rk} = N_{pl,Rd} \times \gamma_{M0} = 1823 \times 1.0 = 1823 \text{ kN}$ 5.3.2(11)

$\gamma_{M1} = 1.0$ 6.1

$n_{pl} = 137.2/(1823/1.0) = 0.0753$ TA.1

$C_{yy} = 1 + (1.112 - 1)\left[\left(2 - \dfrac{1.6}{1.112} \times 0.785^2 \times 0.647 - \dfrac{1.6}{1.112}\right.\right.$

$\qquad \left.\left. \times 0.785^2 \times 0.647^2\right) \times 0.0753 - 0\right] = 1.009$ TA.1

$k_{yy} = 0.785 \times 1.0 \times \dfrac{0.994}{1 - 137.2/4360} \times \dfrac{1}{1.009} = 0.798$ TA.1

Substituting into equation 6.61 of EC3,

$\dfrac{137.2}{0.813 \times 1823} + 0.798 \times \dfrac{144.8}{1.0 \times 155.9} = 0.834 < 1.0$ 6.3.3(4)

and so the member resistance is adequate.

Alternatively, using Annex B,

$$C_{my} = 0.9 \hspace{8cm} \text{TB.3}$$

$$k_{yy} = 0.9 \left\{ 1 + (0.647 - 0.2) \frac{137.2}{0.813 \times 1823/1.0} \right\} = 0.937 \hspace{2cm} \text{TB.1}$$

Substituting into equation 6.61 of EC3,

$$\frac{137.2}{0.813 \times 1823} + 0.937 \times \frac{144.8}{1.0 \times 155.9} = 0.963 < 1.0 \hspace{2cm} 6.3.3(4)$$

and so the member resistance is adequate.

8.5.12   Example 12 – plastic design of a braced frame

Problem. Determine the plastic design resistance of the braced frame shown in Figure 8.3.

Solution. Using the solution of Section 8.5.2,

$$\alpha_{cr} = 11.7 < 15 \hspace{6cm} 5.2.1(3)$$

and so EC3 does not allow first-order plastic analysis to be used for this frame, unless it is clad. The restriction of Clause 5.2.1(3) of EC3 appears to be unnecessarily severe in this case, since the second-order effects are dominated by the buckling of the column 123, and will have little effect on the plastic collapse mechanism of the beam 345.

If first-order plastic analysis were allowed for this frame, then using the solution of Section 8.5.9,

$\alpha_p = 1.248 > 1.0$, and the structure would appear to be satisfactory.

8.5.13   Example 13 – plastic design of an unbraced frame

Problem. Determine the plastic design resistance of the unbraced frame shown in Figure 8.4.

Solution. Using the solution of Section 8.5.3,

$$\alpha_{cr} = 6.31 < 10 \hspace{6cm} 5.2.1(3)$$

and so EC3 does not allow first-order plastic analysis to be used for this frame.

8.6   Unworked examples

8.6.1   Example 14 – truss design

The members of the welded truss shown in Figure 8.15a are all UC sections of S275 steel with their webs perpendicular to the plane of the truss. The member

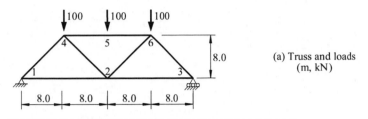

(a) Truss and loads
(m, kN)

Member	Section	I_z(cm^4)	A(cm^2)	f_y(N/mm^2)
14, 36	305 × 305 UC 97	7308	123	275
45, 65	305 × 305 UC 158	12569	201	265
12, 32	305 × 305 UC 97	7308	123	275
42, 62	305 × 305 UC 97	7308	123	275

(b) Member
properties

Member	M_L(kNm)	M_R(kNm)	N (kN)
14	16.8	−68.8	−217.4
45	−150.2	249.8	−172.7
56	249.8	−150.2	−172.7
12	−16.8	8.9	159.1
42	81.5	−40.6	8.5

(c) Elastic
actions

Figure 8.15 Example 14.

properties and the results of a first-order elastic analysis of the truss under the design loads are shown in Figure 8.15b and c.

Determine:

(a) the effective length factors k_{cr} of the compression members and the frame elastic buckling load factor α_{cr},
(b) the amplified first-order design moments M_{Ed} for the compression members,
(c) the adequacy of the members for the actions determined by elastic analysis,
(d) the plastic collapse load factor α_p, and
(e) the adequacy of the truss for plastic design.

8.6.2   Example 15 – unbraced frame design

The members of the unbraced rigid-jointed two-storey frame shown in Figure 8.16a are all of S275 steel. The member properties and the results of a first-order elastic analysis of the frame under its design loads are shown in Figure 8.16b and c.

Determine:

(a) the column effective length factors k_{cr}, the storey buckling load factors $\alpha_{cr,s}$, and the frame buckling load factor α_{cr},
(b) the amplified first-order design moments M_{Ed} for the columns,
(c) the adequacy of the members for the actions determined by elastic analysis,

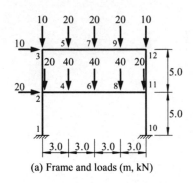

(a) Frame and loads (m, kN)

Member	M_L(kNm)	M_R(kNm)	N (kN)
1–2	−18.8	−13.8	−110.4
10–11	−65.4	79.5	−129.7
2–3	54.6	−48.8	−37.7
11–12	−77.4	75.9	−42.3
2–4	−68.3	89.5	1.7
4–6	89.5	127.4	1.7
6–8	127.4	45.2	1.7
8–11	45.2	−156.9	1.7
3–5	−48.8	34.5	−30.7
5–7	34.5	57.7	−30.7
7–9	57.7	20.9	−30.7
9–12	20.9	−75.9	−30.7

(b) Elastic actions

Member	Section	I_y (cm⁴)	A (cm²)	f_y(N/mm²)
1–3, 10–12	152×152 UC 37	2210	47.1	275
2–11	305×165 UB 40	8503	51.3	275
3–12	254×102 UB 25	3415	32.0	275

(c) Member properties

Figure 8.16 Example 15.

(d) the plastic collapse load factor α_p, and
(e) the adequacy of the frame for plastic design.

References

1. British Standards Institution (2005) *Eurocode 3: Design of Steel Structures- Part 1-8: Design of Joints*, BSI, London.
2. Gardner, L. and Nethercot, D.A. (2005) *Designers' Guide to EN 1993-1-1 Eurocode 3: Design of Steel Structures, General Rules and Rules for Buildings*, Thomas Telford Publishing, London.
3. British Standards Institution (2000) *BS5950-1: 2000 Structural Use of Steelwork in Building – Part 1: Code of Practice for Design – Rolled and Welded Sections*, BSI, London.
4. British Standards Institution (1969) *PD3343 Recommendations for Design (Supplement No. 1 to BS 449)*, BSI, London.
5. Nethercot, D.A. (1986) The behaviour of steel frame structures allowing for semi-rigid joint action, *Steel Structures – Recent Research Advances and Their Applications to Design*, (ed. MN Pavlovic), Elsevier Applied Science Publishers, London, pp. 135–51.
6. Nethercot, D.A. and Chen, W-F. (1988) Effect of connections of columns, *Journal of Constructional Steel Research*, **10**, pp. 201–39.
7. Anderson, D.A, Reading, S.J., and Kavianpour, K. (1991) Wind moment design for unbraced frames, *SCI Publication P-082*, The Steel Construction Institute, Ascot.

8. Couchman, G.H. (1997) Design of semi-continuous braced frames, SCI Publication P-183, The Steel Construction Institute, Ascot.
9. Anderson, D. (ed.) (1996) *Semi-rigid Behaviour of Civil Engineering Structural Connections*, COST-C1, Brussels.
10. Faella, C., Piluso, V., and Rizzano, G. (2000) *Structural Steel Semi-rigid Connections*, CRC Press, Boca Raton.
11. Pippard, A.J.S. and Baker, J.F. (1968) *The Analysis of Engineering Structures*, 4th edition, Edward Arnold, London.
12. Norris, C.H., Wilbur, J.B., and Utku, S. (1976) *Elementary Structural Analysis*, 3rd edition, McGraw-Hill, New York.
13. Coates, R.C., Coutie, M.G., and Kong, F.K. (1988) *Structural Analysis*, 3rd edition, Van Nostrand Reinhold (UK), Wokingham, England.
14. Kleinlogel, A. (1931) *Mehrstielige Rahmen*, Ungar, New York.
15. Steel Construction Institute (2005) *Steel Designers' Manual*, 6th edition, Blackwell Scientific Publications, Oxford.
16. Harrison, H.B. (1973) *Computer Methods in Structural Analysis*, Prentice-Hall, Englewood Cliffs, New Jersey.
17. Harrison, H.B. (1990) *Structural Analysis and Design, Parts 1 and 2*, 2nd ed., Pergamon Press, Oxford.
18. Hancock, G.J., Papangelis, J.P., and Clarke, M.J. (1995) *PRFSA User's Manual*, Centre for Advanced Structural Engineering, University of Sydney.
19. Computer Service Consultants (2006) *S-Frame Enterprise*, CSC (UK) Limited, Leeds.
20. Research Engineers Europe Limited (2005) *STAAD Pro + QSE*, RSE, Bristol.
21. Hancock, G.J. (1984) Structural buckling and vibration analyses on microcomputers, Civil Engineering Transactions, *Institution of Engineers, Australia*, **CE24**, No. 4, pp. 327–32.
22. Bridge, R.Q. and Fraser, D.J. (1987) Improved G-factor method for evaluating effective lengths of columns, *Journal of Structural Engineering*, ASCE, **113**, No. 6, June, pp. 1341–56.
23. Column Research Committee of Japan (1971) *Handbook of Structural Stability*, Corona, Tokyo.
24. Davies, J.M. (1990) In-plane stability of portal frames, *The Structural Engineer*, **68**, No. 8, April, pp. 141–7.
25. Davies, J.M. (1991) The stability of multi-bay portal frames, *The Structural Engineer*, **69**, No. 12, June, pp. 223–9.
26. Trahair, N.S. (1993) Steel structures – elastic in-plane buckling of pitched roof portal frames, *AS4100 DS04*, Standards Australia, Sydney.
27. Silvestre, N. and Camotim, D. (2007) Elastic buckling and second-order behaviour of pitched roof steel frames, *Journal of Constructional Steel Research*, **63**, No. 6, pp. 804–18.
28. Bridge, R.Q. and Trahair, N.S. (1987) Limit state design rules for steel beam-columns, *Steel Construction*, Australian Institute of Steel Construction, **21**, No. 2, September, pp. 2–11.
29. Lai, S.-M.A. and MacGregor, J.G. (1983) Geometric non-linearities in unbraced multistory frames, *Journal of Structural Engineering, ASCE*, **109**, No. 11, pp. 2528–45.
30. Wood, B.R., Beaulieu, D., and Adams, P.F. (1976) Column design by P-delta method, *Journal of the Structural Division, ASCE*, **102**, No. ST2, February, pp. 411–27.

31. Wood, B.R., Beaulieu, D., and Adams, P.F. (1976) Further aspects of design by P-delta method, *Journal of the Structural Division, ASCE*, **102**, No. ST3, March, pp. 487–500.
32. Merchant, W. (1954) The failure load of rigid jointed frameworks as influenced by stability, *The Structural Engineer*, **32**, No. 7, July, pp. 185–90.
33. Levi, V., Driscoll, G.C., and Lu, L-W. (1965) Structural subassemblages prevented from sway, *Journal of the Structural Division, ASCE*, **91**, No. ST5, pp. 103–27.
34. Levi, V., Driscoll, G.C., and Lu, L-W. (1967) Analysis of restrained columns permitted to sway, *Journal of the Structural Division, ASCE*, **93**, No. ST1, pp. 87–108.
35. Hibbard, W.R. and Adams, P.F. (1973) Subassemblage technique for asymmetric structures, *Journal of the Structural Division, ASCE*, **99**, No. ST11, pp. 2259–68.
36. Driscoll, G.C., et al. (1965) *Plastic Design of Multi-storey Frames*, Lehigh University, Pennsylvania.
37. American Iron and Steel Institute (1968) *Plastic Design of Braced Multi-storey Steel Frames*, AISI, New York.
38. Driscoll, G.C., Armacost, J.O., and Hansell, W.C. (1970) Plastic design of multistorey frames by computer, *Journal of the Structural Division, ASCE*, **96**, No. ST1, pp. 17–33.
39. Harrison, H.B. (1967) Plastic analysis of rigid frames of high strength steel accounting for deformation effects, *Civil Engineering Transactions, Institution of Engineers, Australia*, **CE9**, No. 1, April, pp. 127–36.
40. El-Zanaty, M.H. and Murray, D.W. (1983) Nonlinear finite element analysis of steel frames, *Journal of Structural Engineering, ASCE*, **109**, No. 2, February, pp. 353–68.
41. White, D.W. and Chen, W.F. (editors) (1993) *Plastic Hinge Based Methods for Advanced Analysis and Design of Steel Frames*, Structural Stability Research Council, Bethlehem, Pa.
42. Clarke, M.J., Bridge, R.Q., Hancock, G.J., and Trahair, N.S. (1993) Australian trends in the plastic analysis and design of steel building frames, *Plastic Hinge Based Methods for Advanced Analysis and Design of Steel Frames*, Structural Stability Research Council, Bethlehem, Pa, pp. 65–93.
43. Clarke, M.J. (1994) Plastic-zone analysis of frames, Chapter 6 of *Advanced Analysis of Steel Frames: Theory, Software, and Applications*, Chen, W.F. and Toma, S., eds, CRC Press, Inc., Boca Raton, Florida, pp. 259–319.
44. White, D.W., Liew, J.Y.R., and Chen, W.F. (1993) Toward advanced analysis in LRFD, *Plastic Hinge Based Methods for Advanced Analysis and Design of Steel Frames*, Structural Stability Research Council, Bethlehem, Pa, pp. 95–173.
45. Vacharajittiphan, P. and Trahair, N.S. (1975) Analysis of lateral buckling in plane frames, *Journal of the Structural Division, ASCE*, **101**, No. ST7, pp. 1497–516.
46. Vacharajittiphan, P. and Trahair, N.S. (1974) Direct stiffness analysis of lateral buckling, *Journal of Structural Mechanics*, **3**, No. 1, pp. 107–37.
47. Trahair, N.S. (1993) *Flexural-Torsional Buckling of Structures*, E & FN Spon, London.
48. Papangelis, J.P., Trahair, N.S., and Hancock, G.J. (1995) Elastic flexural-torsional buckling of structures by computer, *Proceedings, 6th International Conference on Civil and Structural Engineering Computing*, Cambridge, pp. 109–119.
49. Bradford, M.A., Cuk, P.E., Gizejowski, M.A., and Trahair, N.S. (1987) Inelastic buckling of beam-columns, *Journal of Structural Engineering, ASCE*, **113**, No. 11, pp. 2259–77.

50. Pi, Y.L. and Trahair, N.S. (1994) Nonlinear inelastic analysis of steel beam-columns – Theory, *Journal of Structural Engineering, ASCE*, **120**, No. 7, pp. 2041–61.
51. Pi, Y.L. and Trahair, N.S. (1994) Nonlinear inelastic analysis of steel beam-columns – Applications, *Journal of Structural Engineering, ASCE*, **120**, No. 7, pp. 2062–85.
52. Wongkaew, K. and Chen, W.F. (2002) Consideration of out-of-plane buckling in advanced analysis for planar steel frame design, *Journal of Constructional Steel Research*, **58**, pp. 943–65.
53. Trahair, N.S. and Chan, S.-L. (2003) Out-of-plane advanced analysis of steel structures, *Engineering Structures*, **25**, pp. 1627–37.
54. Trahair, N.S. and Hancock, G.J. (2004) Steel member strength by inelastic lateral buckling, *Journal of Structural Engineering, ASCE*, **130**, No. 1, pp. 64–9.
55. Horne, M.R. (1956) The stanchion problem in frame structures designed according to ultimate carrying capacity, *Proceedings of the Institution of Civil Engineers, Part III*, **5**, pp. 105–46.
56. Horne, M.R. (1964) Safe loads on I-section columns in structures designed by plastic theory, *Proceedings of the Institution of Civil Engineers*, **29**, September, pp. 137–50.
57. Horne, M.R. (1964) The plastic design of columns, *Publication No. 23*, BCSA, London.
58. Institution of Structural Engineers and Institute of Welding (1971) *Joint Committee Second Report on Fully Rigid Multi-Storey Steel Frames*, ISE, London.
59. Wood, R.H. (1973) Rigid-jointed multi-storey steel frame design: a state-of-the-art report, *Current Paper, CP 25/73*, Building Research Establishment, Watford, September.
60. Wood, R.H. (1974) *A New Approach to Column Design*, HMSO, London.

Chapter 9

Joints

9.1 Introduction

Connections or joints are used to transfer the forces supported by a structural member to other parts of the structure or to the supports. They are also used to connect braces and other members which provide restraints to the structural member. Although the terms connections and joints are often regarded as having the same meaning, the definitions of EC3-1-8 [1] are slightly different, as follows. A *connection* consists of fasteners such as bolts, pins, rivets, or welds, and the local member elements connected by these fasteners, and may include additional plates or cleats. A *joint* consists of the zone in which the members are connected, and includes the connection as well as the portions of the member or members at the joint needed to facilitate the action being transferred.

The arrangement of a joint is usually chosen to suit the type of action (force and/or moment) being transferred and the type of member (tension or compression member, beam, or beam-column) being connected. The arrangement should be chosen to avoid excessive costs, since the design, detailing, manufacture, and assembly of a joint is usually time consuming; in particular the joint type has a significant influence on costs. For example, it is often better to use a heavier member rather than stiffeners since this will reduce the number of processes required for its manufacture.

A joint is designed by first identifying the force transfers from the member through the components of the joint to the other parts of the structure. Each component is then proportioned so that it has sufficient strength to resist the force that it is required to transmit. General guidance on joints is given in [2–11].

9.2 Joint components

9.2.1 Bolts

Several different types of bolts may be used in structural joints, including ordinary structural bolts (i.e. commercial or precision bolts and black bolts), and high-strength bolts. Turned close tolerance bolts are now rarely used. Bolts may transfer

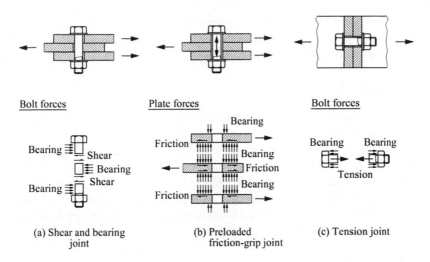

(a) Shear and bearing
joint

(b) Preloaded
friction-grip joint

(c) Tension joint

Figure 9.1 Use of bolts in joints.

loads by shear and bearing, as shown in Figure 9.1a, by friction between plates clamped together as shown in the preloaded friction-grip joint of Figure 9.1b, or by tension as shown in Figure 9.1c. The shear, bearing, and tension capacities of bolts and the slip capacities of preloaded friction-grip joints are discussed in Section 9.6.

The use of bolts often facilitates the assembly of a structure, as only very simple tools are required. This is important in the completion of site joints, especially where the accessibility of a joint is limited, or where it is difficult to provide the specialised equipment required for other types of fasteners. On the other hand, bolting usually involves a significant fabrication effort to produce the bolt holes and the associated plates or cleats. In addition, special but not excessively expensive procedures are required to ensure that the clamping actions required for preloaded friction-grip joints are achieved. Precautions may need to be taken to ensure that the bolts do not become undone, especially in situations where fluctuating or impact loads may loosen them. Such precautions may involve the provision of special locking devices or the use of preloaded high-strength bolts. Guidance on bolted (and on riveted) joints is given in [2–11].

9.2.2 Pins

Pin joints used to be provided in some triangulated frames where it was thought to be important to try to realise the common design assumption that these frames are pin-jointed. The cost of making a pin joint is high because of the machining required for the pin and its holes, and also because of difficulties in assembly. Pins are used in special architectural features where it is necessary to allow relative rotation to occur between the members being connected. In addition, a joint which requires a very large diameter fastener often uses a pin instead of a bolt. Guidance on the design of pin joints is given in Clause 3.13 of EC3-1-8 [1].

9.2.3   Rivets

In the past, hot-driven rivets were extensively used in structural joints. They were often used in the same way as ordinary structural bolts are used in shear and bearing and in tension joints. There is usually less slip in a riveted joint because of the tendency for the rivet holes to be filled by the rivets when being hot-driven. Shop-riveting was cheaper than site riveting, and for this reason shop-riveting was often combined with site-bolting. However, riveting has been replaced by welding or bolting, except in some historical refurbishments. Rivets are treated in a similar fashion to bolts in EC3-1-8 [1].

9.2.4   Welds

Structural joints between steel members are often made by arc-welding techniques, in which molten weld metal is fused with the parent metal of the members or component plates being connected at a joint. Welding is used extensively in fabricating shops where specialised equipment is available and where control and inspection procedures can be readily exercised, ensuring the production of satisfactory welds. Welding is often cheaper than bolting because of the great reduction in the preparation required, while greater strength can be achieved, the members or plates no longer being weakened by bolt holes, and the strength of the weld metal being superior to that of the material connected. In addition, welds are more rigid than other types of load-transferring fasteners. On the other hand, welding often produces distortion and high local residual stresses, and may result in reduced ductility, while site welding may be difficult and costly.

Butt welds, such as that shown in Figure 9.2a, may be used to splice tension members. A full penetration weld enables the full strength of the member to be developed, while the butting together of the members avoids any joint eccentricity. Butt welds often require some machining of the elements to be joined. Special welding procedures are usually needed for full strength welds between thick members to control the weld quality and ductility, while special inspection procedures may be required for critical welds to ensure their integrity. These butt welding limitations often lead to the selection of joints which use fillet welds.

Fillet welds (see Figure 9.2b) may be used to connect lapped plates, as in the tension member splice (Figure 9.2c), or to connect intersecting plates (Figure 9.2d). The member force is transmitted by shear through the weld, either longitudinally or transversely. Fillet welds, although not as efficient as butt welds, require little if any preparation, which accounts for their extensive use. They also require less testing to demonstrate their integrity.

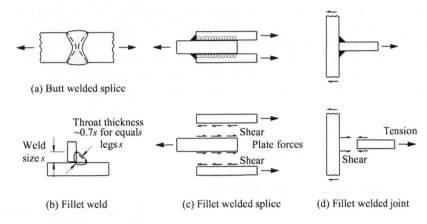

(a) Butt welded splice

(b) Fillet weld   (c) Fillet welded splice   (d) Fillet welded joint

Figure 9.2 Use of welds in joints.

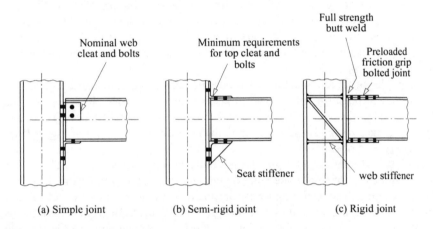

(a) Simple joint   (b) Semi-rigid joint   (c) Rigid joint

Figure 9.3 Beam-to-column joints.

9.2.5 Plates and cleats

Intermediate plates (or gussets), fin (or side) plates, end plates, and angle or tee cleats are frequently used in structural joints to transfer the forces from one member to another. Examples of flange cleats and plates are shown in the beam-to-column joints of Figure 9.3, and an example of a gusset plate is shown in the truss joint of Figure 9.4b. Stiffening plates, such as the seat stiffener (Figure 9.3b) and the column web stiffeners (Figure 9.3c), are often used to help transfer the forces in a joint.

Plates are comparatively strong and stiff when they transfer the forces by in-plane actions, but are comparatively weak and flexible when they transfer the forces by out-of-plane bending. Thus the angle cleat and seat shown in Figure 9.3a are flexible, and allow the relative rotation of the joint members, while the flange plates and web stiffeners (Figure 9.3c) are stiff, and restrict the relative rotation.

The simplicity of welded joints and their comparative rigidity has often resulted in the omission of stiffening plates when they are not required for strength purposes. Thus the rigid joint of Figure 9.3c can be greatly simplified by butt welding the beam directly onto the column flange and by omitting the column web stiffeners. However, this omission will make the joint more flexible since local distortions of the column flange and web will no longer be prevented.

9.3   Arrangement of joints

9.3.1   Joints for force transmission

In many cases, a joint is only required to transmit a force, and there is no moment acting on the group of connectors. While the joint may be capable of also transmitting a moment, it will be referred to as a force joint.

Force joints are generally of two types. For the first, the force acts in the connection plane formed by the interface between the two plates connected, and the connectors between these plates act in shear, as in Figure 9.1a. For the second type, the force acts out of the connection plane and the connectors act in tension, as in Figure 9.1c.

Examples of force joints include splices in tension and compression members, truss joints, and shear splices and joints in beams. A simple shear and bearing bolted tension member splice is shown in Figure 9.1a, and a friction-grip bolted splice in Figure 9.1b. These are simpler than the tension bolt joint of Figure 9.1c, and are typical of site joints.

Ordinary structural and high-strength bolts are used in clearance holes (often 2 or 3 mm oversize to provide erection tolerances) as shown in Figure 9.1a. The hole clearances lead to slip under service loading, and when this is undesirable, a preloaded friction-grip joint such as that shown in Figure 9.1b may be used. In this joint, the transverse clamping action produced by preloading the high-strength bolts allows high frictional resistances to develop and transfer the longitudinal force. Preloaded friction-grip joints are often used to make site joints which need to be comparatively rigid.

The butt and fillet welded splices of Figure 9.2a and c are typical of shop joints, and are of high rigidity. While they are often simpler to manufacture under shop conditions than the corresponding bolted joints of Figure 9.1, special care may need to be taken during welding if these are critical joints.

The truss joint shown in Figure 9.4b uses a gusset plate in order to provide sufficient room for the bolts. The use of a gusset plate is avoided in the joint of Figure 9.4a, while end plates are used in the joint of Figure 9.4c to facilitate the field-bolted connection of shop-welded assemblies.

The beam web shear splice of Figure 9.5a shows a typical shop-welded and site-bolted arrangement. The common simple joint between a beam and column shown in Figure 9.3a is often considered to transmit only shear from the beam to

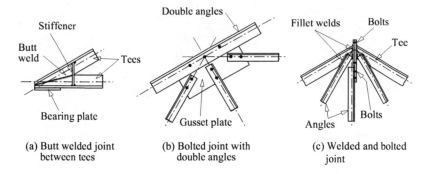

Figure 9.4 Joints between axially loaded members.

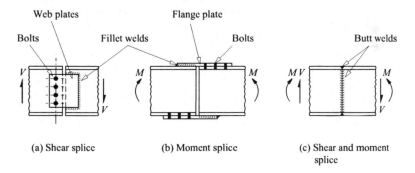

Figure 9.5 Beam splices.

the column flange. Other examples of joints of this type are shown by the beam-to-beam joints of Figure 9.6a and b. Common arrangements of beam splices are given in [2–9] and discussed in Sections 9.5.8 and 9.5.9.

9.3.2   Joints for moment transmission

While it is rare that a real joint transmits only a moment, it is not uncommon that the force transmitted by the joint is sufficiently small for it to be neglected in design. Examples of joints which may be used when the force to be transmitted is negligible include the beam moment splice shown in Figure 9.5b which combines site-bolting with shop-welding, and the welded moment joint of Figure 9.6c. A moment joint is often capable of transmitting moderate forces, as in the case of the beam-to-column joint shown in Figure 9.3c.

9.3.3   Force and moment joints

A force and moment joint is required to transmit both force and moment, as in the beam-to-column joint shown in Figure 9.7. Other examples include the

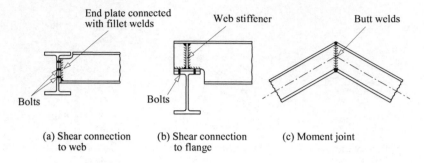

(a) Shear connection to web
(b) Shear connection to flange
(c) Moment joint

Figure 9.6 Joints between beams.

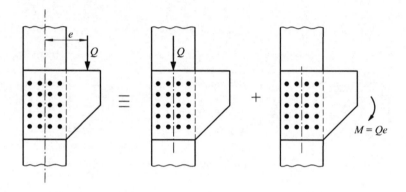

Figure 9.7 Analysis of a force and moment joint.

semi-rigid beam-to-column connection of Figure 9.3b, the full-strength beam splice of Figure 9.5c, and the beam joint of Figure 9.6c. Common beam-to-column joints are given in [2–11], and are discussed in Section 9.5.

In some instances, joints may actually transfer forces or moments which are not intended by the designer. For example, the cleats shown in Figure 9.3a may transfer some horizontal forces to the column, even though the beam is designed as if simply supported, while a similar situation occurs in the shear splice shown in Figure 9.5a.

9.4   Behaviour of joints

9.4.1   Joints for force transmission

When shear and bearing bolts in clearance holes are used in an in-plane force joint, there is an initial slip when the shear is first applied to the joint as some of the clearances are taken up (Figure 9.8). The joint then becomes increasingly stiff as

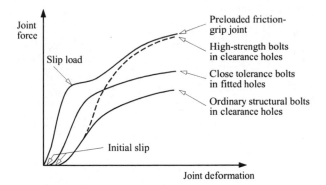

Figure 9.8 Behaviour of bolted joints.

more of the bolts come into play. At higher forces, the more highly loaded bolts start to yield, and the joint becomes less stiff. Later, a state may be reached in which each bolt is loaded to its maximum capacity.

It is not usual to analyse this complex behaviour, and instead it is commonly assumed that equal size bolts share equally in transferring the force as shown in Figure 9.9b (except if the joint is long), even in the service load range. It is shown in Section 9.9.1 that this is the case if there are no clearances and all the bolts fit perfectly, and if the members and connection plates act rigidly and the bolts elastically. If, however, the flexibilities of the joint component members and plates are taken into account, then it is found that the forces transferred are highest in the end bolts of any line of bolts parallel to the joint force and lowest in the centre bolts, as shown in Figure 9.9c. In long bolted joints, the end bolt forces may be so high as to lead to premature failure (before these forces can be redistributed by plastic action) and the subsequent 'unbuttoning' of the joint.

Shear and bearing joints using close tolerance bolts in fitted holes behave in a similar manner to connections with clearance holes, except that the bolt slips are greatly reduced. (It was noted earlier that fitted close tolerance bolts are now rarely used.) On the other hand, slip is not reduced in preloaded friction-grip bolted shear joints, but is postponed until the frictional resistance is overcome at the slip load as shown in Figure 9.8. Again, it is commonly assumed that equal size bolts share equally in transferring the force.

This equal sharing of the force transfer is also often assumed for tension force joints in which the applied force acts out of the connection plane. It is shown in Section 9.9.2 that this is the case for joints in which the plates act rigidly and the bolts act elastically.

Welded force joints do not slip, but behave as if almost rigid. Welds are often assumed to be uniformly stressed, whether loaded transversely or longitudinally. However, there may be significantly higher stresses at the ends of long welds parallel to the joint, just as there are in the case of long bolted joints.

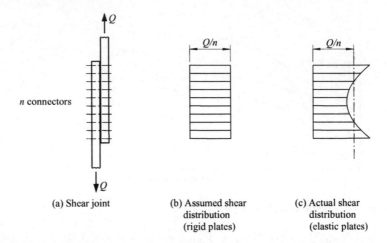

Figure 9.9 Shear distribution in a force connection.

Due to the great differences in the stiffnesses of joints using different types of connectors, it is usual not to allow the joint force to be shared between slip and non-slip connectors, and instead, the non-slip connectors are required to transfer all the force. For example, when ordinary site-bolting is used to hold two members in place at a joint which is subsequently site welded, the bolts are designed for the erection conditions only, while the welds are designed for the final total force. However, it is rational and generally permissible to share the joint force between welds and preloaded friction-grip bolts which are designed against slip, and this is allowed in EC3-1-8 [1].

On the other hand, it is always satisfactory to use one type of connector to transfer the complete force at one part of a joint and a different type at another part. Examples of this are shown in Figure 9.5a and b, where shop welds are used to join each connection plate to one member and field bolts to join it to the other.

9.4.2   Joints for moment transmission

9.4.2.1   General

Although a real connection is rarely required to transmit only a moment, the behaviour of a moment joint may usefully be discussed as an introduction to the consideration of joints which transmit both force and moment and to the component method of joint design used in EC3-1-8 [1].

The idealised behaviour of a moment joint is shown in Figure 9.10, which ignores any initial slip or taking up of clearances. The characteristics of the joint are the design moment resistance $M_{j,Rd}$, the design rotational stiffness S_j, and the rotation capacity ϕ_{Cd}. These characteristics are related to those of the individual components of the joint. For most joints, it is difficult to determine their moment-rotation

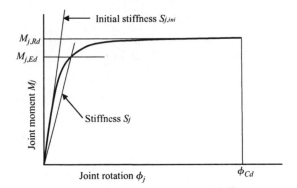

Figure 9.10 Characteristics of a moment joint.

characteristics theoretically, owing to the uncertainties in modelling the various components of the joint, even though a significant body of experimental data is available.

EC3-1-8 uses a *component method* of moment joint design, in which the characteristics of a joint can be determined from the properties of its basic components, including the fasteners or connectors, beam end plates acting in bending, column web panels in shear (Section 4.3), or column webs in transverse compression caused by bearing (Section 4.6). The use of this method for moment joints between I-section members is given in Section 6 of EC3-1-8 and discussed in the following subsections, while the method for moment joints between hollow section members is given in Section 7 of EC3-1-8.

9.4.2.2 Design moment resistance

Each component of a joint must have sufficient resistance to transmit the actions acting on it. Thus the design moment resistance of the joint is governed by the component which has the highest value of its design action to resistance. It is therefore necessary to analyse the joint under its design moment to determine the distribution of the design actions on the joint components.

Joints where the moment acts in the plane of the connectors (as in Figure 9.11a) so that the moment is transferred by connector shear are often analysed elastically [12], by assuming that all the connectors fit perfectly and that each plate acts as if rigid, so that the relative rotation between them is $\delta\theta_x$. In this case, each connector transfers a shear force V_{vi} from one plate to the other. This shear force acts perpendicular to the radius r_i to the axis of rotation, and is proportional to the relative displacement $r_i\delta\theta_x$ of the two plates at the connector, whence

$$V_{vi} = k_v A_i r_i \delta\theta_x,\tag{9.1}$$

where A_i is the shear area of the connector, and the constant k_v depends on the shear stiffness of that type of connector. It is shown in Section 9.9.1 by considering the

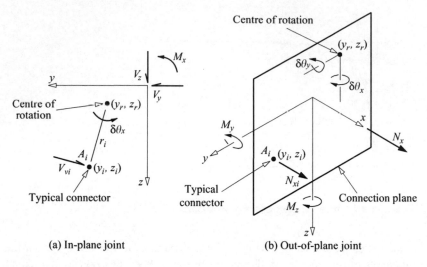

(a) In-plane joint   (b) Out-of-plane joint

Figure 9.11 Joint forces.

equilibrium of the connector forces V_{vi} that the axis of rotation lies at the centroid of the connector group. The moment exerted by the connector force is $V_{vi}r_i$, and so the total moment M_x is given by

$$M_x = k_v \delta\theta_x \sum_i A_i r_i^2. \tag{9.2}$$

If this is substituted into equation 9.1, the connector force can be evaluated as

$$V_{vi} = \frac{M_x A_i r_i}{\sum_i A_i r_i^2}. \tag{9.3}$$

However, the real behaviour of in-plane moment joints is likely to be somewhat different, just as it is for the force joints discussed in Section 9.4.1. This difference is due to the flexibility of the plates, the inelastic behaviour of the connectors at higher moments, and the slip due to the clearances between any bolts and their holes.

Welded moment joints may be analysed by making the same assumptions as for bolted joints [12]. Thus equations 9.2 and 9.3 may be used, with the weld size substituted for the bolt area, and the summations replaced by integrals along the weld.

The simplifying assumptions of elastic connectors and rigid plates are sometimes also made for joints with moments acting normal to the plane of the connectors, as shown in Figure 9.12b. It is shown in Section 9.9.2 that the connector tension

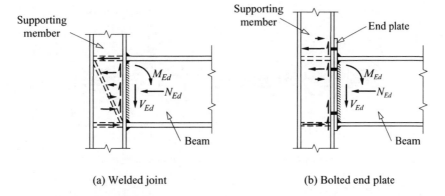

(a) Welded joint   (b) Bolted end plate

Figure 9.12 Common rigid end joints.

forces N_{xi} can be determined from

$$N_{xi} = \frac{M_y A_i z_i}{\sum A_i z_i^2} - \frac{M_z A_i y_i}{\sum A_i y_i^2} \qquad (9.4)$$

in which y_i, z_i are the principal axis coordinates of the connector measured from the centroid of the connector group.

This result implies that the compression forces are transmitted only by the connectors, but in real bolted joints these are transmitted primarily through those portions of the connection plates which remain in contact. Thus the connector tension forces determined from equation 9.4 will be inaccurate when there is a significant difference between the centroid of the contact area and that of the assumed compression connectors. Clause 6.2.7 of EC3-1-8 [1] corrects this defect by defining the centres of compression for a number of bolted beam-column joints, and by requiring the lever arms used in determining the bolt tensions to be measured from these centres.

The result of equation 9.4 is also defective when prying forces are introduced by flexure of any T-stubs (real or idealised) used to transfer bolt tensions, as shown in Figure 9.13 [13–15]. This shows two T-stubs connected through their flanges (or tables) by rows of bolts, which are used to model the tension zone of bolted joints in Clause 6.2.4 of EC3-1-8. Isolating a T-stub simplifies the analysis of a joint. For example, the prying forces in the joint in Figure 9.13b that are produced by flexure of the flange of the T-stub and which must be included in the tensile action $N_{t,Ed}$ on the bolt can be determined from bending theory (Chapter 5). Examples of tension zones which include idealised T-stub assemblies are shown in Figures 9.12 and 9.14.

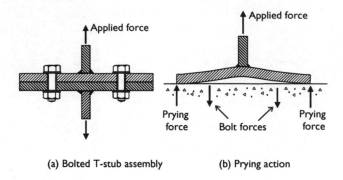

(a) Bolted T-stub assembly   (b) Prying action

Figure 9.13 T-stub elements.

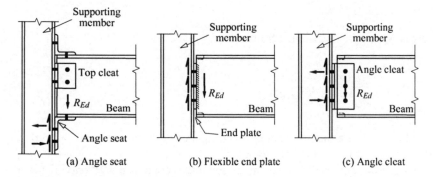

(a) Angle seat   (b) Flexible end plate   (c) Angle cleat

Figure 9.14 Common flexible end joints.

9.4.2.3   Design moment stiffness

Joints need to be classified for frame analysis purposes (Chapter 8) as being either effectively pinned (low moment stiffness), semi-rigid, or effectively rigid (high-moment stiffness). When a joint cannot be assumed to be either effectively pinned or rigid, then its moment stiffness needs to be determined so that it can be classified. If a frame is to be analysed as if semi-rigid (Section 8.3.3), then the moment stiffnesses of its semi-rigid joints need to be used in the analysis.

In the component method used in Clause 6.3 of EC3-1-8 [1], the moment stiffness of a joint is determined from the inverse of the sum of the flexibilities of the components used at each link in the chain of force transfer through the joint. Thus the flexibilities of any plate or cleat components used must be included with those of the fasteners. Plates are comparatively stiff (and often assumed to be rigid) when loaded in their planes, but are comparatively flexible when bent out of their planes. In general, the overall behaviour of a joint can be assessed by determining the path by which force is transferred through the joint, and by synthesising the responses of all the components to their individual loads.

9.4.2.4   Rotation capacity

When rigid plastic analysis is used in the design of a frame (Section 8.5.3.7), each plastic hinge location must have sufficient rotation capacity to ensure that the plastic moment can be maintained until the collapse mechanism is developed. Clause 6.4 of EC3-1-8 [1] provides methods of determining whether joints between I-section members have sufficient rotation capacity.

9.4.2.5   Joint classification

The behaviour of a structure is influenced as much by the behaviour of its joints as by the behaviour of its individual members. Flexural frames are therefore analysed as being simple, semi-continuous, or continuous (Section 8.3.1), depending on the resistance and stiffness classifications of their joints.

Clause 5.2.3 of EC3-1-8 [1] classifies joints according to strength as being nominally pinned, of partial strength, or of full strength. A joint may be classified as full strength if its design moment resistance is not less than that of the connected members. A joint may be classified as nominally pinned if its design moment resistance is not greater than 0.25 times that required for a full strength joint, provided it has sufficient rotation capacity. A joint which cannot be classified as either nominally pinned or full strength is classified as partial strength.

Clause 5.2.2 of EC3-1-8 classifies joints according to stiffness as being nominally pinned, semi-rigid, or rigid. A joint in a frame that is braced against significant sway may be classified as rigid if $S_j \geq 8EI_b/L_b$ where EI_b is the flexural rigidity of the beam connected and L_b is its length. A joint may be classified as nominally pinned if $S_j \leq 0.5EI_b/L_b$. A joint which cannot be classified as either nominally pinned or rigid is classified as semi-rigid.

9.4.3   Force and moment joints

Joints which are required to transfer both force and moment may be analysed elastically by using the method of superposition, as shown in Section 9.9. Thus, the individual bolt forces in the eccentrically loaded in-plane plate connections shown in Figure 9.7 can be determined as the vector sum of the components caused by a concentric force Q and a moment Qe. Similarly, the individual connector forces in a joint loaded out of its plane as shown in Figure 9.11b may be determined from the sum of the forces due to the out-of-plane force N_x and the principal axis moments M_y and M_z in which the centre of compression is used to determine the connector lever arms.

The connector forces in joints subjected to combined loadings may be analysed elastically by using superposition to combine the separate effects of in-plane and out-of-plane loading.

The following section provides simplified descriptions of the response of common joints.

9.5 Common joints

9.5.1 General

There are many different joint arrangements used for force and moment transmission between members and supports, depending on the actions which are to be transferred. Fabrication and erection procedures may be simplified by standardising a number of joints for the more frequently occurring situations. Examples of some common joints are shown in Figures 9.12 and 9.14–9.17.

Flexible joints for simple construction (for which any moment transfer can be neglected) are described in Sections 9.5.2–9.5.4, a semi-rigid joint in Section 9.5.5, and rigid joints for continuous structures (which are able to transfer moment as well as force) in Sections 9.5.6 and 9.5.7. Welded and bolted splices are described in Sections 9.5.8 and 9.5.9, and seats and base plates in Sections 9.5.10 and 9.5.11, respectively.

Guidance on the arrangement and analysis of some of these joints is given in [2–11], while EC3-1-8 [1] provides extensive and detailed methods of analysis for a large number of joints, by considering them as assemblies of basic components whose properties are known. Worked examples of a web side plate (fin plate) and

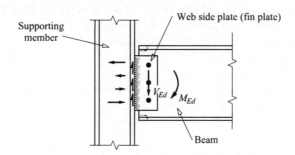

Figure 9.15 Common semi-rigid end joint.

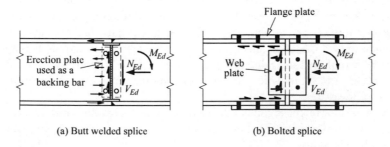

(a) Butt welded splice   (b) Bolted splice

Figure 9.16 Common splices.

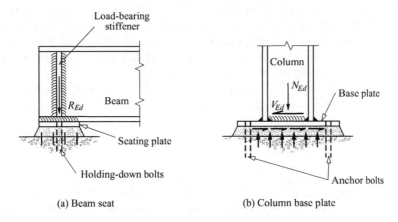

Figure 9.17 Common beam seat and column base plate arrangements.

a flexible end plate connection designed in accordance with EC3-1-8 are given in Section 9.10.

9.5.2 Angle seat joint

An angle seat joint (Figure 9.14a) transfers a beam reaction force R_{Ed} to the supporting member through the angle seat. The top cleat is for lateral restraint only, and may be bolted either to the top of the web or to the top flange. The angle seat may be bolted or fillet welded to the supporting member. The joint has very little moment capacity, and is classified as a nominally pinned joint. It may be used for simple construction.

The beam reaction is transferred by bearing, shear, and bending of the horizontal leg of the angle, by vertical shear through the connectors, and by horizontal forces in the connectors and between the vertical leg and the supporting member.

In this joint, the beam is designed for zero end moment, and the supporting member for the eccentric beam reaction. The beam web may need to be stiffened to resist shear (Section 4.7.4) and bearing (Section 4.7.6). Although this joint is easily designed, its use is often discouraged because of erection difficulties associated with the close depth tolerances required at the top and bottom of the beam.

9.5.3 Flexible end plate joint

A flexible end plate joint (Figures 9.6a and 9.14b) also transfers a beam reaction R_{Ed} to the supporting member. The end plate is fillet welded to the beam web, and bolted to the supporting member. The flanges may be notched or coped, if required. This joint also has very little moment capacity, as there may be significant flexibility in the end plate, and is classified as a nominally pinned joint. It may be used for simple construction.

The beam reaction is transferred by weld shear to the end plate, by shear and bearing to the bolts, and by shear and bearing to the supporting member. In modelling this joint, the idealised T-stub assembly consists of the end plate and the beam web to which it is welded as one T-stub, which is bolted to the column flange which with the column web forms the other T-stub.

The beam is designed for zero end moment, with the end plate augmenting the web shear and bearing capacity, while the supporting member is designed for the eccentric beam reaction.

9.5.4 Angle cleat joint

An angle cleat joint (Figure 9.14c) also transfers a beam reaction R_{Ed} to the supporting member. One or two angle cleats may be used, and bolted to the beam web and to the supporting member. The flanges may be notched or coped, if required. This joint also has very little moment capacity, as there may be significant flexibility in the angle legs connected to the supporting member and in the bolted connection to the beam web. The joint is classified as a nominally pinned joint, and may be used for simple construction.

The beam reaction is transferred by shear and bearing from the web to the web bolts and to the angle cleats. These actions are transferred by the cleats to the supporting member bolts, and by these to the supporting member by shear, tension and compression. If two angle cleats are used, the idealised T-stub assembly consists of the column flange and web as one T-stub, and angle cleats bolted to the beam web as the other T-stub.

The beam is designed for zero end moment, with the angles augmenting the web shear and bearing capacity, while the supporting member is designed for the eccentric beam reaction.

9.5.5 Web side plate (fin plate) joint

A web side plate (fin plate) joint (Figure 9.15) transfers a beam reaction R_{Ed} to the supporting member, and can also transfer a moment M_{Ed}. The fin plate is fillet welded to the supporting member, and bolted to the beam web. The flanges may be notched or coped, if required. This joint has limited flexibility, and is classified as semi-rigid. It may be used for simple construction, and also for semi-continuous construction provided that the degree of interaction between the members can be established.

The beam reaction R_{Ed} and moment M_{Ed} are transferred by shear through the bolts to the fin plate, by in-plane bending and shear to the welds, and by vertical and horizontal shear to the supporting member.

The beam and the supporting member are designed for the reaction R_{Ed} and moment M_{Ed}.

9.5.6 Welded moment joint

A fully welded moment joint (Figure 9.12a) transfers moment M_{Ed}, axial force N_{Ed}, and shear V_{Ed} from one member to another. The welds may be fillet or butt welds, and erection cleats may be used to facilitate field welding. The concentrated flange forces resulting from the moment M_{Ed} and axial force N_{Ed} may require the other member in the joint to be stiffened (Section 4.5) in order to transfer these forces. When suitably stiffened, such a joint acts as if rigid, and may be used in continuous construction.

The shear V_{Ed} is transferred by shear through the welds from one member to the other. The flange forces are transferred by tension through the flange welds and by compression to the other member, while the web moment and the axial force are transferred by tension through the web welds and by compression to the other member.

9.5.7 Bolted moment end plate joint

A bolted moment end plate (extended end plate) joint (Figure 9.12b) also transfers moment M_{Ed}, axial force N_{Ed}, and shear V_{Ed} from one member to another. The end plate is fillet welded to the web and flanges of one member, and bolted to the other member. Concentrated flange forces may require the other member to be stiffened in order to transfer these forces. This joint is also classified as rigid, although it is less stiff than a welded moment joint, due to flexure of the end plate. EC3-1-8 [1] allows bolted moment end plate joints to be used in continuous construction.

The beam moment, axial force, and shear are transferred by tension through the flange and web welds and compression, and by shear through the web welds to the end plate, by bending and shear through the end plate to the bolts, and by bolt shear and tension and by horizontal plate reaction to the other member.

The bolt tensions are increased by prying actions resulting from bending of the end plate, as in Figure 9.13. They can be determined by considering an analysis of the idealised T-stub assembly of the end plate and beam web, and of the column flange and web to which it is bolted. These prying actions must be considered in the design of bolts subjected to tension, and are discussed in Section 9.6.2.

9.5.8 Welded splice

A fully welded splice (Figure 9.16a) transfers moment M_{Ed}, axial force N_{Ed}, and shear V_{Ed} from one member to another concurrent member. The flange welds are butt welds, while the web welds may be fillet welds, and erection plates may be used to facilitate site welding. A welded splice acts as a rigid joint between the members, and may be used in continuous construction.

The shear V_{Ed} is transferred by shear through the welds. The flange forces resulting from the moment M_{Ed} and axial force N_{Ed} are transferred by tension or compression through the flange welds, while the web moment and the axial force are transferred by tension or compression (or shear) through the web welds.

9.5.9 Bolted splice

A fully bolted splice (Figure 9.16b) also transfers moment M_{Ed}, axial force N_{Ed}, and shear V_{Ed} from one member to another concurrent member. Flange and web plates may be provided on one or both sides. This joint may be used in continuous construction.

The moment, axial force, and shear are transferred from one member by bearing and shear through the bolts to the plates, to the other bolts, and then to the other member. The flange plates only transfer the flange axial force components of the moment M_{Ed}, axial force N_{Ed}, while the web plates transfer all of the shear V_{Ed} together with the web components of M_{Ed} and N_{Ed}.

9.5.10 Beam seat

A beam seat (Figure 9.17a) transfers a beam reaction R_{Ed} to its support. A seating plate may be fillet welded to the bottom flange to increase the support bearing area, while holding-down bolts provide positive connections to the support. If a load-bearing stiffener is provided to increase the web bearing capacity, then this will effectively prevent lateral deflection of the top flange (see Figure 6.19).

The beam reaction R_{Ed} is transferred by bearing through the seating plate to the support.

9.5.11 Base plate

A base plate (Figure 9.17b) transfers a column axial force N_{Ed} and shear V_{Ed} to a support or a concrete foundation, and may also transfer a moment M_{Ed}. The base plate is fillet welded to the flanges and web of the column, while the anchor or holding-down bolts provide connections to the support. Pinned bases used in simple construction often use four anchor bolts, and the flexibility of these and the base plate limits the effective moment resistance.

An axial compression N_{Ed} is transferred from the column by end bearing or weld shear to the base plate, and then by plate bending, shear, and bearing to the support. An axial tension force N_{Ed} is transferred from the column by weld shear to the base plate, by plate bending and shear to the holding-down anchor bolts, and then by bolt tension to the support. The shear force V_{Ed} is transferred from the column by weld shear to the base plate, and then by shear and bearing through the holding-down bolts to the support, or by friction. Specific guidance on the design of holding-down bolts is given in Clause 6.2.6.12 of EC3-1-8 [1], while more general guidance on the design of base plates as idealised T-stub assemblies is given in Clause 6.2.

9.6 Design of bolts

9.6.1 Bearing bolts in shear

The resistance F_v of a bolt in shear (Figure 9.1a) depends on the shear strength of the bolt (of tensile strength f_{ub}) and the area A of the bolt in a particular shear plane (either the gross area, or the tensile stress area through the threads A_s, as appropriate). It can be expressed in the form

$$F_v = \alpha_v f_{ub} A \tag{9.5}$$

in which $\alpha_v = 0.6$ generally (except Grade 10.9 bolts when the shear plane passes through the threaded portion of the bolt, in which case $\alpha_v = 0.5$). The factor $\alpha_v = 0.6$ is close to the theoretical yield value of $\tau_y/f_y \approx 0.577$ (Section 1.3.1) and the experimental value of 0.62 reported in [9].

EC3-1-8 [1] therefore requires the design shear force $F_{v,Ed}$ to be limited by

$$F_{v,Ed} \leq F_{v,Rd} = \frac{f_{ub}}{\gamma_{M2}} \sum_n \alpha_{vn} A_n, \tag{9.6}$$

where $\gamma_{M2} = 1.25$ is the partial factor for the connector resistance (1.50 for Grade 4.6 bolts), n is the number of shear planes, A_n and α_{vn} are the appropriate values for the nth shear plane. It is common and conservative to determine the area A_n at a shear plane by substituting the tensile stress area A_s for the shank area A of the bolt.

For bolts in long joints with a distance L_j between the end bolts in the joint, the shear resistance of all bolts is reduced by using the factor

$$\beta_{Lf} = 1 - \frac{L_j - 15d}{200d} \tag{9.7}$$

provided that $0.75 \leq \beta_{Lf} \leq 1$, where d is the diameter of the bolt.

9.6.2   Bolts in tension

The resistance of a bolt in tension (Figure 9.1c) depends on the tensile strength f_{ub} of the bolt and the minimum cross-sectional area of the threaded length of the bolt. The design force $F_{t,Ed}$ is limited by EC3-1-8 [1] to

$$F_{t,Ed} \leq F_{t,Rd} = 0.9 f_{ub} A_s / \gamma_{M2}, \tag{9.8}$$

where $\gamma_{M2} = 1.25$ (1.50 for Grade 4.6 bolts) and A_s is the tensile stress area of the bolt. The use of $0.9 f_{ub}$ for the limit state of bolt fracture in addition to the partial factor γ_{M2} for the connector resistance reflects the reduced ductility at tensile fracture compared with shear failure.

If any of the connecting plates is sufficiently flexible, then additional prying forces may be induced in the bolts, as in Figure 9.13b. However, EC3-1-8 does not provide specific guidance on how to determine the prying force in the bolt, even in the T-stub component based method of design. Under some circumstances which minimise plate flexure in a joint component so that the additional tensile forces caused by prying are small, it has been suggested that the prying forces need not be calculated, provided that $F_{t,Rd}$ is determined from

$$F_{t,Rd} = 0.72f_{ub}A_s/\gamma_{M2}, \tag{9.9}$$

which results from the insertion of a reduction factor of 0.8 into equation 9.8. Some methods for determining prying forces in T-stub connections are given in [13–15].

9.6.3 Bearing bolts in shear and tension

Test results [9] for bearing bolts in shear and tension suggest a circular interaction relationship (Figure 9.18) for the strength limit state. EC3-1-8 [1] uses a more conservative interaction between shear and tension, which can be expressed in the form of equations 9.6, 9.8, and

$$\frac{F_{v,Ed}}{F_{v,Rd}} + \frac{F_{t,Ed}}{1.4F_{t,Rd}} \leq 1 \tag{9.10}$$

where $F_{v,Rd}$ is the shear resistance when there is no tension, and $F_{t,Rd}$ is the tension resistance when there is no shear.

9.6.4 Bolts in bearing

It is now commonly the case that bolt materials are of much higher strengths than those of the steel plates or elements through which the bolts pass. As a result of

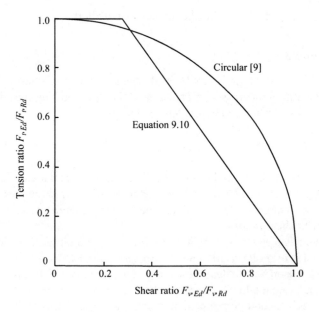

Figure 9.18 Bearing bolts in shear and tension.

this, bearing failure usually takes place in the plate material rather than in the bolt. The design of plates against bearing failure is discussed in Section 9.7.2.

For the situation in which bearing failure occurs in the bolt rather than the plate, EC3-1-8 [1] requires the design bearing force $F_{b,Ed}$ on a bolt whose ultimate tensile strength f_{ub} is less than that of the plate material f_u to be limited to

$$F_{b,Ed} \leq F_{b,Rd} = \frac{f_{ub}dt}{\gamma_{M2}} \tag{9.11}$$

in which $\gamma_{M2} = 1.25$ (1.50 for Grade 4.6 bolts), t is the thickness of the connected element, and d is the diameter of the bolt. Equation 9.11 ignores the enhancement of the bearing strength caused by the triaxial stress state that exists in the bearing area of the bolt, but it seldom governs except for very high-strength plates.

9.6.5 Preloaded friction-grip bolts

9.6.5.1 Design against bearing

High-strength bolts may be used as bearing bolts, or as preloaded bolts in friction-grip joints (Figure 9.1b) which are designed against joint slip under service or factored loads. When high-strength bolts are used as bearing bolts, they should be designed as discussed in Sections 9.6.1–9.6.4.

9.6.5.2 Design against slip

For design against slip in a friction-grip joint for either serviceability or at ultimate loading, the shear force on a preloaded bolt $F_{v,Ed,ser}$ or $F_{v,Ed}$ determined from the appropriate loads must satisfy

$$F_{v,Ed} \leq F_{s,Rd} = nk_s\mu F_{p,C}/\gamma_{M3} \tag{9.12}$$

in which $\gamma_{M3} = 1.25$ for ultimate loading (for serviceability $\gamma_{M3,ser} = 1.1$), n is the number of friction surfaces, k_s is a coefficient given in Table 3.6 of EC3-1-8 [1] which allows for the shape and size of the hole, μ is a slip factor given in Table 3.7 of EC3-1-8, and the preload $F_{p,C}$ is taken as

$$F_{p,C} = 0.7f_{ub}A_s. \tag{9.13}$$

When the joint loading induces bolt tensions, these tend to reduce the friction clamping forces. EC3-1-8 requires the shear serviceability or ultimate design load $F_{v,Ed,ser}$ or $F_{v,Ed}$ to satisfy

$$F_{v,Ed} \leq F_{s,Rd} = nk_s\mu \left(F_{p,C} - 0.8F_{t,Ed}\right)/\gamma_{M3} \tag{9.14}$$

in which $F_{t,Ed,ser}$ or $F_{t,Ed}$ is the total applied tension at service loading or ultimate loading, including any prying forces, and the slip resistance partial safety factor is $\gamma_{M3} = 1.1$ for service loading and 1.25 for ultimate loading.

9.7 Design of bolted plates

9.7.1 General

A connection plate used in a joint is required to transfer actions which may act in the plane of the plate, or out of it. These actions include axial tension and compression forces, shear forces, and bending moments, and may include components induced by prying actions. The presence of bolt holes often weakens the plate, and failure may occur very locally by the bearing of a bolt on the surface of the bolt hole through the plate (punching shear), or in an overall mode along a path whose position is determined by the positions of several holes and the actions transferred by the plate, such as that considered in Section 2.2.3 for staggered connectors in tension members.

Generally, the proportions of the plates should be such as to ensure that there are no instability effects, in which case it will be conservative for the strength limit state to design against general yield, and satisfactory to design against local fracture.

The actual failure stress distributions in bolted connection plates are both uncertain and complicated. When design is to be based on general yielding, it is logical to take advantage of the ductility of the steel and to use a simple plastic analysis. A combined yield criterion such as

$$\sigma_y^2 + \sigma_z^2 - \sigma_y\sigma_z + 3\tau_{yz}^2 \leq f_y^2 \tag{9.15}$$

(Section 1.3.1) may then be used, in which σ_y and σ_z are the design normal stresses and τ_{yz} is the design shear stress.

When design is based on local fracture, an elastic analysis may be made of the stress distribution. Often approximate bending and shear stresses may be determined by elastic beam theory (Chapter 5). The failure criterion should then be tested at all potentially critical locations.

9.7.2 Bearing and tension

Bearing failure of a plate may occur where a bolt bears against part of the surface of the bolt hole through the plate, as shown in Figure 9.19a. After local yielding, the plate material flows plastically, increasing the circumference and thickness of the bearing area, and redistributing the contact force exerted by the bolt.

EC3-1-8 [1] requires the plate-bearing force $F_{b,Ed}$ due to the design loads to be limited by

$$F_{b,Ed} \leq F_{b,Rd} = k_1\alpha_d f_u dt/\gamma_{M2} \tag{9.16}$$

in which f_u is the ultimate tensile strength of the plate material, d is the bolt diameter, t is the plate thickness and $\gamma_{M2} = 1.25$ is the partial factor for connector resistance (1.50 for Grade 4.6 bolts).

The coefficient α_d in equation 9.16 is associated with plate tear-out by shearing in the direction of load transfer, commonly when the bolt is close to the end of a plate, as shown in Figure 9.19b. In this case, EC3-1-8 requires α_d to be determined from

$$\alpha_d = \frac{e_1}{3d_0} \leq 1 \qquad (9.17)$$

for an end bolt, or from

$$\alpha_d = \frac{p_1}{3d_0} - 0.25 \leq 1 \qquad (9.18)$$

for internal bolts if these bolts are closely spaced, in which d_0 is the bolt diameter and the dimensions e_1 and p_1 are shown in Figure 2.7.

The coefficient k_1 in equation 9.16 is associated with tension fracture perpendicular to the direction of load transfer, commonly when the bolt is close to an edge, as shown in Figure 9.19e. Hence, EC3-1-8 uses

$$k_1 = 2.8e_2/d_0 - 1.7 \leq 2.5 \qquad (9.19)$$

if the bolts are at the edge of the plate, and

$$k_1 = 1.4p_2/d_0 - 1.7 \leq 2.5 \qquad (9.20)$$

for internal bolts, where p_2 is the distance between rows of holes (p in Figure 2.5).

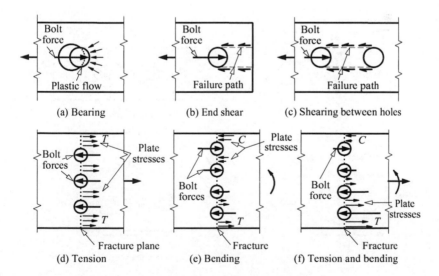

Figure 9.19 Bolted plates in bearing, shear, tension, or bending.

Tension failure may be caused by tension actions as shown in Figure 9.19d. Tension failures of tension members are treated in Section 2.2, and it is logical to apply the EC3 tension members method discussed in Section 2.6.2 to the design of plates in tension. However, there are no specific limits given in EC3-1-8.

9.7.3   Shear and tension

Plate sections may be subjected to simultaneous normal and shear stress, as in the case of the splice plates shown in Figure 9.20a and b. These may be designed conservatively against general yield by using the shear and bending stresses determined by elastic analyses of the gross cross-section in the combined yield criterion of equation 9.15, and against fracture by using the stresses determined by elastic analyses of the net section in an ultimate strength version of equation 9.15.

Block failure may occur in some connection plates as shown in Figure 9.20c and d. In these failures, it may be assumed that the total resistance is provided partly by the tensile resistance across one section of the failure path, and partly by the shear resistance along another section of the failure path. This assumption implies considerable redistribution from the elastic stress distribution, which is likely to be very non-uniform. Hence EC3-1-8 [1] limits the block-tearing resistance for situations of concentric loading such as in Figure 9.20c to

$$N_{Ed} \le V_{eff,1,Rd} = f_u A_{nt}/\gamma_{M2} + \left(f_y/\sqrt{3}\right) A_{nv}/\gamma_{M0}, \tag{9.21}$$

in which A_{nt} and A_{nv} are the net areas subjected to tension and shear, respectively, $f_y/\sqrt{3}$ is the yield stress of the plate in shear (Section 1.3), $\gamma_{M0} = 1.0$ is the partial

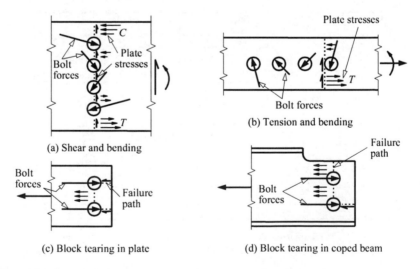

(a) Shear and bending

(b) Tension and bending

(c) Block tearing in plate

(d) Block tearing in coped beam

Figure 9.20 Bolted plates in shear and tension.

factor for plate resistance to yield and $\gamma_{M2} = 1.1$ is the partial factor for fracture. Equation 9.21 corresponds to an elastic analysis of the net section subjected to uniform shear and tensile stresses.

For situations of eccentric loading such as in Figure 9.20d, the block-tearing resistance is limited to

$$N_{Ed} \le V_{eff,2,Rd} = 0.5 f_u A_{nt} / \gamma_{M2} + \left(f_y / \sqrt{3} \right) A_{nv} / \gamma_{M0}, \qquad (9.22)$$

which corresponds to an elastic analysis of the net section subjected to a uniform shear stress and a triangularly varying tensile stress.

9.8   Design of welds

9.8.1   Full penetration butt welds

Full penetration butt welds are so made that their thicknesses and widths are not less than the corresponding lesser values for the elements joined. When the weld metal is of higher strength than those of the elements joined (and this is usually the case), the static capacity of the weld is greater than those of the elements joined. Hence, the design is controlled by these elements, and there are no design procedures required for the weld. EC3-1-8 [1] requires butt welds to have equal or superior properties to those of the elements joined, so that the design resistance is taken as the weaker of the parts connected.

Partial penetration butt welds have effective (throat) thicknesses which are less than those of the elements joined. EC3-1-8 requires partial penetration butt welds to be designed as deep-penetration fillet welds.

9.8.2   Fillet welds

Each of the fillet welds connecting the two plates shown in Figure 9.21 is of length L, and the throat thickness a of the weld is inclined at $\alpha = \tan^{-1}(s_2/s_1)$. Each weld transfers a longitudinal shear V_L and transverse forces or shears V_{Ty} and V_{Tz} between the plates. The average normal and shear stresses σ_w and τ_w on the weld throat may be expressed in terms of the forces

$$\sigma_w L a = V_{Ty} \sin \alpha + V_{Tz} \cos \alpha \qquad (9.23)$$

$$\tau_w L a = \sqrt{\left[\left(V_{Ty} \cos \alpha - V_{Tz} \sin \alpha \right)^2 + V_L^2 \right]}. \qquad (9.24)$$

In addition to these stresses, there are local stresses in the weld arising from bending effects, shear lag, and stress concentrations, together with longitudinal normal stresses induced by compatibility between the weld and each plate.

It is customary to assume that the static strength of the weld is determined by the average throat stresses σ_w and τ_w alone, and that the ultimate strength of the

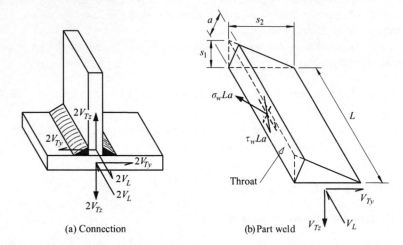

(a) Connection   (b) Part weld

Figure 9.21 Fillet welded joint.

weld is reached when

$$\sqrt{(\sigma_w^2 + 3\tau_w^2)} = f_{uw} \tag{9.25}$$

in which f_{uw} is the ultimate tensile strength of the weld, which is assumed to be greater than that of the plate. This equation is similar to equation 1.1 used in Section 1.3.1 to express the distortion energy theory for yield under combined stresses. Substituting equations 9.23 and 9.24 into equation 9.25 and rearranging leads to

$$3\left(V_{Ty}^2 + V_{Tz}^2 + V_L^2\right) - 2\left(V_{Ty}\sin\alpha + V_{Tz}\cos\alpha\right)^2 = \left(f_{uw}La\right)^2. \tag{9.26}$$

This is often simplified conservatively to

$$\frac{V_R}{La} = \frac{f_{uw}}{\sqrt{3}} \tag{9.27}$$

in which V_R is the vector resultant

$$V_R = \sqrt{\left(V_{Ty}^2 + V_{Tz}^2 + V_L^2\right)} \tag{9.28}$$

of the applied forces V_{Ty}, V_{Tz}, and V_L.

In the simple design method of EC3-1-8 [1] based on equations 9.27 and 9.28, the design weld forces $F_{Ty,Ed}$, $F_{Tz,Ed}$, and $F_{L,Ed}$ per unit length due to the factored loads are limited by

$$F_{w,Ed} \leq F_{w,Rd} \tag{9.29}$$

where $F_{w,Ed}$ is the resultant of all of the forces transmitted by the weld per unit length given by

$$F_{w,Ed} = \sqrt{\left(F_{Ty,Ed}^2 + F_{Tz,Ed}^2 + F_{L,Ed}^2 \right)} \tag{9.30}$$

and

$$F_{w,Rd} = f_{vw,d}a \tag{9.31a}$$

is the design weld resistance per unit length, in which

$$f_{vw,d} = \frac{f_u/\sqrt{3}}{\beta_w \gamma_{M2}} \tag{9.31b}$$

is the design shear strength of the weld, f_u is the ultimate tensile strength of the plate which is less than that of the weld, β_w is a correlation factor for the steel type between 0.8 and 1.0 given in Table 4.1 of EC3-1-8, and $\gamma_{M2} = 1.25$ is the connector resistance partial safety factor.

EC3-1-8 also provides a less conservative directional method which is based on equations 9.23 to 9.25, and which makes some allowance for the dependence of the weld strength on the direction of loading by assuming that the normal stress parallel to the axis of the weld throat does not influence the design resistance. For this method, the normal stress perpendicular to the throat $\sigma_\perp$ is determined from equation 9.23 as

$$\sigma_\perp a = F_{Ty,Ed} \sin\alpha + F_{Tz,Ed} \cos\alpha, \tag{9.32}$$

the shear stress perpendicular to the throat $\tau_\perp$ from equation 9.24 as

$$\tau_\perp a = F_{Ty,Ed} \cos\alpha - F_{Tz,Ed} \sin\alpha, \tag{9.33}$$

and the shear stress parallel to the throat $\tau_{||}$ from equation 9.24 as

$$\tau_{||} a = F_{L,Ed}. \tag{9.34}$$

The stresses in equations 9.32 to 9.34 are then required to satisfy

$$\sqrt{\left[\sigma_\perp^2 + 3 \left(\tau_\perp^2 + \tau_{||}^2 \right) \right]} \leq \frac{f_u}{\beta_w \gamma_{M2}} \tag{9.35}$$

and

$$\sigma_\perp \leq 0.9 f_u / \gamma_{M2} \tag{9.36}$$

where $\gamma_{M2} = 1.25$ and β_w is the correlation factor used in equation 9.31b.

9.9   Appendix – elastic analysis of joints

9.9.1   In-plane joints

The in-plane joint shown in Figure 9.11a is subjected to a moment M_x and to forces V_y, V_z acting through the centroid of the connector group (which may consist of bolts or welds) defined by (see Section 5.9)

$$\left.\begin{array}{l} \sum A_i y_i = 0 \\ \sum A_i z_i = 0 \end{array}\right\} \tag{9.37}$$

in which A_i is the area of the ith connector and y_i, z_i its coordinates.

It is assumed that the joint undergoes a rigid body relative rotation $\delta\theta_x$ between its plate components about a point whose coordinates are y_r, z_r. If the plate components are rigid and the connectors elastic, then it may be assumed that each connector transfers a force V_{vi}, which acts perpendicular to the line r_i to the centre of rotation (Figure 9.11a) and which is proportional to the distance r_i, so that

$$V_{vi} = k_v A_i r_i \delta\theta_x \tag{9.1}$$

in which k_v is a constant which depends on the elastic shear stiffness of the connector.

The centroidal force resultants V_y, V_z of the connector forces are

$$V_y = -\Sigma V_{vi}\,(z_i - z_r)/r_i = k_v \delta\theta_x z_r \Sigma A_i$$
$$V_z = \Sigma V_{vi}\,(y_i - y_r)/r_i = -k_v \delta\theta_x y_r \Sigma A_i \tag{9.38}$$

after using equations 9.37 and 9.1. The moment resultant M_x of the connector forces about the centroid is

$$M_x = \Sigma V_{vi}(z_i - z_r)\,z_i/r_i + \Sigma V_{vi}(y_i - y_r)\,y_i/r_i,$$

whence

$$M_x = k_v \delta\theta_x \sum A_i \left(y_i^2 + z_i^2\right), \tag{9.39}$$

after using equations 9.37 and 9.1. This can be used to eliminate $k_v \delta\theta_x$ from equations 9.38 and these can then be rearranged to find the coordinates of the centre of rotation as

$$y_r = \frac{-V_z \sum A_i \left(y_i^2 + z_i^2\right)}{M_x \sum A_i}, \tag{9.40}$$

$$z_r = \frac{V_y \sum A_i \left(y_i^2 + z_i^2\right)}{M_x \sum A_i}. \tag{9.41}$$

The term $k_v \delta \theta_x$ can also be eliminated from equation 9.1 by using equation 9.39 so that the connector force can be expressed as

$$V_{vi} = \frac{M_x A_i r_i}{\sum A_i \left(y_i^2 + z_i^2 \right)}. \tag{9.42}$$

Thus the greatest connector shear stress V_{vi}/A_i occurs at the connector at the greatest distance r_i from the centre of rotation y_r, z_r.

When there are n bolts of equal area, these equations simplify to

$$y_r = \frac{-V_z \sum \left(y_i^2 + z_i^2 \right)}{n M_x}, \tag{9.40a}$$

$$z_r = \frac{V_y \sum \left(y_i^2 + z_i^2 \right)}{n M_x}, \tag{9.41a}$$

$$V_{vi} = \frac{M_x r_i}{\sum \left(y_i^2 + z_i^2 \right)}. \tag{9.42a}$$

When the connectors are welds, the summations in equations 9.40–9.42 should be replaced by integrals, so that

$$y_r = \frac{-V_z \left(I_y + I_z \right)}{M_x A}, \tag{9.40b}$$

$$z_r = \frac{V_y \left(I_y + I_z \right)}{M_x A}, \quad \text{and} \tag{9.41b}$$

$$\tau_w = \frac{M_x r}{\left(I_y + I_z \right)} \tag{9.42b}$$

in which A is the area of the weld group given by the sum of the products of the weld lengths L and throat thicknesses a, I_y, and I_z are the second moments of area of the weld group about its centroid, and r is the distance to the weld where the shear stress is τ_w. The properties of some specific weld groups are given in [5].

In the special case of a moment joint with V_y, V_z equal to zero, the centre of rotation is at the centroid of the connector group (equations 9.41 and 9.42), and the connector forces are given by

$$V_{vi} = \frac{M_x A_i r_i}{\sum\limits_i A_i r_i^2} \tag{9.3}$$

with $r_i = \sqrt{\left(y_i^2 + z_i^2 \right)}$. In the special case of a force joint with M_x, V_z equal to zero, the centre of rotation is at $z_r = \infty$, and the connector forces are given by

$$V_{vi} = \frac{V_y A_i}{\sum A_i}. \tag{9.43}$$

In the special case of a force joint with M_x, V_y equal to zero, the centre of rotation is at $y_r = -\infty$, and the connector forces are given by

$$V_{vi} = \frac{V_z A_i}{\sum A_i}. \tag{9.44}$$

Equations 9.43 and 9.44 indicate that the connector stresses V_{vi}/A_i caused by forces V_y, V_z are constant throughout the joint.

It can be shown that the superposition by vector addition of the components of V_{vi} obtained from equations 9.3, 9.43, and 9.44 for the separate connection actions of V_y, V_z, and M_x leads to the general result given in equation 9.42.

9.9.2   Out-of-plane joints

The out-of-plane joint shown in Figure 9.11b is subjected to a normal force N_x and moments M_y, M_z acting about the principal axes y, z of the connector group (which may consist of bolts or welds) defined by (Section 5.9)

$$\left.\begin{array}{l} \sum A_i y_i = 0 \\ \sum A_i z_i = 0 \\ \sum A_i y_i z_i = 0 \end{array}\right\} \tag{9.45}$$

in which A_i is the area of the ith connector and y_i, z_i are its coordinates.

It is assumed that the plate components of the joint undergo rigid body relative rotations $\delta\theta_y$, $\delta\theta_z$ about axes which are parallel to the y, z principal axes and which pass through a point whose coordinates are y_r, z_r and that only the connectors transfer forces. If the plates are rigid and the connectors elastic, then it may be assumed that each connector transfers a force N_{xi} which acts perpendicular to the plane of the joint, and which has components which are proportional to the distances $(y_i - y_r)$, $(z_i - z_r)$ from the axes of rotation, so that

$$N_{xi} = k_t A_i (z_i - z_r)\, \delta\theta_y - k_t A_i (y_i - y_r)\, \delta\theta_z \tag{9.46}$$

in which k_t is a constant which depends on the axial stiffness of the connector.

The centroidal force resultant N_x of the connector forces is

$$N_x = \sum N_{xi} = k_t \left(-z_r\delta\theta_y + y_r\delta\theta_z\right) \sum A_i, \tag{9.47}$$

after using equations 9.45. The moment resultants M_y, M_z of the connector forces about the centroidal axes are

$$M_y = \sum N_{xi} z_i = k_t\delta\theta_y \Sigma A_i z_i^2, \tag{9.48}$$

$$M_z = -\sum N_{xi} y_i = k_t\delta\theta_z \Sigma A_i y_i^2, \tag{9.49}$$

after using equations 9.46.

Equations 9.47–9.49 can be used to eliminate k_t, $\delta\theta_y$, $\delta\theta_z$, y_r, z_r from equation 9.46, which then becomes

$$N_{xi} = \frac{N_x A_i}{\sum A_i} + \frac{M_y A_i z_i}{\sum A_i z_i^2} - \frac{M_z A_i y_i}{\sum A_i y_i^2}. \tag{9.50}$$

This result demonstrates that the connector force can be obtained by superposition of the separate components due to the centroidal force N_x and the principal axis moments M_y, M_z. When there are n bolts of equal area, this equation simplifies to

$$N_{xi} = \frac{N_x}{n} + \frac{M_y z_i}{\sum z_i^2} - \frac{M_z y_i}{\sum y_i^2}. \tag{9.50a}$$

When the connectors are welds, the summations in equation 9.47 should be replaced by integrals, so that

$$\sigma_w = \frac{N_x}{A} + \frac{M_y z}{I_y} - \frac{M_z y}{I_z} \tag{9.50b}$$

in which y, z are the coordinates of the weld where the stress is σ_w.

The assumption that only the connectors transfer compression forces through the connection may be unrealistic when portions of the plates remain in contact. In this case the analysis above may still be used, provided that the values A_i, y_i, z_i used for the compression regions represent the actual contact areas between the plates.

9.10 Worked examples

9.10.1 Example 1 – in-plane analysis of a bolt group

Problem. The semi-rigid web side plate (fin plate) joint shown in Figure 9.22a is to transmit factored design actions equivalent to a vertical downwards force of Q kN acting at the centroid of the bolt group and a clockwise moment of $0.2Q$ kNm. Determine the maximum bolt shear force.

Solution.
For the bolt group, $\Sigma(y_i^2 + z_i^2) = 4 \times (70^2 + 105^2) + 4 \times (70^2 + 35^2) = 88\,200\,\text{mm}^2$.

$$\text{Using equation } 9.40, y_r = \frac{-Q \times 10^3 \times 88\,200}{-0.2Q \times 10^6 \times 8} = 55.1\,\text{mm}$$

and so the most heavily loaded bolts are at the top and bottom of the right-hand row.

For these, $r_i = \sqrt{\{(55.1 + 70)^2 + 105^2\}} = 163.3\,\text{mm}$,
and the maximum bolt force may be obtained from equation 9.3 as

$$F_{v,Ed} = \frac{-0.2Q \times 10^6}{88200}\,\text{N} = -0.370\,Q\,\text{kN}.$$

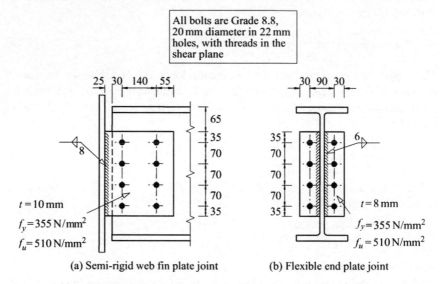

(a) Semi-rigid web fin plate joint   (b) Flexible end plate joint

Figure 9.22 Examples 1–9.

9.10.2   Example 2 – in-plane design resistance of a bolt group

Problem. Determine the design resistance of the bolt group in the web fin plate joint shown in Figure 9.22a.

Solution. For 20 mm Grade 8.8 bolts, $A_s = A_t = 245 \text{ mm}^2$,

$$\alpha_v = 0.6, \qquad\qquad\qquad\qquad \text{EC3-1-8 T3.4}$$

$$f_{ub} = 800 \text{ N/mm}^2, \qquad\qquad\qquad \text{EC3-1-8 T3.1}$$

and so

$$F_{v,Rd} = \frac{0.6 \times 800 \times 245}{1.25} \text{ N} = 94.1 \text{ kN}. \qquad \text{EC3-1-8 T3.4}$$

Using this with the solution of example 1,

$$0.370 \, Q \leq 94.1 \text{ so that } Q_v \leq 254.0.$$

9.10.3   Example 3 – plate-bearing resistance

Problem. Determine the bearing resistance of the web fin plate shown in Figure 9.22a.

Solution.

$$\alpha_d = 35/(3 \times 22) = 0.530, f_{ub}/f_u = 800/510 = 1.569 > 0.530,$$

EC3-1-8 T3.4

and so

$$\alpha_b = 0.530.$$ EC3-1-8 T3.4

$$2.8e_2/d_0 - 1.7 = 2.8 \times 55/22 - 1.7 = 5.3 > 2.5, \text{and so } k_1 = 2.5.$$

EC3-1-8 T3.4

$$F_{b,Rd} = 2.5 \times 0.530 \times 510 \times 20 \times 10/1.1 \, \text{N} = 122.9 \, \text{kN}.$$ EC3-1-8 T3.4

Using this with the solution of example 1, $Q_b \leq 122.9/0.370 = 331.9 > 254.0(\geq Q_v)$.

9.10.4   Example 4 – plate shear and tension resistance

Problem. Determine the shear and tension resistance of the web fin plate shown in Figure 9.22a.

Solution. If the yield criterion of equation 9.15 is used with the elastic stresses on the gross cross-section of the plate, then the elastic bending stress may be determined using an elastic section modulus of $bd^2/6$, so that

$$\sigma_y = 0.2 Q \times 10^6 \times 6/(10 \times 280^2) \, \text{N/mm}^2 = 1.531 \, Q \, \text{N/mm}^2$$

and the average shear stress is

$$\tau_{yz} = Q \times 10^3/(280 \times 10) \, \text{N/mm}^2 = 0.357 \, Q \, \text{N/mm}^2.$$

Substituting into equation 9.15, $(1.531 \, Q)^2 + 3 \times (0.357Q)^2 \leq 355^2$

so that $Q_{vty} \leq 215.0 < 254.0 (\geq Q_v)$.

If the equivalent fracture criterion derived from equation 9.15 is used with the elastic stresses on the net cross-section of the plate, then

$$I_p = 280^3 \times 10/12 - 2 \times 22 \times 10 \times 105^2 - 2 \times 22 \times 10 \times 35^2$$
$$= 12.90 \times 10^6 \, \text{mm}^4$$

so that

$$W_{el,p} = 12.90 \times 10^6/140 = 92.2 \times 10^3 \text{mm}^3, \text{and}$$
$$\sigma_y = 0.2 Q \times 10^6/(92.2 \times 10^3) \, \text{N/mm}^2 = 2.170 \, Q \, \text{N/mm}^2$$

and the average shear stress is

$$\tau_{yz} = Q \times 10^3/\{(280 - 4 \times 22) \times 10\} \, \text{N/mm}^2 = 0.521 \, Q \, \text{N/mm}^2.$$

Using the equivalent of equation 9.15 for fracture with $f_u = 510 \, \text{N/mm}^2$ (EN10025-2),

$$(2.170 \, Q)^2 + 3 \times (0.521 \, Q)^2 \leq 510^2$$

so that $Q_{vtu} \leq 228.5 > 215.0 (\geq Q_{vty})$.

However, these calculations are conservative, since the maximum bending stresses occur at the top and bottom of the plate, where the shear stresses are zero. If the shear stresses are ignored, then $Q_{ty} \leq 355/1.531 = 231.9 < 254.3 (\geq Q_v)$ and

$$Q_{tu} \leq 510/2.170 = 235.0 > 231.9 (\geq Q_{ty}).$$

9.10.5   Example 5 – fillet weld resistance

Problem. Determine the resistance of the two fillet welds shown in Figure 9.22a.
Weld forces per unit length.
At the welds, the design actions consist of a vertical shear of Q kN and a moment of

$$(-0.2 \, Q - Q \times (25 + 30 + 70)/1000) \, \text{kNm} = -0.325 \, Q \, \text{kNm}.$$

$$l_{eff} = 280 - 2 \times 8/\sqrt{2} = 268.7 \, \text{mm} \hspace{2cm} \text{EC3-1-8 4.5.1(1)}$$

The average shear force per unit weld length can be determined as

$$F_{L,Ed} = (Q \times 10^3)/(2 \times 268.7) \, \text{N/mm} = 1.861 \, Q \, \text{N/mm},$$

and the maximum bending force per unit weld length from equation 9.50b as

$$F_{Ty,Ed} = \frac{(-0.325 \, Q \times 10^6) \times (268.7/2)}{2 \times 268.7^3/12} \text{N/mm} = -13.51 \, Q \, \text{N/mm},$$

and using equation 9.30, the resultant of these forces is

$$F_{w,Ed} = \sqrt{[(1.861 \, Q)^2 + (13.51 \, Q)^2]} = 13.63 \, Q \, \text{N/mm}.$$

Simplified method of EC3-1-8.

For Grade S355 steel, $\beta_w = 0.9$ \hspace{2cm} EC3-1-8 T4.1

$$f_{vw,d} = \frac{510/\sqrt{3}}{0.9 \times 1.25} = 261.7 \, \text{N/mm}^2 \hspace{2cm} \text{EC3-1-8 4.5.3.3}$$

$$F_{w,Rd} = 261.7 \times (8/\sqrt{2}) = 1481 \, \text{N/mm} \hspace{2cm} \text{EC3-1-8 4.5.3.3}$$

Hence $13.63 \, Q \leq 1481$

so that

$$Q_w \leq 108.6 < 231.9 (\geq Q_{ty}).$$

Directional method of EC3-1-8.

Using equations 9.32–9.34,

$$\sigma_\perp = \frac{(-0.325\,Q \times 10^6) \times (268.7/2) \sin 45^\circ}{2 \times (268.7^3/12) \times (8/\sqrt{2})} = -1.688\,Q\,\mathrm{N/mm^2}$$

$$\tau_\perp = \frac{(-0.325\,Q \times 10^6) \times (268.7/2) \cos 45^\circ}{2 \times (268.7^3/12) \times (8/\sqrt{2})} = -1.688\,Q\,\mathrm{N/mm^2}$$

$$\tau_\parallel = \frac{Q \times 10^3}{2 \times 268.7 \times (8/\sqrt{2})} = 0.329\,Q\,\mathrm{N/mm^2}$$

The strength condition is

$$\{(-1.688\,Q)^2 + 3 \times [(-1.688\,Q)^2 + (0.329Q)^2]\}^{0.5} \leq \frac{510}{0.9 \times 1.25}$$
$$\text{EC3-1-8 } 4.5.3.2(6)$$

so that

$$Q_w \leq 132.4 < 231.9 (\geq Q_{ty})$$

and

$$1.688Q \leq 0.9 \times 510/1.25 \qquad\qquad \text{EC3-1-8 } 4.5.3.2(6)$$

so that

$$Q_w \leq 217.5 > 132.4$$

and so the joint resistance is governed by the shear and bending capacity of the welds.

9.10.6 *Example 6 – bolt slip*

Problem. If the bolts of the semi-rigid web fin plate joint shown in Figure 9.22a are preloaded, then determine the value of Q at which the first bolt slip occurs, if the joint is designed to be non-slip in service and the friction surface is Class B.

Solution.

$$F_{p,C} = 0.7 \times 800 \times 245 \,\text{N} = 137.2 \,\text{kN} \qquad\qquad \text{EC3-1-8 3.9.1(2)}$$

$$k_s = 1.0 \qquad\qquad\qquad\qquad\qquad\qquad\qquad\qquad \text{EC3-1-8 T3.6}$$

$$\mu = 0.4 \qquad\qquad\qquad\qquad\qquad\qquad\qquad\qquad \text{EC3-1-8 T3.7}$$

$$\gamma_{M3,ser} = 1.1 \qquad\qquad\qquad\qquad\qquad\qquad\qquad \text{EC3-1-8 2.2(2)}$$

$$F_{s,Rd} = 1.0 \times 0.4 \times 137.2/1.1 = 49.9 \,\text{kN} \qquad \text{EC3-1-8 3.9.1(1)}$$

and using the solution of example 1,

$$Q_{sL} = 49.9/0.370 = 134.7.$$

9.10.7   Example 7 – out-of-plane resistance of a bolt group

Problem. The flexible end plate joint shown in Figure 9.22b is to transmit factored design actions equivalent to a downwards force of Q kN acting at the centroid of the bolt group and an out-of-plane moment of $M_x = -0.05 \, Q$ kNm. Determine the maximum bolt forces, and the design resistance of the bolt group.

Bolt group analysis.

$$\Sigma z_i^2 = 4 \times 105^2 + 4 \times 35^2 \,\text{mm}^2 = 49\,000 \,\text{mm}^2,$$

and using equation 9.50a, the maximum bolt tension is

$$F_{t,Ed} = \frac{(-0.05 \, Q \times 10^6) \times (-35 - 70)}{49\,000} \text{N} = 0.107 \, Q \,\text{kN}.$$

(A less-conservative value of $F_{t,Ed} = 0.0824 \, Q$ kN is obtained if the centre of compression is assumed to be at the lowest pair of bolts so that the lever arm to the highest pair of bolts is 175 mm).
  The average bolt shear is $F_{v,Ed} = (Q \times 10^3)/8 \,\text{N} = 0.125 \, Q \,\text{kN}$.

Plastic resistance of plate.
If the web thickness is 8.8 mm,

$$m = 90/2 - 8.8/2 - 0.8 \times 6 = 35.8 \,\text{mm} \qquad\qquad \text{EC3-1-8 F6.8}$$

$$l_{eff} = 2 \times 35.8 + 0.625 \times 30 + 35 = 125.4 \,\text{mm} \qquad \text{EC3-1-8 T6.4}$$

$$M_{pl,1,Rd} = 0.25 \times 125.4 \times 8^2 \times 355/1.0 \,\text{Nmm} = 0.712 \,\text{kNm} \quad \text{EC3-1-8 T6.2}$$

$$F_{T,1,Rd} = 4 \times 0.712 \times 10^6/30 \,\text{N} = 94.9 \,\text{kN} \qquad\qquad \text{EC3-1-8 T6.2}$$

and so $F_{T,1,Rd} \geq 2 \times 0.107 \, Q$, whence $Q_p \leq 443.0 \,\text{kN}$.

Bolt resistance.
At plastic collapse of the plate, the prying force causes a plastic hinge at the bolt line, so that the prying force $= 0.712 \times 10^6/30$ N $= 23.7$ kN.

Bolt tension $= 0.107 \times 443.0 + 23.7 = 70.6$ kN.

Using the solution of example 2, $F_{v,Rd} = 94.1$ kN.
For 20 mm Grade 8.8 bolts, $A_t = 245$ mm^2 and $f_{ub} = 800$ N/mm^2

EC3-1-8 T3.1

$F_{t,Rd} = 0.9 \times 800 \times 245/1.25$ N $= 141.1$ kN.   EC3-1-8 T3.4

Hence, $\dfrac{0.125 \times 443.0}{94.1} + \dfrac{70.6}{1.4 \times 141.1} = 0.946 < 1.0$ EC3-1-8 T3.1

and so the bolt resistance does not govern.
 Thus $Q \leq 443.0$.

9.10.8 Example 8 – plate-bearing resistance

Problem. Determine the bearing resistance of the end plate of example 7 shown in Figure 9.22b.

Solution.

$\alpha_d = 35/(3 \times 22) = 0.530$, $f_{ub}/f_u = 800/510 = 1.569 > 0.530$,   EC3-1-8 T3.4

and so $\alpha_b = 0.530$.   EC3-1-8 T3.4

$k_1 = 2.8 \times 30/22 = 3.82 > 2.5$ so that $k_1 = 2.5$ EC3-1-8 T3.4

$F_{b,Rd} = 2.5 \times 0.530 \times 510 \times 20 \times 8/1.1$ N $= 98.3$ kN   EC3-1-8 T3.4

Using this with the analysis solution of example 7, $0.125\,Q \leq 98.3$
so that $Q_b \leq 786.8 > 443.0\,(\geq Q_p)$.

9.10.9 Example 9 – fillet weld resistance

Problem. Determine the resistance of the two fillet welds shown in Figure 9.22b.

Solution.
 $l_w = 268.7$ mm and $F_{L,Ed} = 1.861\,Q$ N/mm, as in example 5.

The maximum bending force per unit weld length can be determined from equation 9.50b as

$$F_{Ty,Ed} = \frac{(-0.05\,Q \times 10^6) \times (268.7/2)}{2 \times 268.7^3/12} = 2.078\,Q\,\text{N/mm}.$$

Using equation 9.30, the resultant of these is

$$F_{w,Ed} = \sqrt{[(1.861\,Q)^2 + (2.078\,Q)^2]} = 2.789\,Q\,\text{N/mm}.$$

For S355 Grade steel, $\beta_w = 0.9$ EC3-1-8 T4.1

$$f_{vw,d} = \frac{510/\sqrt{3}}{0.9 \times 1.25} = 261.7\,\text{N/mm}^2 \qquad\qquad \text{EC3-1-8 4.5.3.3}$$

$$F_{w,Rd} = 261.7 \times (6/\sqrt{2}) = 1110\,\text{N/mm}. \qquad\qquad \text{EC3-1-8 4.5.3.3}$$

Hence $2.789\,Q \le 1110$,
so that $Q_w \le 398.1 < 443.0\ (\ge Q_p)$
   Using the directional method of EC3-1-8 with equations 9.32–9.34,

$$\sigma_\perp = \frac{(-0.05\,Q \times 10^6) \times (268.7/2)\sin 45^\circ}{2 \times (268.7^3/12) \times (6/\sqrt{2})} = -0.346\,Q\,\text{N/mm}^2$$

$$\tau_\perp = \frac{(-0.05\,Q \times 10^6) \times (268.7/2)\cos 45^\circ}{2 \times (268.7^3/12) \times (6/\sqrt{2})} = -0.346\,Q\,\text{N/mm}^2$$

$$\tau_\parallel = \frac{Q \times 10^3}{2 \times 268.7 \times (6/\sqrt{2})} = 0.439\,\text{N/mm}^2$$

The strength condition is

$$\{(-0.346\,Q)^2 + 3 \times [(-0.346\,Q)^2 + (0.439\,Q)^2]\}^{0.5} \le \frac{510}{0.9 \times 1.25}$$
EC3-1-8 4.5.3.2(6)

so that

$$Q_w \le 441.0 < 443.0\ (\ge Q_p)$$

and

$$0.346\,Q \le 0.9 \times 510/1.25$$

so that

$$Q_w \le 1060 > 441.0.$$

Thus the connection capacity is governed by the shear and bending resistance of the welds.

9.11 Unworked examples

9.11.1 Example 10 – bolted joint

Arrange and design a bolted joint to transfer a design force of 800 kN from the inclined member to the truss joint shown in Figure 9.23a.

9.11.2 Example 11 – tension member splice

Arrange and design the tension member splice shown in Figure 9.23b so as to maximise the tension resistance.

9.11.3 Example 12 – tension member welds

Proportion the weld lengths shown in Figure 9.23c to transfer the design tension force concentrically.

9.11.4 Example 13 – channel section bracket

Determine the design resistance Q of the fillet welded bracket shown in Figure 9.23d.

9.11.5 Example 14 – tee-section bracket

Design the fillet welds for the bracket shown in Figure 9.23e.

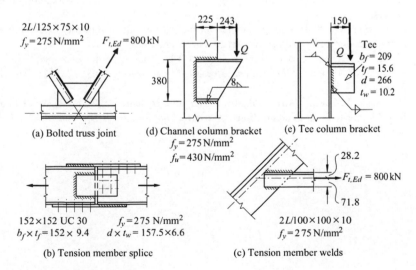

(a) Bolted truss joint

$2L/125 \times 75 \times 10$
$f_y = 275 \, \text{N/mm}^2$ $F_{t,Ed} = 800 \, \text{kN}$

(d) Channel column bracket
$f_y = 275 \, \text{N/mm}^2$
$f_u = 430 \, \text{N/mm}^2$

(e) Tee column bracket

Tee
$b_f = 209$
$t_f = 15.6$
$d = 266$
$t_w = 10.2$

(b) Tension member splice

152×152 UC 30 $f_y = 275 \, \text{N/mm}^2$
$b_f \times t_f = 152 \times 9.4$ $d \times t_w = 157.5 \times 6.6$

(c) Tension member welds

$2L/100 \times 100 \times 10$
$f_y = 275 \, \text{N/mm}^2$

$F_{t,Ed} = 800 \, \text{kN}$

Figure 9.23 Examples 10–14.

References

1. British Standards Institute (2006) *Eurocode 3: Design of Steel Structures – Part 1–8: Design of Joints,* BSI, London.
2. The Steel Construction Institute and the British Constructional Steelwork Association (2002) *Joints in Steel Construction: Simple Connections,* SCI and BCSA, Ascot.
3. The Steel Construction Institute and the British Constructional Steelwork Association (1995) *Joints in Steel Construction: Moment Connections,* SCI and BCSA, Ascot.
4. Jaspart, J.P., Renkin, S., and Guillaume, M.L. (2003) *European Recommendations for the Design of Simple Joints in Steel Structures,* University of Liège, Liège.
5. Faella, C., Piluso, V., and Rizzano, G. (2000) *Structural Steel Semi-rigid Connections,* CRC Press, Boca Raton.
6. Owens, G.W. and Cheal, B.D. (1989) *Structural Steelwork Connections,* Butterworths, London.
7. Davison, B. and Owens, G.W. (eds) (2003) *Steel Designer's Manual,* 6th edition, Blackwell Publishing, Oxford.
8. Morris, L.J. (1988) Connection design in the UK, *Steel Beam-to-Column Building Connections,* (ed. Chen, W.F.), Elsevier Applied Science, pp. 375–414.
9. Kulak, G.L., Fisher, J.W., and Struik, J.H.A. (1987) *Guide to Design Criteria for Bolted and Rivetted Joints,* 2nd edition, John Wiley and Sons, New York.
10. CIDECT (1992) *Design Guide Recommendations for Rectangular Hollow Section (RHS) Joints Under Predominantly Static Loading,* Comité International pour le Développment et l'Étude de la Construction Tubulaire, Cologne.
11. CIDECT (1991) *Design Guide Recommendations for Circular Hollow Section (CHS) Joints Under Predominantly Static Loading,* Comité International pour le Développment et l'Étude de la Construction Tubulaire, Cologne.
12. Harrison, H.B. (1980) *Structural Analysis and Design, Parts 1 and 2,* Pergamon Press, Oxford.
13. Zoetemeijer, P. (1974) A design method for the tension side of statically loaded bolted beam-to-column connections, *Heron,* **20**, No. 1, pp. 1–59.
14. Piluso, V., Faella, C., and Rizzano G. (2001) Ultimate behavior of bolted T-stubs. I: Theoretical model, *Journal of Structural Engineering, ASCE,* **127**, No. 6, pp. 686–93.
15. Witteveen, J., Stark, J.W.B., Bijlaard, F.S.K., and Zoetemeijer, P. (1982) Welded and bolted beam-to-column connections, *Journal of the Structural Division, ASCE,* **108**, No. ST2, pp. 433–55.

Chapter 10

Torsion members

10.1   Introduction

The resistance of a structural member to torsional loading may be considered to be the sum of two components. When the rate of change of the angle of twist rotation is constant along the member (see Figure 10.1a), it is in a state of uniform (or St Venant) torsion [1, 2], and the longitudinal warping deflections are also constant along the member. In this case, the torque acting at any cross-section is resisted by a single set of shear stresses distributed around the cross-section. The ratio of the torque acting to the twist rotation per unit length is defined as the torsional rigidity GI_t of the member.

The second component of the resistance to torsional loading may act when the rate of change of the angle of twist rotation varies along the member (see Figures 10.1b and c), so that it is in a state of non-uniform torsion [1]. In this case the warping deflections vary along the member, and an additional set of shear stresses may act in conjunction with those due to uniform torsion to resist the torque acting. The stiffness of the member associated with these additional shear stresses is proportional to the warping rigidity EI_w.

When the first component of the resistance to torsional loading completely dominates the second, the member is in a state of uniform torsion. This occurs when the torsion parameter $K = \sqrt{(\pi^2 EI_w/GI_t L^2)}$ is very small, as indicated in Figure 10.2, which is adapted from [1]. Thin-walled closed-section members whose torsional rigidities are very large behave in this way, as do members with narrow rectangular sections and angle and tee-sections, whose warping rigidities are negligible. If, on the other hand, the second component of the resistance to torsional loading completely dominates the first, the member is in a limiting state of non-uniform torsion referred to as warping torsion. This may occur when the torsion parameter K is very large, as indicated in Figure 10.2, which is the case for some very thin-walled open sections (such as light gauge cold-formed sections) whose torsional rigidities are very small. Between these two extremes, the torsional loading is resisted by a combination of the uniform and warping torsion components, and the beam is in the general state of non-uniform torsion. This occurs for intermediate values of the parameter K, as shown in

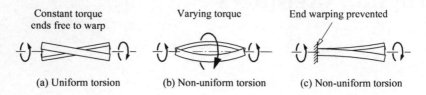

(a) Uniform torsion   (b) Non-uniform torsion   (c) Non-uniform torsion

Figure 10.1 Uniform and non-uniform torsion of an I-section member.

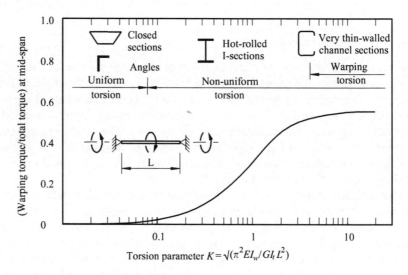

Figure 10.2 Effect of cross-section on torsional behaviour.

Figure 10.2, which are appropriate for most hot-rolled I- or channel-section members.

Whether a member is in a state of uniform or non-uniform torsion also depends on the loading arrangement and the warping restraints. If the torque resisted is constant along the member and warping is unrestrained (as in Figure 10.la), then the member will be in uniform torsion, even if the torsional rigidity is very small. If, however, the torque resisted varies along the length of the member (Figure 10.1b), or if the warping displacements are restrained in any way (Figure 10.1c), then the rate of change of the angle of twist rotation will vary, and the member will be in non-uniform torsion. In general, these variations must be accounted for, but in some cases they can be ignored, and the member analysed as if it were in uniform torsion. This is the case for members of very low warping rigidity

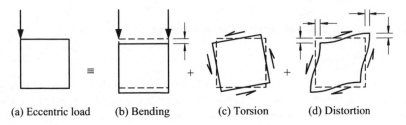

(a) Eccentric load   (b) Bending   (c) Torsion   (d) Distortion

Figure 10.3 Eccentric loading of a thin-walled closed section.

and for members of high-torsional rigidity, for which the rates of change of the angle of twist rotation vary only locally near points of concentrated torque and warping restraint. This simple method of analysis usually leads to satisfactory predictions of the angles of twist rotation, but may produce underestimates of the local stresses.

In this chapter, the behaviour and design of torsion members are discussed. Uniform torsion is treated in Section 10.2, and non-uniform torsion in Section 10.3, while the design of torsion members for strength and serviceability is treated in Section 10.4.

Structural members, however, are rarely used to resist torsion alone, and it is much more common for torsion to occur in conjunction with bending and other effects. For example, when the box section beam shown in Figure 10.3a is subjected to an eccentric load, this causes bending (Figure 10.3b), torsion (Figure 10.3c), and distortion (Figure 10.3d). There may also be interactions between bending and torsion. For example, the longitudinal stresses due to bending may cause a change in the effective torsional rigidity. This type of interaction is of some importance in the flexural–torsional buckling of thin-walled open-section members, as discussed in Sections 3.7.5 and 6.10. Also, significant twist rotations of the member may cause increases in the bending stresses in thin-walled open-section members. However, these interactions are usually negligible when the torsional rigidity is very high, as in thin-walled closed-section members. The behaviour of members under combined bending and torsion is discussed in Section 10.5.

The effects of distortion of the cross-section are only significant in very thin-walled open-sections, and in thin-walled closed sections with high-distortional loadings. Thus, for the box-section beam shown in Figure 10.3, the relatively flexible plates may bend out of their planes as indicated in Figure 10.3d, causing the cross-section to distort. Because of this, the in-plane plate bending and shear stress distributions are changed, and significant out-of-plane plate bending stresses may be induced. The distortional behaviour of structural members is discussed briefly in Section 10.6.

10.2   Uniform torsion

10.2.1   Elastic deformations and stresses

10.2.1.1   General

In elastic uniform torsion, the twist rotation per unit length is constant along the length of the member. This occurs when the torque is constant and the ends of the member are free to warp, as shown in Figure 10.1a. When a member twists in this way,

(a) lines which were originally parallel to the axis of twist become helices,
(b) cross-sections rotate ϕ as rigid bodies about the axis of twist, and
(c) cross-sections warp out of their planes, the warping deflections (u) being constant along the length of the member (see Figure 10.4).

The uniform torque T_t acting at any section induces shear stresses τ_{xy}, τ_{xz} which act in the plane of the cross-section. The distribution of these shear stresses can be visualised by using Prandtl's membrane analogy, as indicated in Figure 10.5 for a rectangular cross-section. Imagine a thin uniformly stretched membrane which is fixed to the cross-section boundary and displaced by a uniform transverse pressure. The contours of the displaced membrane coincide with the shear stress

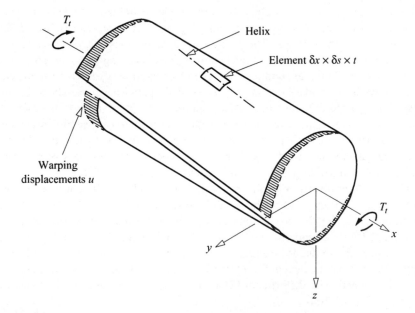

Figure 10.4 Warping displacements u due to twisting.

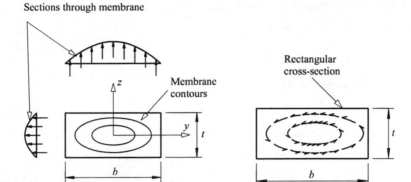

Sections through membrane

Membrane contours

Rectangular cross-section

(a) Transversely loaded membrane

(b) Distribution of shear stress

Figure 10.5 Prandtl's membrane analogy for uniform torsion.

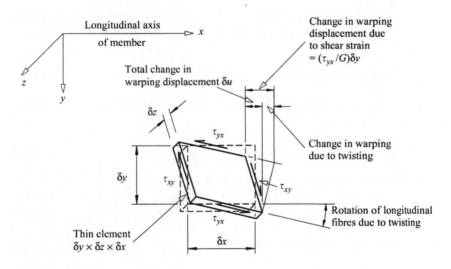

Longitudinal axis of member

Change in warping displacement due to shear strain $= (\tau_{yx}/G)\delta y$

Total change in warping displacement δu

Change in warping due to twisting

Rotation of longitudinal fibres due to twisting

Thin element $\delta y \times \delta z \times \delta x$

Figure 10.6 Warping displacements.

trajectories, the slope of the membrane is proportional to the shear stress, and the volume under the membrane is proportional to the torque. The shear strains caused by these shear stresses are related to the twisting and warping deformations, as indicated in Figure 10.6.

The stress distribution can be found by solving the equation [2]

$$\frac{\partial^2 \theta}{\partial y^2} + \frac{\partial^2 \theta}{\partial z^2} = -2G\frac{d\phi}{dx}, \tag{10.1}$$

(in which G is the shear modulus of elasticity) for the stress function θ (which corresponds to the membrane displacement in Prandtl's analogy) defined by

$$\left.\begin{aligned} \tau_{xz} = \tau_{zx} = -\partial\theta/\partial y \\[2mm] \tau_{xy} = \tau_{yx} = \partial\theta/\partial z \end{aligned}\right\}, \tag{10.2}$$

subject to the condition that

$$\theta = \text{constant} \tag{10.3}$$

around the section boundary. The uniform torque T_t, which is the static equivalent of the shear stresses τ_{xy}, τ_{xz}, is given by

$$T_t = 2 \iint\limits_{\text{section}} \theta \, dy \, dz. \tag{10.4}$$

10.2.1.2   Solid cross-sections

Closed-form solutions of equations 10.1–10.3 are not generally available, except for some very simple cross-sections. For a solid circular section of radius R, the solution is

$$\theta = \frac{G}{2}\frac{d\phi}{dx}(R^2 - y^2 - z^2). \tag{10.5}$$

If this is substituted in equations 10.2, the circumferential shear stress τ_t at a radius $r = \sqrt{(y^2 + z^2)}$ is found to be

$$\tau_t = Gr\frac{d\phi}{dx}. \tag{10.6}$$

The torque effect of these shear stresses is

$$T_t = \int_0^R \tau_t(2\pi r)dr \tag{10.7}$$

which can be expressed in the form

$$T_t = GI_t\frac{d\phi}{dx} \tag{10.8}$$

where the torsion section constant I_t is given by

$$I_t = \pi R^4/2, \tag{10.9}$$

which can also be expressed as

$$I_t = \frac{A^4}{4\pi^2\,(I_y + I_z)}. \tag{10.10}$$

The same result can be obtained by substituting equation 10.5 directly into equation 10.4. Equation 10.6 indicates that the maximum shear stress occurs at the boundary and is equal to

$$\tau_{t,max} = \frac{T_t R}{I_t}.$$ (10.11)

For other solid cross-sections, equation 10.10 gives a reasonably accurate approximation for the torsion section constant, but the maximum shear stress depends on the precise shape of the section. When sections have re-entrant corners, the local shear stresses may be very large (as indicated in Figure 10.7), although they may be reduced by increasing the radius of the fillet at the re-entrant corner.

10.2.1.3   Rectangular cross-sections

The stress function θ for a very narrow rectangular section of width b and thickness t is approximated by

$$\theta \approx G\frac{d\phi}{dx}\left(\frac{t^2}{4} - z^2\right).$$ (10.12)

The shear stresses can be obtained by substituting equation 10.12 into equations 10.2, whence

$$\tau_{xy} = -2zG\frac{d\phi}{dx},$$ (10.13)

which varies linearly with z as shown in Figure 10.8c. The torque effect of the τ_{xy} shear stresses can be obtained by integrating their moments about the x axis,

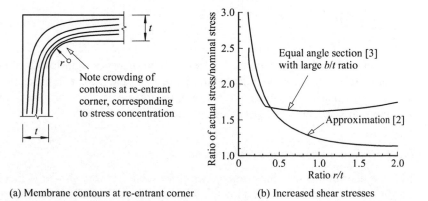

(a) Membrane contours at re-entrant corner   (b) Increased shear stresses

Figure 10.7 Stress concentrations at re-entrant corners.

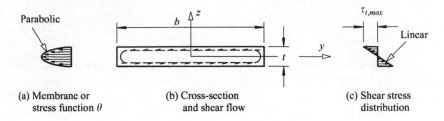

(a) Membrane or   (b) Cross-section   (c) Shear stress
    stress function θ and shear flow   distribution

Figure 10.8 Uniform torsion of a thin rectangular section.

whence

$$-\int_{-t/2}^{t/2} \tau_{xy} bz \, \mathrm{d}z = G\frac{bt^3}{6}\frac{\mathrm{d}\phi}{\mathrm{d}x} \qquad (10.14)$$

which is one half of the total torque effect. The other half arises from the shear stresses τ_{xz}, which have their greatest effects near the ends of the thin rectangle. Although they act there over only a small length of the order of t (compared with b for the τ_{xy} stresses), they have a large lever arm of the order of $b/2$ (instead of $t/2$), and they make an equal contribution. The total torque can therefore be expressed in the form of equation 10.8 with

$$I_t \approx bt^3/3. \qquad (10.15)$$

The same result can be obtained by substituting equation 10.12 directly into equation 10.4. Equation 10.13 indicates that the maximum shear stress occurs at the centre of the long boundary, and is given by

$$\tau_{t,max} \approx \frac{T_t t}{I_t}. \qquad (10.16)$$

For stockier rectangular sections, the stress function θ varies significantly along the width b of the section, as indicated in Figure 10.5, and these approximations may not be sufficiently accurate. In this case, the torsion section constant I_t and the maximum shear stress $\tau_{t,max}$ can be obtained from Figure 10.9.

10.2.1.4 Thin-walled open cross-sections

The stress distribution in a thin-walled open cross-section is very similar to that in a narrow rectangular section, as shown in Figure 10.10a. Thus, the shear stresses are parallel to the walls of the section, and vary linearly across the thickness t, this pattern remaining constant around the section except at the ends. Because of this

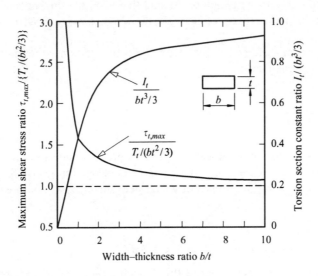

Figure 10.9 Torsion properties of rectangular sections.

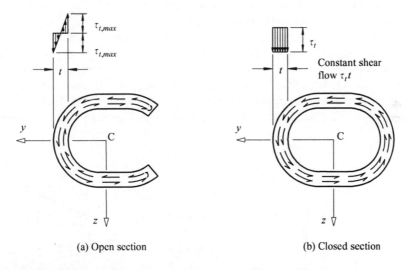

(a) Open section   (b) Closed section

Figure 10.10 Shear stresses due to uniform torsion of thin-walled sections.

similarity, the torsion section constant I_t can be closely approximated by

$$I_t \approx \sum bt^3/3, \qquad\qquad (10.17)$$

where b is the developed length of the mid-line and t the thickness of each thin-walled element of the cross-section, while the summation is carried out for all such elements.

The maximum shear stress away from the concentrations at re-entrant corners can also be approximated as

$$\tau_{t,max} \approx \frac{T_t t_{max}}{I_t},\tag{10.18}$$

where t_{max} is the maximum thickness. When more accurate values for the torsion section constant I_t are required, the formulae developed in [3] can be used. The values given in [4–6] for British hot-rolled steel sections have been calculated in this way.

10.2.1.5   Thin-walled closed cross-sections

The uniform torsional behaviour of thin-walled closed-section members is quite different from that of open-section members, and there are dramatic increases in the torsional stiffness and strength. The shear stress no longer varies linearly across the thickness of the wall, but is constant as shown in Figure 10.10b, and there is a constant shear flow around the closed section.

This shear flow is required to prevent any discontinuities in the longitudinal warping displacements of the closed section. To show this, consider the slit rectangular tube shown in Figure 10.11a. The mid-thickness surface of this open section is unstrained, and so the warping displacements of this surface are entirely due to the twisting of the member. The distribution of the warping displacements caused by twisting (see Section 10.3.1.2) is shown in Figure 10.11b. The relative

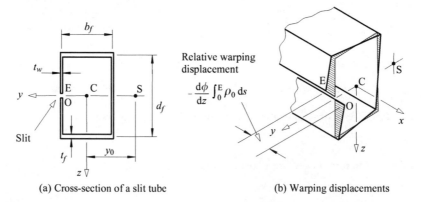

(a) Cross-section of a slit tube   (b) Warping displacements

Figure 10.11 Warping of a slit tube.

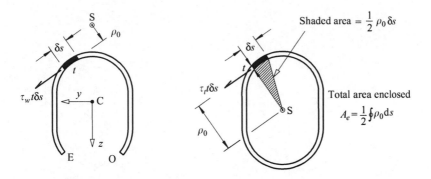

(a) Warping torque in an open section   (b) Uniform torque in a closed section

Figure 10.12 Torques due to shear flows in thin-walled sections.

warping displacement at the slit is equal to

$$u_E - u_0 = -\frac{d\phi}{dx}\int_0^E \rho_0 ds, \tag{10.19}$$

in which ρ_0 is the distance (see Figure 10.12a) from the tangent at the mid-thickness line to the axis of twist (which passes through the shear centre of the section, as discussed in Sections 10.3.1.2 and 5.4.3).

However, when the tube is not slit, this relative warping displacement cannot occur. In this case, the relative warping displacement due to twisting shown in Figure 10.13a is exactly balanced by the relative warping displacement caused by shear straining (see Figure 10.6) of the mid-thickness surface shown in Figure 10.13b, so that the total relative warping displacement is zero. It is shown in Section 10.7.1 that the required shear straining is produced by a constant shear flow $\tau_t t$ around the closed section which is given by

$$\tau_t t = \frac{T_t}{2A_e} \tag{10.20}$$

where A_e is the area enclosed by the section, and that the torsion section constant is given by

$$I_t = \frac{4A_e^2}{\oint (1/t)ds}. \tag{10.21}$$

The maximum shear stress can be obtained from equation 10.20 as

$$\tau_{t,max} = \frac{T_t}{2A_e t_{min}} \tag{10.22}$$

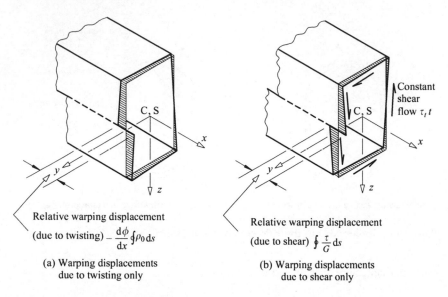

Relative warping displacement
(due to twisting) $-\dfrac{d\phi}{dx}\oint \rho_0 \, ds$

Relative warping displacement
(due to shear) $\oint \dfrac{\tau}{G} \, ds$

(a) Warping displacements
due to twisting only

(b) Warping displacements
due to shear only

Figure 10.13 Warping of a closed section.

The membrane analogy can also be used for thin-walled closed sections, by imagining that the inner-section boundary is fixed to a weightless horizontal plate supported at a distance equivalent to $\tau_t t$ above the outer-section boundary by the transverse pressures and the forces in the membrane stretched between the boundaries, as shown in Figure 10.14b. Thus the slope $(\tau_t t)/t$ of the membrane is substantially constant across the wall thickness, and is equivalent to the shear stress τ_t, while twice the volume $A_e \tau_t t$ under the membrane is equivalent to the uniform torque T_t given by equation 10.20.

The membrane analogy is used in Figure 10.14 to illustrate the dramatic increases in the stiffnesses and strengths of thin-walled closed sections over those of open sections. It can be seen that the torque (which is proportional to the volume under the membrane) is much greater for the closed section when the maximum shear stress is the same. The same conclusion can be reached by considering the effective lever arms of the shear stresses, which are of the same order as the overall dimensions of the closed section, compared with those of the same order as the wall thickness of the open section.

The behaviour of multi-cell closed-section members in uniform torsion can be determined by using the warping displacement continuity condition for each cell, as indicated in Section 10.7.1. A general matrix method of carrying out this analysis by computer has been given in [7], and a computer program has been developed [8]. Alternatively, the membrane analogy can be extended by considering a set of horizontal plates at different heights, one for each cell of the cross-section.

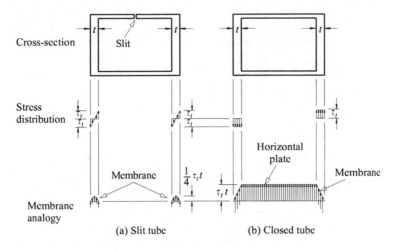

Figure 10.14 Comparison of uniform torsion of open and closed sections.

10.2.2   Elastic analysis

10.2.2.1   Statically determinate members

Some members in uniform torsion are statically determinate, such as the cantilever shown in Figure 10.15. In these cases, the distribution of the total torque $T_x = T_t$ can be determined from statics, and the maximum stresses can be evaluated from equations 10.11, 10.16, 10.18, or 10.20, or from Figures 10.7 or 10.9. The twisted shape of the member can be determined by integrating equation 10.8, using the sign conventions of Figure 10.16a. Thus, the angle of twist rotation in a region of constant torque T_x will vary linearly in accordance with

$$\phi = \phi_0 + \frac{T_x x}{GI_t}, \tag{10.23}$$

as indicated in Figure 10.15c.

10.2.2.2   Statically indeterminate members

Many torsion members are statically indeterminate, such as the beam shown in Figure 10.17, and the distribution of torque cannot be determined from statics alone. The redundant quantities can be determined by substituting the corresponding compatibility conditions into the solution of equation 10.8. Once these have been found, the maximum torque can be evaluated by statics, and the maximum stress can be determined. The maximum angle of twist can also be derived from the solution of equation 10.8.

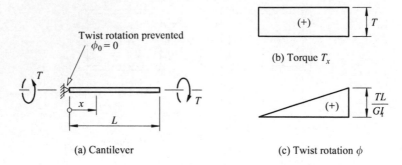

(a) Cantilever

(b) Torque T_x

(c) Twist rotation ϕ

Figure 10.15 Uniform torsion of a statically determinate cantilever.

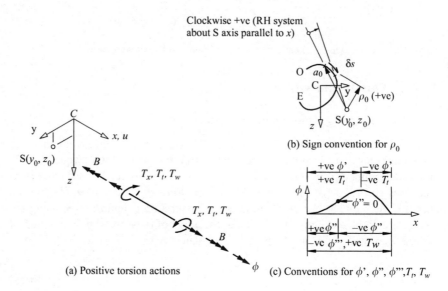

(a) Positive torsion actions

(b) Sign convention for ρ_0

(c) Conventions for ϕ', ϕ'', ϕ''', T_t, T_w

Figure 10.16 Torsion sign conventions.

As an example of this procedure, the indeterminate beam shown in Figure 10.17 is analysed in Section 10.7.2. The theoretical twisted shape of this beam is shown in Figure 10.17c, and it can be seen that there is a jump discontinuity in the twist per unit length $d\phi/dx$ at the loaded point. This discontinuity is a consequence of the use of the uniform torsion theory to analyse the behaviour of the member. In practice such a discontinuity does not occur, because an additional set of local warping stresses is induced. These additional stresses, which may have high values at the loaded point, have been investigated in [9, 10]. High local stresses can usually be tolerated in static loading situations when the material is ductile, whether the stresses are induced by concentrated torques or by other concentrated loads. In

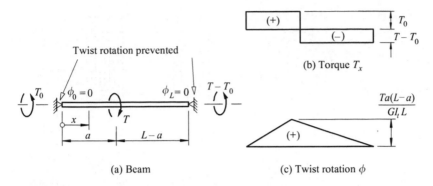

(b) Torque T_x

Twist rotation prevented

(a) Beam

(c) Twist rotation ϕ

Figure 10.17 Uniform torsion of a statically indeterminate beam.

members for which the use of the uniform torsion theory is appropriate, such as those with high torsional rigidity and low warping rigidity, these additional stresses decrease rapidly as the distance from the loaded point increases. Because of this, the angles of twist predicted by the uniform torsion analysis are of sufficient accuracy.

10.2.3 Plastic collapse analysis

10.2.3.1 Fully plastic stress distribution

The elastic shear stress distributions described in the previous sub-sections remain valid while the shear yield stress τ_y is not exceeded. The uniform torque at nominal first yield T_{ty} may be obtained from equations 10.11, 10.16, 10.18, or 10.22 by using $\tau_{t,max} = \tau_y$. Yielding commences at T_{ty} at the most highly stressed regions, and then generally spreads until the section is fully plastic.

The fully plastic shear stress distribution can be visualised by using the 'sand heap' modification of the Prandtl membrane analogy. Because the fully plastic shear stress distribution is constant, the slope of the Prandtl membrane is also constant, and its contours are equally spaced in the same way as are those of a heap formed by pouring sand on to a base area of the same shape as the member cross-section until the heap is fully formed.

This is demonstrated in Figure 10.18a for a circular cross-section, for which the sand heap is conical. Its fully plastic uniform torque may be obtained from

$$T_{tp} = \int_0^R \tau_y(2\pi r)r \, dr \qquad (10.24)$$

whence

$$T_{tp} = (2\pi R^3/3)\,\tau_y \qquad (10.25)$$

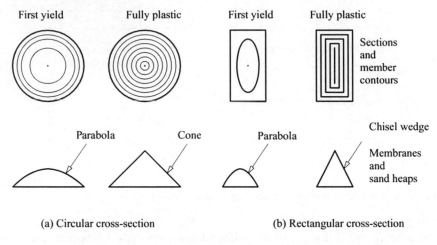

Figure 10.18 First yield and fully plastic uniform torsion shear stress distributions.

The corresponding first yield uniform torque may be obtained from equations 10.9 and 10.11 as

$$T_{ty} = \left(\pi R^3 / 2 \right) \tau_y \tag{10.26}$$

and so the uniform torsion plastic shape factor is $T_{tp}/T_{ty} = 4/3$.

The sand heap for a fully plastic rectangular section $b \times t$ is chisel-shaped as shown in Figure 10.18b. In this case, the fully plastic uniform torque is given by

$$T_{tp} = \frac{bt^2}{2} \left(1 - \frac{t}{3b} \right) \tau_y, \tag{10.27}$$

and for a narrow rectangular section this becomes $T_{tp} \approx (bt^2/2)\tau_y$. The corresponding first yield uniform torque is

$$T_{ty} \approx \left(bt^2 / 3 \right) \tau_y \tag{10.28}$$

and so the uniform torsion plastic shape factor is $T_{tp}/T_{ty} = 3/2$.

For thin-walled open sections, the fully plastic uniform torque is approximated by

$$T_{tp} \approx \Sigma \left(bt^2 / 2 \right) \tau_y \tag{10.29}$$

and the first yield torque is obtained from equation 10.18 as

$$T_{ty} = \frac{\tau_y I_t}{t_{max}} \tag{10.30}$$

For a hollow section, the equilibrium condition of constant shear flow limits the maximum shear flow to the shear flow $\tau_y t_{min}$ in the thinnest wall (see equation 10.22), so that the first yield uniform torque is given by

$$T_{ty} = 2A_e t_{min} \tau_y \tag{10.31}$$

This is the maximum torque that can be resisted, even though walls with $t > t_{min}$ are not fully yielded, and so this value of T_{ty} should also be used for the fully plastic uniform torque T_{tp}.

10.2.3.2 Plastic analysis

A member in uniform torsion will collapse plastically when there is a sufficient number of fully plastic cross-sections to transform the member into a mechanism, as shown for example in Figure 10.19. Each fully plastic cross-section can be thought of as a shear 'hinge' which allows increasing torsional rotations under the uniform plastic torque T_{tp}. The plastic shear hinges usually form at the supports where there are reaction torques, as indicated in Figure 10.19. In general, the plastic shear hinges develop progressively until the collapse mechanism is formed. Examples of uniform torsion plastic collapse mechanisms are shown in Figures 10.20 and 10.21.

At plastic collapse, the mechanism is statically determinate, and can be analysed by using statics to determine the plastic collapse torques. For the member shown in Figure 10.19, the mechanism forms when there are plastic shear hinges at each support. The total applied torque $\alpha_t T$ at plastic collapse must be in equilibrium with these plastic uniform torques, and so

$$\alpha_t T = 2T_{tp} \tag{10.32}$$

so that the uniform torsion plastic collapse load factor is given by

$$\alpha_t = 2T_{tp}/T \tag{10.33}$$

Values of the uniform torsion plastic collapse load factor α_t for a number of example torsion members can be obtained from Figures 10.20 and 10.21.

10.3 Non-uniform torsion

10.3.1 Elastic deformations and stresses

10.3.1.1 General

In elastic non-uniform torsion, both the rate of change of the angle of twist rotation $d\phi/dx$ and the longitudinal warping deflections u vary along the length of the

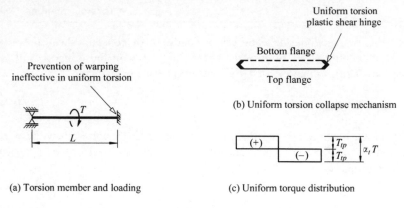

Prevention of warping ineffective in uniform torsion

(a) Torsion member and loading

Uniform torsion plastic shear hinge

Bottom flange

Top flange

(b) Uniform torsion collapse mechanism

(c) Uniform torque distribution

Figure 10.19 Uniform torsion plastic collapse.

Member and loading	Plastic collapse				Rotations	
	Uniform torsion		Warping torsion		Uniform tm	Warping
	Mechanism	$\alpha_t T/T_{tp}$	Mechanism	$\alpha_w TL/M_{fp}d_f$	$\phi_{tm}GI_t/TL$	$\phi_{wm}EI_w/TL^3$
		1.0		0	1.0	∞
		1.0		1.0	1.0	0.333
		1.0		1.0	1.0	0.667
		1.0		1.0	1.0	0.583
		2.0		4.0	0.25	0.0208
		2.0		6.0	0.25	0.00931
		2.0		8.0	0.25	0.00521
		2.0		6.0	0.25	0.0150
		2.0		6.0	0.25	0.0142
		2.0		8.0	0.25	0.00751
		2.0		8.0	0.25	0.00721

× Free to warp

⌇ Warping prevented

—— Top flange

······· Bottom flange

○ Warping torsion frictionless hinge

• Warping torsion plastic hinge

❮ ❯ Uniform torsion shear hinges

Figure 10.20 Plastic collapse and maximum rotation – concentrated torque.

Member and loading	Plastic collapse				Rotations	
	Uniform torsion		Warping torsion		Uniform	Warping
	Mechanism	$\alpha_t T/T_{tp}$	Mechanism	$\alpha_w TL/M_{fp}d_f$	$\phi_{tm}GI_t/TL$	$\phi_{wm}EI_w/TL^3$
		1.0		0	0.5	∞
		1.0		2.0	0.5	0.139
		1.0		2.0	0.5	0.292
		1.0		2.0	0.5	0.250
		2.0		8.0	0.125	0.0130
		2.0		11.66	0.125	0.00541
		2.0		16.0	0.125	0.00260
		2.0		11.66	0.125	0.00915
		2.0		11.66	0.125	0.00860
		2.0		16.0	0.125	0.00417
		2.0		16.0	0.125	0.00397

×   Free to warp   o   Warping torsion frictionless hinge

⌐   Warping prevented   •   Warping torsion plastic hinge

——   Top flange   ❮ ❯   Uniform torsion shear hinges

········   Bottom flange

Figure 10.21 Plastic collapse and maximum rotation – uniformly distributed torque.

member. The varying warping deflections induce longitudinal strains and stresses σ_w. When these warping normal stresses also vary along the member, there are associated warping shear stresses τ_w distributed around the cross-section, and these may act in conjunction with the shear stresses due to uniform torsion to resist the torque acting at the section [1, 11]. In the following sub-sections, the warping deformations, stresses, and torques in thin-walled open-section members of constant cross-section are treated. The local warping stresses induced by the non-uniform torsion of closed sections are beyond the scope of this book, but are treated in [9, 10], while some examples of the non-uniform torsion of tapered open-section members are discussed in [12, 13].

10.3.1.2 Warping deflections and stresses

In thin-walled open-section members, the longitudinal warping deflections u due to twist are much greater than those due to shear straining (see Figure 10.6), and the latter are usually neglected. The warping deflections due to twist arise because the lines originally parallel to the axis of twist become helical locally, as indicated in Figures 10.4, 10.22 and 10.23. In non-uniform torsion, the axis of twist is the locus of the shear centres (see Section 5.4.3), since otherwise the shear centre axis would be deformed, and the member would be bent as well as twisted. It is shown in Section 10.8.1 that the warping deflections due to twist (see Figure 10.23) can be expressed as

$$u = (\alpha_n - \alpha)\frac{d\phi}{dx}, \tag{10.34}$$

where

$$\alpha = \int_0^s \rho_0 ds, \tag{10.35}$$

and

$$\alpha_n = \frac{1}{A}\int_0^E \alpha t \, ds \tag{10.36}$$

where ρ_0 is the perpendicular distance from the shear centre S to the tangent to the mid-line of the section wall (see Figures 10.16b and 10.23).

In non-uniform torsion, the warping displacements u vary along the length of the member, as shown for example in Figure 10.1b and c. Because of this, longitudinal

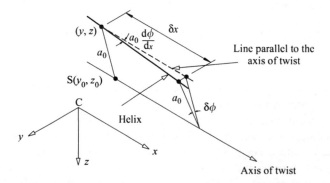

Figure 10.22 Rotation of a line parallel to the axis of twist.

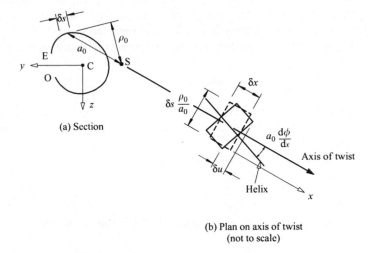

(a) Section

(b) Plan on axis of twist
(not to scale)

Figure 10.23 Warping of an element $\delta x \times \delta s \times t$ due to twisting.

strains are induced, and the corresponding longitudinal warping normal stresses are

$$\sigma_w = E(\alpha_n - \alpha)\frac{d^2\phi}{dx^2}.$$ (10.37)

These stresses define the bimoment stress resultant

$$B = -\int_0^E \sigma_w(\alpha_n - \alpha)\,t ds.$$ (10.38)

Substituting equation 10.37 leads to

$$B = -EI_w\frac{d^2\phi}{dx^2}$$ (10.39)

in which

$$I_w = \int_0^E (\alpha_n - \alpha)^2 t ds$$ (10.40)

is the warping section constant. Substituting equation 10.39 into equation 10.37 allows the warping normal stresses to be expressed as

$$\sigma_w = -\frac{B(\alpha_n - \alpha)}{I_w}.$$ (10.41)

Values of the warping section constant I_w and of the warping normal stresses σ_w for structural sections are given in [5, 6] while a general computer program for their evaluation has been developed [8].

When the warping normal stresses σ_w vary along the length of the member, warping shear stresses τ_w are induced, in the same way that variations in the bending normal stresses in a beam induce bending shear stresses (see Section 5.4). The warping shear stresses can be expressed in terms of the warping shear flow

$$\tau_w t = -E \frac{\mathrm{d}^3 \phi}{\mathrm{d}x^3} \int_0^s (\alpha_n - \alpha) \, t \mathrm{d}s. \tag{10.42}$$

Information for calculating the warping shear stresses in structural sections is also given in [5, 6], while a general computer program for their evaluation has been developed [8].

The warping shear stresses exert a warping torque

$$T_w = \int_0^E \rho_0 \tau_w t \mathrm{d}s \tag{10.43}$$

about the shear centre, as shown in Figure 10.12a. It can be shown [14, 15] that this reduces to

$$T_w = -EI_w \frac{\mathrm{d}^3 \phi}{\mathrm{d}x^3}, \tag{10.44}$$

after substituting for τ_w and integrating by parts. The warping torque T_w given by equation 10.44 is related to the bimoment B given by equation 10.39 through

$$T_w = \frac{\mathrm{d}B}{\mathrm{d}x}. \tag{10.45}$$

Sign conventions for T_w and B are illustrated in Figure 10.16.

A somewhat simpler explanation of the warping torque T_w and bimoment B can be given for the I-section illustrated in Figure 10.24. This is discussed in Section 10.8.2 where it is shown that for an equal flanged I-section, the bimoment B is related to the flange moment M_f through

$$B = d_f M_f \tag{10.46}$$

and the warping section constant I_w is given by

$$I_w = \frac{I_z d_f^2}{4} \tag{10.47}$$

in which d_f is the distance between the flange centroids.

10.3.2 Elastic analysis

10.3.2.1 Uniform torsion

Some thin-walled open-section members, such as thin rectangular sections and angle and tee sections, have very small warping section constants. In such cases it is usually sufficiently accurate to ignore the warping torque T_w, and to analyse the member as if it were in uniform torsion, as discussed in Section 10.2.2.

10.3.2.2   Warping torsion

Very thin-walled open-section members have very small torsion-section constants I_t, while some of these, such as I- and channel sections, have significant warping section constants I_w. In such cases, it is usually sufficiently accurate to analyse the member as if the applied torque were resisted entirely by the warping torque, so that $T_x = T_w$. Thus, the twisted shape of the member can be obtained by solving

$$-EI_w\frac{d^3\phi}{dx^3} = T_x, \tag{10.48}$$

as

$$-EI_w\phi = \int_0^x \int_0^x \int_0^x T_x dx dx dx + \frac{A_1 x^2}{2} + A_2 x + A_3, \tag{10.49}$$

where A_1, A_2, and A_3 are constants of integration whose values depend on the boundary conditions.

For statically determinate members, there are three boundary conditions from which the three constants of integration can be determined. As an example of the three commonly assumed boundary conditions, the statically determinate can-tilever shown in Figures 10.1c and 10.25 is analysed in Section 10.8.3. The twisted shape of the cantilever is shown in Figure 10.25c, and it can be seen that the maximum angle of twist rotation occurs at the loaded end and is equal to $TL^3/3EI_w$.

For statically indeterminate members such as the beam shown in Figure 10.26, there is an additional boundary condition for each redundant quantity involved. These redundants can be determined by substituting the additional boundary

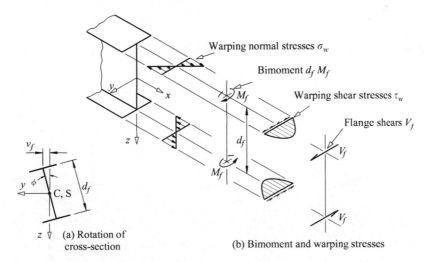

(a) Rotation of cross-section   (b) Bimoment and warping stresses

Figure 10.24 Bimoment and stresses in an I-section member.

conditions into equation 10.49. As an example of this procedure, the beam shown in Figure 10.26 is analysed in Section 10.8.4.

The warping torsion analysis of equal flanged I-sections can also be carried out by analysing the flange bending model shown in Figure 10.27. Thus the warping

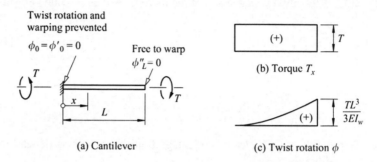

Figure 10.25 Warping torsion of a statically determinate cantilever.

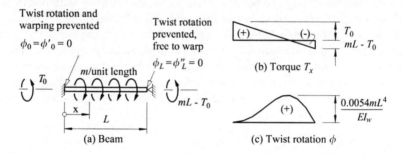

Figure 10.26 Warping torsion of a statically indeterminate beam.

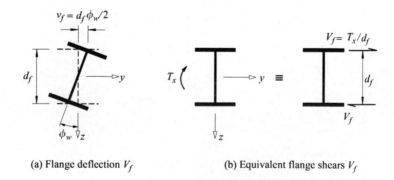

Figure 10.27 Flange bending model of warping torsion.

torque $T_w = T_x$ acting on the I-section is replaced by two transverse shears

$$V_f = T_x/d_f,$$ (10.50)

one on each flange. The bending deflections v_f of each flange can then be determined by elastic analysis or by using available solutions [4], and the twist rotations can be calculated from

$$\phi_w = 2v_f/d_f$$ (10.51)

10.3.2.3   Non-uniform torsion

When neither the torsional rigidity nor the warping rigidity can be neglected, as in hot-rolled steel I- and channel sections, the applied torque is resisted by a combination of the uniform and warping torques, so that

$$T_x = T_t + T_w.$$ (10.52)

In this case, the differential equation of non-uniform torsion is

$$T_x = GI_t \frac{d\phi}{dx} - EI_w \frac{d^3\phi}{dx^3},$$ (10.53)

the general solution of which can be written as

$$\phi = p(x) + A_1 e^{x/a} + A_2 e^{-x/a} + A_3,$$ (10.54)

where

$$a^2 = \frac{EI_w}{GI_t} = \frac{K^2 L^2}{\pi^2}.$$ (10.55)

The function $p(x)$ in this solution is a particular integral whose form depends on the variation of the torque T_x along the beam. This can be determined by the standard techniques used to solve ordinary linear differential equations (see [16] for example). The values of the constants of integration $A_1, A_2,$ and A_3 depend on the form of the particular integral and on the boundary conditions.

Complete solutions for a number of torque distributions and boundary conditions have been determined, and graphical solutions for the twist ϕ and its derivatives $d\phi/dx$, $d^2\phi/dx^2$, and $d^3\phi/dx^3$ are available [5, 6]. As an example, the non-uniform torsion of the cantilever shown in Figure 10.28 is analysed in Section 10.8.5.

Sometimes the accuracy of the method of solution described above is not required, in which case a much simpler method may be used. The maximum angle of twist rotation ϕ_m may be approximated by

$$\phi_m = \frac{\phi_{tm}\phi_{wm}}{\phi_{tm} + \phi_{wm}}$$ (10.56)

in which ϕ_{tm} is the maximum uniform torsion angle of twist rotation obtained by solving equation 10.8 with $T_t = T_x$, and ϕ_{wm} is the maximum warping torsion

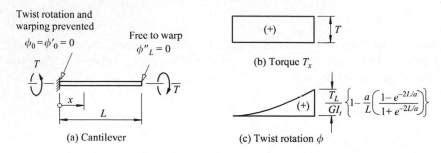

Figure 10.28 Non-uniform torsion of a cantilever.

angle of twist rotation obtained by solving equation 10.48. Values of ϕ_{tm} and ϕ_{wm} can be obtained from Figures 10.20 and 10.21 for a number of different torsion members.

This approximate method tends to overestimate the true maximum angle of twist rotation, partly because the maximum values ϕ_{tm} and ϕ_{wm} often occur at different locations along the member. In the case of the cantilever of Figure 10.28, the approximate solution

$$\phi_m = \frac{TL/GI_t}{1 + 3EI_w/GI_tL^2} \tag{10.57}$$

obtained using Figures 10.15c and 10.25c in equation 10.56 has a maximum error of about 12% [17].

10.3.3   Plastic collapse analysis

10.3.3.1   Fully plastic bimoment

The warping normal stresses σ_w developed in elastic warping torsion are usually much larger than the warping shear stresses τ_w. Thus the limit of elastic behaviour is often reached when the bimoment is close to its nominal first yield value for which the maximum value of σ_w is equal to the yield stress f_y.

For the equal flanged I-section shown in Figure 10.24, the first yield bimoment B_y corresponds to the flange moment M_f reaching the yield value

$$M_{fy} = f_y b_f^2 t_f /6 \tag{10.58}$$

so that

$$B_y = f_y d_f b_f^2 t_f /6 \tag{10.59}$$

in which b_f is the flange width, t_f is the flange thickness, and d_f is the distance between flange centroids. As the bimoment increases above B_y, yielding spreads from the flange tips into the cross-section, until the flanges become fully yielded at

$$M_{fp} = f_y b_f^2 t_f / 4 \qquad\qquad (10.60)$$

which corresponds to the fully plastic bimoment

$$B_p = f_y d_f b_f^2 t_f / 4 \qquad\qquad (10.61)$$

The plastic shape factor for warping torsion of the I-section is therefore given by $B_p / B_y = 3/2$.

Expressions for the full plastic bimoments of a number of monosymmetric thin-walled open sections are given in [18].

10.3.3.2   Plastic analysis of warping torsion

An equal flanged I-section member in warping torsion will collapse plastically when there is a sufficient number of warping hinges (frictionless or plastic) to transform the member into a mechanism, as shown for example in Figure 10.29. In general, the warping hinges develop progressively until the collapse mechanism forms.

A frictionless warping hinge occurs at a point where warping is unrestrained ($\phi'' = 0$) so that there is a frictionless hinge in each flange, while at a plastic warping hinge, the bimoment is equal to the fully plastic value B_p (equation 10.61) and the moment in each flange is equal to the fully plastic value M_{fp} (equation 10.60), so that there is a flexural plastic hinge in each flange. Warping torsion plastic collapse therefore corresponds to the simultaneous plastic collapse

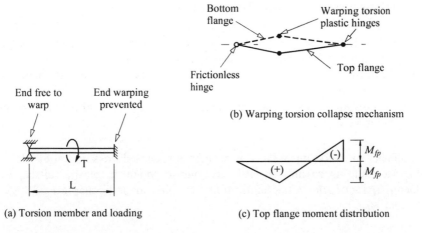

(a) Torsion member and loading

(b) Warping torsion collapse mechanism

(c) Top flange moment distribution

Figure 10.29 Warping torsion plastic collapse.

of the flanges in opposite directions, with a series of flexural hinges (frictionless or plastic) in each flange. Thus warping torsion plastic collapse can be analysed by using the methods discussed in Section 5.5.5 to analyse the flexural plastic collapse of the flanges.

Warping torsion plastic hinges often form at supports or at points of concentrated torque, as indicated in Figure 10.29. Examples of warping torsion plastic collapse mechanisms are shown in Figures 10.20 and 10.21.

At plastic collapse, each flange becomes a mechanism which is statically determinate, so that it can be analysed by using statics to determine the plastic collapse torques. For the member shown in Figure 10.29, each flange forms a mechanism with plastic hinges at the concentrated torque and the right-hand support and a frictionless hinge at the left-hand support where warping is unrestrained. If the concentrated torque $\alpha_w T$ acts at mid-length, then the flanges collapse when

$$\alpha_w(T/d_f)L/4 = M_{fp} + M_{fp}/2 \tag{10.62}$$

so that the warping torsion plastic collapse load factor is given by

$$\alpha_w = 6M_{fp}d_f/TL \tag{10.63}$$

Values of the warping torsion plastic collapse load factor α_w for a number of example-torsion members are given in Figures 10.20 and 10.21.

10.3.3.3   Plastic analysis of non-uniform torsion

It has not been possible to develop a simple but rigorous model for the analysis of the plastic collapse of members in non-uniform torsion where both uniform and warping torsion are important. This is because

(a) different types of stress (shear stresses τ_t, τ_w, and normal stress σ_w) are associated with uniform and warping torsion collapse,
(b) the different stresses τ_t, τ_w, and σ_w are distributed differently across the section, and
(c) the uniform torque T_t and the warping torque T_w are distributed differently along the member.

A number of approximate theories have been proposed the simplest of which is the Merchant approximation [19] according to which the plastic collapse load factor α_x is the sum of the uniform and warping torsion collapse load factors, so that

$$\alpha_x = \alpha_t + \alpha_w \tag{10.64}$$

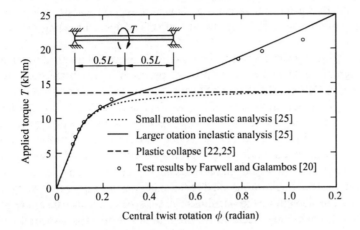

Figure 10.30 Comparison of analysis and test results.

This approximation assumes that there is no interaction between uniform and warping torsion at plastic collapse, so that the separate plastic collapse capacities are additive.

If strain-hardening is ignored and only small twist rotations are considered, then the errors arising from this approximation are on the unsafe side, but are small because the warping torsion shear stresses τ_w are small, because the yield interactions between normal and shear stresses (see Section 1.3.1) are small, and because cross-sections which are fully plastic due to warping torsion often occur at different locations along the member than those which are fully plastic due to uniform torsion. These small errors are more than compensated for by the conservatism of ignoring strain-hardening and second-order longitudinal stresses that develop at large rotations. Test results [20, 21] and numerical studies [22, 23] have shown that these cause significant strengthening at large rotations as indicated in Figure 10.30, and that the approximation of equation 10.64 is conservative.

An example of the plastic collapse analysis of the non-uniform torsion is given in Section 10.9.6.

10.4 Torsion design

10.4.1 General

EC3 gives limited guidance for the analysis and design of torsion members. While both elastic and plastic analysis are permitted generally, only accurate and approximate methods of elastic analysis are specifically discussed for torsion members. Also, while both first yield and plastic design resistances are referred to generally, only the first yield design resistance is specifically discussed for torsion members. Further, there is no guidance on section classification for torsion members, nor on how to allow for the effects of local buckling on the design resistance.

In the following sub-sections, the specific EC3 provisions for torsion analysis and design are extended so as to provide procedures which are consistent with those used in EC3 for the design of beams against bending and shear.

First the member cross-section is classified as Class 1, Class 2, Class 3, or Class 4 in much the same way as is a beam cross-section (see Sections 4.7.2 and 5.6.1.2), and then the effective section resistances for uniform and warping torsion are determined. Following this, an appropriate method of torsion analysis (plastic or elastic) is selected, and then an appropriate method (plastic, first hinge, first yield, or local buckling) is used for strength design. All of the strength design methods suggested ignore the warping shear stresses τ_w because these are generally small, and because they occur at different points in the cross-section and along the member than do the much more significant uniform torsion shear stresses τ_t and the warping bimoment normal stresses σ_w. Finally, a method of serviceability design is discussed.

10.4.2   Section classification

Cross-sections of torsion members need to be classified according to the extent by which local buckling effects may reduce their cross-section resistances. Cross-sections which are capable of reaching and maintaining plasticity while a torsion plastic hinge collapse mechanism develops may be called Class 1, as are the corresponding cross-sections of beams. Class 2 sections are capable of developing a first hinge, but inelastic local buckling may prevent the development of a plastic collapse mechanism. Class 3 sections are capable of reaching the nominal first yield before local buckling occurs, while Class 4 sections will buckle locally before the nominal first yield is reached.

For open cross-sections, the warping shear stresses τ_w are usually very small, and can be neglected without serious error. The uniform torsion shear stresses τ_t change sign and vary linearly across the wall thickness t, and so can be considered to have no effect on local buckling. The flange elastic and plastic warping normal stress distributions are similar to those due to bending in the plane of the flange, and so the section classification may be based on the same width–thickness limits. Thus the flange outstands of a Class 1 open section would satisfy

$$\lambda \leq 9 \tag{10.65}$$

in which λ is the element slenderness given by

$$\lambda = (c/t) \sqrt{(f_y/235)} . \tag{10.66}$$

Class 2 open sections would satisfy

$$9 < \lambda \le 10, \tag{10.67}$$

Class 3 open sections would satisfy

$$10 < \lambda \le 14, \tag{10.68}$$

and Class 4 open sections would satisfy

$$14 < \lambda. \tag{10.69}$$

For closed cross-sections, the only significant stresses that develop during torsion are uniform torsion shear stresses τ_t that are constant across the wall thickness t and along the length b of any element of the cross-section. This stress distribution is very similar to that caused by a bending shear force in the web of an equal-flanged I-section, and so the section classification may be based on the same width–thickness limits [24]. Thus the elements of Class 1, Class 2, and Class 3 rectangular hollow sections would satisfy

$$\lambda \le 60 \tag{10.70}$$

in which λ is the element slenderness given by

$$\lambda = (h_w/t) \sqrt{(f_y/235)}. \tag{10.71}$$

in which h_w is the clear distance between flanges. Rectangular hollow sections which do not satisfy this condition would be classified as Class 4. Circular hollow sections do not buckle under shear, unless they are exceptionally slender, and so they may generally be classified as Class 1.

10.4.3 Uniform torsion section resistance

The effective uniform torsion section resistance of all open sections $T_{t,Rd}$ can be approximated by the plastic section resistance T_{tp} given by equation 10.29 in which the shear yield stress is given by

$$\tau_y = f_y/\sqrt{3} \tag{10.72}$$

The uniform torsion section resistance of Class 1, Class 2, and Class 3 rectangular hollow sections $T_{t,Rd,123}$ can be determined by using the first yield uniform torque T_{ty} of equation 10.31. The effective uniform torsion section resistance of a Class 4 rectangular hollow section $T_{t,Rd,4}$ can be approximated by reducing the first yield uniform torque T_{ty} of equation 10.31 to

$$T_{t,Rd,4} = T_{ty}(60/\lambda) \tag{10.73}$$

The effective uniform torsion section resistance of a circular hollow section $T_{t,Rd}$ can generally be approximated by using equation 10.31 to find the first yield uniform torque T_{ty}.

10.4.4 Bimoment section resistance

The bimoment section resistance of Class 1 and Class 2 equal-flanged I-sections $B_{Rd,12}$ is given by equation 10.61. The bimoment section resistance of a Class 3 equal-flanged I-section can be approximated by using

$$B_{Rd,3} = B_y + (B_p - B_y)(14 - \lambda)/4 \qquad (10.74)$$

in which B_y is the first yield bimoment given by equation 10.59. The bimoment section resistance of a Class 4 equal-flanged I-section can be approximated by using

$$B_{Rd,4} = B_y (14/\lambda) \qquad (10.75)$$

These equations may also be used for the bimoment section resistances of other open sections by using their fully plastic bimoments B_p [18] or first yield bimoments B_y. Alternatively, the bimoment section resistances of other Class 1, Class 2, or Class 3 open sections can be conservatively approximated by using

$$B_{Rd,123} = B_y \qquad (10.76)$$

and the bimoment section resistances of other Class 4 open sections can be approximated using equation 10.75.

It is conservative to ignore bimoments in hollow section members, and so there is no need to consider their bimoment section resistances.

10.4.5 Plastic design

The use of plastic design should be limited to Class 1 members which have sufficient ductility to reach the plastic collapse mechanism. Plastic design should be carried out by using plastic analysis to determine the plastic collapse load factor α_x (see Section 10.3.3.3), and then checking that

$$1 \le \alpha_x \qquad (10.77)$$

10.4.6 First hinge design

The use of first hinge design should be limited to Class 1 and Class 2 members in which the first hinge of a collapse mechanism can form. The design uniform torques T_t and bimoments B can be determined by using elastic analysis, and the member is satisfactory if

$$\left(\frac{T_t}{T_{t,Rd,2}}\right)^2 + \left(\frac{B}{B_{Rd,2}}\right)^2 \le 1 \qquad (10.78)$$

is satisfied at all points along the member. The first hinge method is more conservative than plastic design, and more difficult to use because elastic member analysis is much more difficult than plastic analysis, and also because equation 10.78 often needs to be checked at a number of points along the member.

10.4.7 First yield design

The use of first yield design should be limited to Class 1, Class 2, and Class 3 members which can reach first yield. The maximum design uniform torques T_t and bimoments B can be determined by elastic analysis. The member is satisfactory if

$$\left(\frac{T_t}{T_{t,Rd,3}}\right)^2 + \left(\frac{B}{B_{Rd,3}}\right)^2 \le 1 \tag{10.79}$$

is satisfied at all points along the member. This first yield method is the same as that specified in EC3 except for the neglect of the warping shear stresses τ_w. It is just as difficult to use as the first hinge method.

10.4.8 Local buckling design

Class 4 members should designed against local buckling. The maximum design uniform torques T_t and bimoments B can be determined by elastic analysis. The member is satisfactory if

$$\left(\frac{T_t}{T_{t,Rd,4}}\right)^2 + \left(\frac{B}{B_{Rd,4}}\right)^2 \le 1 \tag{10.80}$$

is satisfied at all points along the member.

10.4.9 Design for serviceability

Serviceability design usually requires the estimation of the deformations under part or all of the serviceability loads. Because these loads are usually significantly less than the factored combined loads used for strength design, the member remains largely elastic, and the serviceability deformations can be evaluated by linear elastic analysis.

There are no specific serviceability criteria for satisfactory deformations of a torsion member given in EC3. However, general recommendations such as 'the deformations of a structure and of its component members should be appropriate to the location, loading and function of the structure and the component members' may be applied to torsion members, so that the onus is placed on the designer to determine what are appropriate magnitudes of the twist rotations.

Because of the lack of precise serviceability criteria, highly accurate predictions of the serviceability rotations are not required. Thus the maximum twist rotations of members in non-uniform torsion may be approximated by using equation 10.56.

Usually, the twist rotations of thin-walled closed-section members are very small, and can be ignored (although in some cases, the distortional deformations may be quite large, as discussed in Section 10.6). On the other hand, thin-walled open-section members are comparatively flexible, and special measures may be required to limit their rotations.

10.5 Torsion and bending

10.5.1 Behaviour

Pure torsion occurs very rarely in steel structures. Most commonly, torsion occurs in combination with bending actions. The torsion actions may be classified as primary or secondary, depending on whether the torsion action is required to transfer load (primary torsion), or whether it arises as a secondary action. Secondary torques may arise as a result of differential twist rotations compatible with the joint rotations of primary frames, as shown in Figure 10.31, and are often predicted by three-dimensional analysis programs. They are not unlike the secondary bending moments which occur in rigid-jointed trusses, but which are usually ignored (a procedure justified by many years of satisfactory experience based on the long-standing practice of analysing rigid-jointed trusses as if pin-jointed). Secondary torques are usually small when there are alternative load paths of high stiffness, and may often be ignored.

Primary torsion actions may be classified as being restrained, free, or destabilising, as shown in Figure 10.32. For restrained torsion, the member applying the torsion action (such as ABC shown in Figure 10.32b) also applies a restraining action to the member resisting the torsion (DEC in Figure 10.32b). In this case, the structure is redundant, and compatibility between the members must be satisfied in the analysis if the magnitudes of the torques and other actions are to be determined correctly. Free torsion occurs when the member applying the torsion action (such as

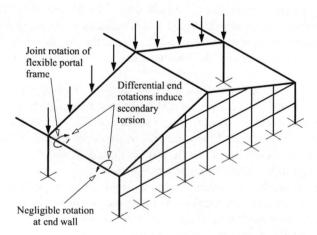

Joint rotation of
flexible portal
frame

Differential end
rotations induce
secondary
torsion

Negligible rotation
at end wall

Figure 10.31 Secondary torsion in an industrial frame.

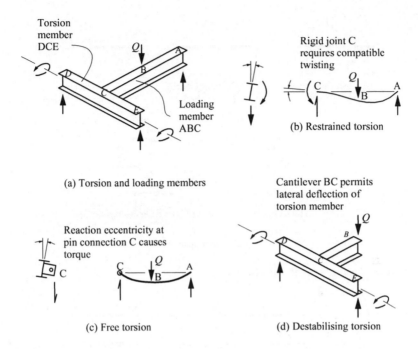

(a) Torsion and loading members

(b) Restrained torsion

(c) Free torsion

(d) Destabilising torsion

Figure 10.32 Torsion actions.

ABC in Figure 10.32c) does not restrain the twisting of the torsion member (DEF), but does prevent its lateral deflection. Destablising torsion may occur when the member applying the torsion action (such as BC in Figure 10.32d) does not restrain either the twisting or the lateral deflection of the torsion member (DCE). In this case, lateral buckling actions (Chapter 6) caused by the in-plane loading of the torsion member amplify the torsion and out-of-plane bending behaviour.

The inelastic non-linear bending and torsion of fully braced, centrally braced, and unbraced I-beams with central concentrated loads (see Figure 10.33) have been analysed [25], and interaction equations developed for predicting their strengths. It was found that while circular interaction equations are appropriate for short length braced beams, these provide unsafe predictions for beams subject to destablising torsion, where lateral buckling effects become important. On the other hand, destablising interactions between lateral buckling and torsion tend to be masked by the favourable effects of the secondary axial stresses that develop at large rotations, and linear interaction equations based on plastic analyses provide satisfactory strength predictions, as shown in Figure 10.34. Proposals based on these findings are made below for the analysis and design of members subject to combined torsion and bending.

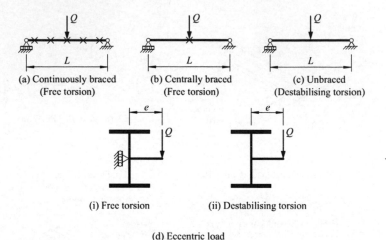

Figure 10.33 Beams in torsion and bending.

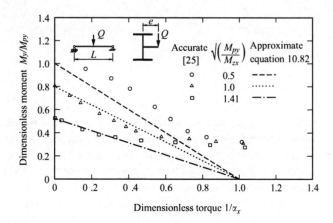

Figure 10.34 Interaction between lateral buckling and torsion for unbraced beams.

10.5.2 Plastic design of fully braced members

The member must be Class 1. Plastic collapse analyses should be used to determine the plastic collapse load factors α_i for in-plane bending (Section 5.5) and α_x for torsion (Section 10.3.3.3).

The member should satisfy the circular interaction equation

$$1/\alpha_i^2 + 1/\alpha_x^2 \leq 1 \tag{10.81}$$

10.5.3   Design of members without full bracing

A member without full bracing has its bending resistance reduced by lateral buckling effects (Chapter 6). Its bending should be analysed elastically, and the resulting moment distribution used to determine its bending resistance $M_{b,Rd}$ according to EC3.

An elastic torsion analysis should be used to find the values of the design uniform torques T_t and bimoments B. An appropriate torsion design method should be selected from Sections 10.4.5–10.4.8 and used to find the load factor α_x by which the design torques and bimoments should be multiplied so that the appropriate torsion design equation (equations 10.77–10.80) is just satisfied.

The member should then satisfy the linear interaction equation

$$\frac{M_{Ed}}{M_{b,Rd}} + \frac{1}{\alpha_x} \leq 1 \tag{10.82}$$

There is no need to amplify the moment because of the finding [25] that linear interaction equations are satisfactory, even for unbraced beams, as shown in Figure 10.34.

10.6   Distortion

Twisting and distortion of a flexural member may be caused by the local distribution around the cross-section of the forces acting, as shown in Figure 10.3. If the member responds significantly to either of these actions, the bending stress distribution may depart considerably from that calculated in the usual way, while additional distortional stresses may be induced by out-of-plane bending of the wall (see Figure 10.3d). To avoid possible failure, the designer must either increase the strength of the member, which may be uneconomic, or must limit both twisting and distortion.

The resistance to distortion of a thin-walled member depends on the arrangement of the cross-section. Members with triangular closed cells have high resistances because of the truss-like action of the triangulated cross-section which causes the distortional loads to be transferred by in-plane bending and shear of the cell walls. On the other hand, the walls of the members with rectangular or trapezoidal cells bend out of their planes when under distortional loading and when the resistances to distortion are low [26], while members of open cross-section are even more more flexible. Concentrated distortional loads can be distributed locally by providing stiff diaphragms which reduce or prevent local distortion. However, isolated diaphragms are not fully effective when the distortional loads are distributed or can move along the member. In such cases, it is necessary to analyse the distortion of the section and the stresses induced by the distortion.

The resistance to torsion of a thin-walled open section is comparatively small, and the dominant mode of deformation is twisting of the member. Because this twisting is usually accounted for, the importance of distortion lies in its modifying influence on the torsional behaviour. Significant distortions occur only in very thin-walled members, for which the distortional rigidity, which varies as the cube of the wall thickness, reduces at a much faster rate than the warping rigidity, which varies directly with the thickness [27]. These distortions may induce significant wall bending stresses, and may increase the angles of twist and change the warping stress distributions [28, 29].

Thin-walled closed-section members are very stiff torsionally, and the twisting deformations due to uniform torsion are not usually of great importance except in curved members. On the other hand, rectangular and trapezoidal members are not very resistant to distortional loads, and the distortional deformations may be large, while the stresses induced by the distortional loads may dominate the design [26].

The distortional behaviour of single-cell rectangular or trapezoidal members can be classified as uniform or non-uniform. Uniform distortion occurs when the applied distortional loading is uniformly distributed along a member of constant cross-section which has no diaphragms. In this case the distortional loading is resisted solely by the out-of-plane bending rigidity of the plate elements of the section. In non-uniform distortion, the distortions vary along the member, and the plate elements bend in their planes, producing in-plane shear stresses which help to resist the distortional loading.

Perhaps the simplest method of analysing non-uniform distortion is by using an analogy to the problem [30] of a beam on an elastic foundation (BEF analogy). In this analogy, which is shown diagrammatically in Figure 10.35, the distortional

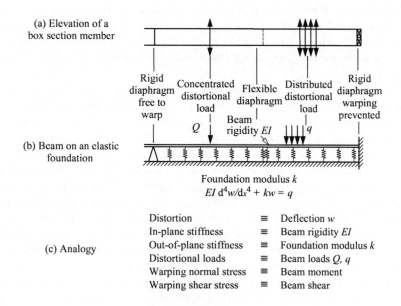

Figure 10.35 BEF analogy for box section distortion.

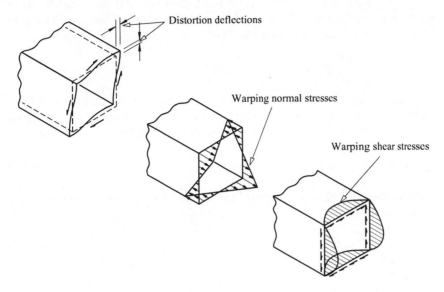

Distortion deflections

Warping normal stresses

Warping shear stresses

Figure 10.36 Warping stresses in non-uniform distortion.

loading corresponds to the applied beam loads in the analogy, and the out-of-plane resistance of the plate elements corresponds to the stiffness k of the elastic foundation, while the in-plane resistance of these elements corresponds to the flexural rigidity EI of the analogous beam. The BEF analogy can account for concentrated and distributed distortional loads, for rigid diaphragms which either permit or prevent warping, and for flexible diaphragms and stiffening rings, as indicated in Figure 10.35. Many solutions for problems of beams on elastic foundations have been obtained [31], and these can be used to find the out-of-plane bending stresses in the plate elements (which are related to the foundation reactions), and the warping and normal shear stresses due to non-uniform distortion (which are related to the bending moment and shear force in the beam on the elastic foundation). The distributions of the warping stresses in a rectangular box section are shown in Figure 10.36.

10.7   Appendix – uniform torsion

10.7.1   Thin-walled closed sections

The shear flow $\tau_t t$ around a thin-walled closed-section member in uniform torsion can be determined by using the condition that the total change in the warping displacement u around the complete closed section must be zero for continuity. The change in the warping displacement due to the twist rotations ϕ is equal to

$-(\mathrm{d}\phi/\mathrm{d}x) \oint \rho_0 \mathrm{d}s$ (see Section 10.2.1.5 and Figure 10.13a), while the change in the warping displacement due to the shear straining of the mid-surface is equal to $\oint (\tau_t/G)\mathrm{d}s$ (see Figures 10.6 and 10.13b). Thus

$$\oint \frac{\tau_t}{G}\mathrm{d}s = \frac{\mathrm{d}\phi}{\mathrm{d}x} \oint \rho_0 \mathrm{d}s.$$

The shear flow $\tau_t t$ is constant around the closed section (otherwise the corresponding longitudinal shear stresses would not be in equilibrium), while $\oint \rho_0 \mathrm{d}s$ is equal to twice the area A_e enclosed by the section, as shown in Figure 10.12b. Thus,

$$\tau_t t \oint \frac{1}{t}\mathrm{d}s = 2GA_e \frac{\mathrm{d}\phi}{\mathrm{d}x}. \tag{10.83}$$

The moment resultant of the shear flow around the section is equal to the torque acting, and so

$$T_t = \oint \tau_t t \rho_0 \mathrm{d}s \tag{10.84}$$

whence

$$\tau_t t = \frac{T_t}{2A_e}. \tag{10.20}$$

Substituting this into equation 10.83 leads to

$$T_t = G\frac{4A_e^2}{\oint (1/t)\mathrm{d}s}\frac{\mathrm{d}\phi}{\mathrm{d}x},$$

and so the torsion section constant is

$$I_t = \frac{4A_e^2}{\oint (1/t)\mathrm{d}s}. \tag{10.21}$$

The torsion section constants and the shear flows in multi-cell closed-section members can be determined by extending this method. If the member consists of n junctions linked by m walls, then there are $(m - n + 1)$ independent cells. The longitudinal equilibrium conditions require there to be m constant shear flows $\tau_t t$, one for each wall, and $(n - 1)$ junction equations of the type

$$\sum_{\text{junction}} (\tau_t t) = 0 \tag{10.85}$$

There are $(m - n + 1)$ independent cell warping continuity conditions of the type

$$\sum_{\text{cell}} \int_{\text{wall}} (\tau_t/G)\mathrm{d}s = 2A_e \frac{\mathrm{d}\phi}{\mathrm{d}x} \tag{10.86}$$

where A_e is the area enclosed by the cell. Equations 10.85 and 10.86 can be solved simultaneously for the m shear flows $\tau_t t$. The total uniform torque T_t can be determined as the torque resultant of these shear flows, and the torsion section constant I_t can then be found from

$$I_t = \frac{T_t}{G\mathrm{d}\phi/\mathrm{d}x}.$$

10.7.2 Analysis of statically indeterminate members

The uniform torsion of the statically indeterminate beam shown in Figure 10.17a may be analysed by taking the left-hand reaction torque T_0 as the redundant quantity. The distribution of torque T_x is therefore as shown in Figure 10.17b, and equation 10.8 becomes

$$GI_t\frac{\mathrm{d}\phi}{\mathrm{d}x} = T_0 - T\langle x - a\rangle^0,$$

where the value of the second term is taken as zero when the value inside the Macaulay brackets $\langle\ \rangle$ is negative.

The solution of this equation which satisfies the condition that the angle of twist at the left-hand support is zero is

$$GI_t\phi = T_0x - T\langle x - a\rangle. \tag{10.87}$$

The other condition which must be satisfied is that the angle of twist at the right-hand support must be zero. Using this in equation 10.87,

$$0 = T_0L - T(L - a),$$

and so

$$T_0 = \frac{T(L - a)}{L}.$$

Thus the maximum torque is T_0 when a is less than $L/2$, while the maximum angle of twist rotation occurs at the loaded point, and is equal to

$$\phi_a = \frac{Ta(L - a)}{GI_tL}.$$

10.8 Appendix – non-uniform torsion

10.8.1 Warping deflections

When a member twists, lines originally parallel to the axis of twist become helical, and any element $\delta x \times \delta s \times t$ of the section wall rotates $a_0\mathrm{d}\phi/\mathrm{d}x$ about the line a_0

from the shear centre S, as shown in Figures 10.4 and 10.22. The increase δu in the warping displacement of the element is shown in Figure 10.23, and is equal to

$$\delta u = -\rho_0 \frac{d\phi}{dx} \delta s$$

where ρ_0 is the perpendicular distance from the shear centre S to the tangent to the mid-line of the wall. The sign convention for ρ_0 is demonstrated in Figure 10.16b, which shows a cross-section viewed by looking in the positive x direction. The value of ρ_0 is positive when the line a_0 from the shear centre rotates in a clockwise sense when the distance s increases, which is the case for Figure 10.16b. Thus the warping displacement is given by

$$u = (\alpha_n - \alpha)\frac{d\phi}{dx}, \tag{10.34}$$

where

$$\alpha = \int_0^s \rho_0 ds, \tag{10.35}$$

and the quantity α_n is chosen to be

$$\alpha_n = \frac{1}{A} \int_0^E \alpha t ds, \tag{10.36}$$

so that the average warping displacement of the section is zero.

10.8.2   Warping torque and bimoment in an I-section

When an equal flanged I-section member twists ϕ as shown in Figure 10.24a, the flanges deflect laterally

$$v_f = \frac{d_f}{2}\phi.$$

Thus the flanges may have curvatures

$$\frac{d^2 v_f}{dx^2} = \frac{d_f}{2}\frac{d^2\phi}{dx^2}$$

in opposite directions as for example in the cantilever shown in Figure 10.1c. These curvatures can be thought of as being produced by the equal and opposite moments

$$I_f \frac{d^2 v_f}{dx^2} = \frac{EI_z}{2} \frac{d_f}{2} \frac{d^2 \phi}{dx^2}$$

shown in Figure 10.24b. This pair of equal and opposite flange moments M_f, spaced d_f apart, form the bimoment

$$B = d_f M_f. \tag{10.46}$$

The bending stresses induced by the bimoment are the warping normal stresses

$$\sigma_w = \mp \frac{M_f y}{I_f} = \mp E \frac{d_f}{2} y \frac{d^2 \phi}{dx^2}$$

shown in Figure 10.24b. These can also be obtained from equation 10.37.

When the flange moments M_f vary along the member, they induce equal and opposite flange shears

$$V_f = -\frac{dM_f}{dx} = -\frac{EI_z}{2} \frac{d_f}{2} \frac{d^3 \phi}{dx^3},$$

as shown in Figure 10.24b. The warping shear stresses

$$\tau_w = \pm \frac{V_f}{b_f t_f} \left(1.5 - \frac{6y^2}{b_f^2} \right)$$

(in which b_f is the flange width) which are statically equivalent to these shear forces are also shown in Figure 10.24b. They can also be obtained from equation 10.42.

The equal and opposite flange shears V_f are statically equivalent to the warping torque

$$T_w = V_f d_f = -\frac{EI_z d_f^2}{4} \frac{d^3 \phi}{dx^3}.$$

If this is compared with equation 10.44, it can be deduced that the warping section constant for the I-section is

$$I_w = \frac{I_z d_f^2}{4}. \tag{10.47}$$

This result can also be obtained from equation 10.40.

10.8.3 Warping torsion analysis of statically determinate members

The twisted shape of a statically determinate member in warping tors~~ion~~ by equation 10.49. For example, the cantilever shown in Figure 10.25a, T_x is constant and equal to the applied torque T, and so the twisted shape is gi~~ven~~ by

$$-EI_w\phi = \frac{Tx^3}{6} + \frac{A_1 x^2}{2} + A_2 x + A_3. \tag{10.88}$$

The constants of integration A_1, A_2, and A_3 depend on the boundary conditions. At the support, it is assumed that twisting is prevented, so that

$$\phi_0 = 0$$

and that warping is also prevented, so that (see equation 10.34)

$$\left(\frac{d\phi}{dx}\right)_0 = 0.$$

At the loaded end, the member is free to warp (i.e. the warping normal stresses σ_w are zero), so that (see equation 10.37)

$$\left(\frac{d^2\phi}{dx^2}\right)_L = 0.$$

By substituting these conditions into equation 10.88, the constants of integration can be determined as

$$A_1 = -TL,$$
$$A_2 = 0,$$
$$A_3 = 0.$$

If these are substituted into equation 10.88, the complete solution for the twisted shape is obtained as

$$-EI_w\phi = \frac{Tx^3}{6} - \frac{TLx^2}{2}.$$

The maximum angle of twist occurs at the loaded end and is equal to

$$\phi_L = \frac{TL^3}{3EI_w}.$$

The warping shear stress distribution (see equation 10.42) is constant along the member, but the maximum warping normal stress (see equation 10.37) occurs at the support where $d^2\phi/dx^2$ reaches its highest value.

10.8.4 Warping torsion analysis of statically indeterminate members

The warping torsion of the statically indeterminate beam shown in Figure 10.26a may be analysed by taking the left-hand reaction torque T_0 as the redundant. The distribution of torque T_x is then given by (see Figure 10.26b)

$$T_x = T_0 - mx$$

and equation 10.49 becomes

$$-EI_w\phi = \frac{T_0 x^3}{6} - \frac{mx^4}{24} + \frac{A_1 x^2}{2} + A_2 x + A_3.$$

After using the boundary conditions $\phi_0 = (d\phi/dx)_0 = 0$, the constants A_2 and A_3 are determined as

$$A_2 = A_3 = 0,$$

while the condition $\phi_L = 0$ requires that

$$A_1 = -\frac{T_0 L}{3} + \frac{mL^2}{12}.$$

The additional boundary condition is $(d^2\phi/dx^2)_L = 0$, and so

$$T_0 = 5mL/8.$$

Thus the complete solution is

$$-EI_w\phi = m\left(\frac{-x^4}{24} + \frac{5Lx^3}{48} - \frac{L^2 x^2}{16}\right).$$

The maximum angle of twist occurs near $x = 0.58L$, and is approximately equal to $0.0054mL^4/EI_w$. The warping shear stresses are greatest at the left-hand support, where the torque T_x has its greatest value of $5mL/8$. The warping normal stresses are greatest at $x = 5L/8$, where $-EI_w d^2\phi/dx^2$ has its maximum value of $9mL^2/128$.

10.8.5 Non-uniform torsion analysis

The twisted shape of a member in non-uniform torsion is given by equation 10.54. For the example of the cantilever shown in Figure 10.28a, the torque T_x is constant and equal to the applied torque T. In this case, a particular integral is

$$p(x) = \frac{Tx}{GI_t},$$

while the boundary conditions require that

$$\left. \begin{array}{l} 0 = A_1 + A_2 + A_3 \\[2mm] 0 = \dfrac{T}{GI_t} + \dfrac{A_1}{a} - \dfrac{A_2}{a} \\[3mm] 0 = \dfrac{A_1}{a^2}e^{L/a} + \dfrac{A_2}{a^2}e^{-L/a} \end{array} \right\},$$

so that

$$\phi = \frac{Tx}{GI_t} - \frac{Ta}{GI_t(1 + e^{-2L/a})}\left[e^{(x-2L)/a} - e^{-x/a} + 1 - e^{-2L/a}\right].$$

At the fixed end, the uniform torque is zero because $d\phi/dx = 0$. Because of this, the torque T_x there is resisted solely by the warping torque, and the maximum warping shear stresses occur at this point. The value of $d^2\phi/dx^2$ is greatest at the fixed end, and therefore the value of the warping normal stress is also greatest there. The value of $d\phi/dx$ is greatest at the loaded end of the cantilever, and therefore the value of the shear stress due to uniform torsion is also greatest there. The warping torque at the loaded end is

$$T_{wL} = T\frac{2e^{-L/a}}{1 + e^{-2L/a}},$$

which decreases from T to zero as L/a increases from 0 to ∞. Thus the warping shear stresses at the loaded end are not zero, but are less than those at the fixed end.

10.9 Worked examples

10.9.1 Example 1 – approximations for the serviceability twist rotation

Problem. The cantilever shown in Figure 10.28 is 5 m long and has the properties shown in Figure 10.37. If the serviceability end torque T is 3 kNm, determine approximate values of the serviceability end twist rotation either by

(a) assuming that the cantilever is in uniform torsion ($EI_w \equiv 0$), or
(b) assuming that the cantilever is in warping torsion ($GI_t \equiv 0$), or
(c) using the approximation of equation 10.56.

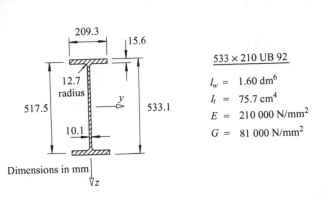

209.3

15.6

12.7
radius

517.5

y

533.1

10.1

Dimensions in mm

z

533 × 210 UB 92

$I_w = 1.60 \text{ dm}^6$

$I_t = 75.7 \text{ cm}^4$

$E = 210\,000 \text{ N/mm}^2$

$G = 81\,000 \text{ N/mm}^2$

Figure 10.37 Examples 1–7.

Solution.

(a) If $EI_w = 0$, then the end twist rotation (see Section 10.2.2.1) is

$$\phi_{um} = \frac{TL}{GI_t} = \frac{3 \times 10^6 \times 5000}{81\,000 \times 75.7 \times 10^4} = 0.245 \text{ radians} = 14.0°.$$

(b) If $GI_t = 0$, then the end twist rotation is (see Section 10.3.2.2)

$$\phi_{wm} = TL^3/3EI_w = \frac{3 \times 10^6 \times 5000^3}{3 \times 210\,000 \times 1.6 \times 10^{12}} = 0.372 \text{ radians} = 21.3°.$$

(c) Using the results of (a) and (b) above in the approximation of equation 10.56

$$\phi_m = \frac{0.245 \times 0.372}{0.245 + 0.372} = 0.148 \text{ radians} = 8.5°$$

10.9.2   Example 2 – serviceability twist rotation

Problem. Use the solution obtained in Section 10.8.5 to determine an accurate value of the serviceability end twist rotation of the cantilever of example 1.

Solution. Using equation 10.51,

$$a = \sqrt{\frac{EI_w}{GI_t}} = \sqrt{\left(\frac{210\,000 \times 1.6 \times 10^{12}}{81\,000 \times 75.7 \times 10^4}\right)} = 2341 \text{ mm}.$$

Therefore $e^{-2L/a} = e^{-2 \times 5000/2341} = 0.01396$

Using Section 10.8.5,

$$\phi_L = \frac{TL}{GI_t}\left[1 - \frac{a}{L}\frac{(1 - e^{-2L/a})}{(1 + e^{-2L/a})}\right]$$

$$= \frac{3\times10^6\times5000}{81\,000\times75.7\times10^4}\left[1 - \frac{2341}{5000}\frac{(1 - 0.01396)}{(1 + 0.01396)}\right] = 0.133 \text{ radians} = 7.6°.$$

which is about 10% less than the approximate value of 8.5° obtained in Section 10.9.1c.

10.9.3 Example 3 – elastic uniform torsion shear stress

Problem. For the cantilever of example 2, determine the maximum elastic uniform torsion shear stress caused by a strength design end torque of 5 kNm.

Solution. An expression for the nominal maximum elastic uniform torsion shear stress can be obtained by combining equations 10.8 and 10.18, whence

$$\tau_{t,max} \approx G\left(\frac{d\phi}{dx}\right)_{max} t_{max}.$$

An expression for $d\phi/dx$ can be obtained from the solution for ϕ in Section 10.8.5, whence

$$\frac{d\phi}{dx} = \frac{T}{GI_t}\left\{1 - \frac{[e^{(x-2L)/a} + e^{-x/a}]}{(1 + e^{-2L/a})}\right\}.$$

The maximum value of this occurs at the free end $x = L$ and is given by

$$\left(\frac{d\phi}{dx}\right)_{max} = \frac{T}{GI_t}\left[1 - \frac{2e^{-L/a}}{(1 + e^{-2L/a})}\right].$$

Adapting the solution of example 2,

$$\left(\frac{d\phi}{dx}\right)_{max} = \frac{T}{GI_t}\left[1 - \frac{2\times\sqrt{(0.01396)}}{1.01396}\right] = 0.767\frac{T}{GI_t}.$$

Thus

$$\tau_{t,max} \approx 0.767\frac{Tt_{max}}{I_t} \approx \frac{0.767\times5\times10^6\times15.6}{75.7\times10^4} \text{ N/mm}^2 \approx 79.0 \text{ N/mm}^2.$$

This nominal maximum shear stress occurs at the centre of the long edge of the flange. A similar value may be calculated for the opposite edge, but the actual stresses near this point are increased because of the re-entrant corner at the junction of the web and flange. An approximate estimate of the stress concentration factor can be made by using Figure 10.7b with

$$\frac{r}{t} = \frac{12.7}{15.6} = 0.81.$$

Thus the actual local maximum stress is approximately $\tau_{t,max} = 1.7 \times 79.0 = 134 \text{ N/mm}^2$.

10.9.4   Example 4 – elastic warping shear and normal stresses

Problem. For the cantilever of example 3, determine the maximum elastic warping shear and normal stresses.

Solution. The maximum warping shear stress τ_w occurs at the fixed end $(x = 0)$ of the cantilever (see Section 10.8.5) where the uniform torque is zero. Thus the total torque there is resisted only by the warping torque, and so

$$V_f = T/d_f,$$

where V_f is the flange shear (see Section 10.8.2). The shear stresses in the flange vary parabolically (see Figure 10.24b), and so

$$\tau_{w,max} = \frac{1.5\, V_f}{b_f t_f}.$$

Thus

$$\tau_{w,max} = \frac{1.5 \times (5 \times 10^6)/517.5}{209.3 \times 15.6} \text{ N/mm}^2 = 4.4 \text{ N/mm}^2.$$

The maximum warping normal stress σ_w occurs where $d^2\phi/dx^2$ is a maximum (see equation 10.37). An expression for $d^2\phi/dx^2$ can be obtained from the solution in Section 10.8.5 for ϕ,

whence
$$\frac{d^2\phi}{dx^2} = \frac{T}{aGI_t}\left[\frac{(-e^{(x-2L)/a} + e^{-x/a})}{(1 + e^{-2L/a})}\right].$$

The maximum value of this occurs at the fixed end $x = 0$ and is given by

$$\left(\frac{d^2\phi}{dx^2}\right)_{max} = \frac{T}{aGI_t}\frac{(1 - e^{-2L/a})}{(1 + e^{-2L/a})}$$

The corresponding maximum warping normal stress (see Section 10.8.2) is

$$\sigma_{w,max} = \frac{E d_f b_f}{2 \times 2} \left(\frac{d^2\phi}{dx^2} \right)_{max}$$

$$= \frac{210\,000 \times 517.5 \times 209.3 \times 5 \times 10^6}{2 \times 2 \times 2341 \times 81\,000 \times 75.7 \times 10^4} \times \frac{(1 - 0.01396)}{(1 + 0.01396)} \text{ N/mm}^2$$

$$= 192.6 \text{ N/mm}^2.$$

10.9.5　Example 5 – first yield design torque capacity

Problem. If the cantilever of example 2 has a yield stress of 275 N/mm^2, then use Section 10.4.7 to determine the torque resistance which is consistent with EC3 and a first yield strength criterion.

Solution. In this case, the maximum uniform torsion shear and warping torsion shear and normal stresses occur either at different points along the cantilever, or else at different points around the cross-section. Because of this, each stress may be considered separately.

For first yield in uniform torsion, equation 10.79 is equivalent to

$$\tau_t / \tau_y \le 1$$

in which $\tau_y \approx 275/\sqrt{3} = 159$ N/mm^2. Thus

$$\tau_{t,max} \le 159 \text{ N/mm}^2.$$

Using Section 10.9.3, a design torque of $T = 5$ kNm causes a maximum shear stress of $\tau_{t,max} = 79.0$ N/mm^2, if the stress concentration at the flange–web-junction is ignored. Thus the design uniform torsion torque resistance is

$$T_t = 5 \times 159/79.0 = 10.05 \text{ kNm}.$$

The maximum warping shear stress $\tau_{w,max}$ caused by a design torque of $T = 5$ kNm (see Section 10.9.4) is 4.4 N/mm^2, which is much less than the corresponding maximum uniform torsion shear stress $\tau_{t,max} = 79.0$ N/mm^2. Because of this, the design torque resistance is not limited by the warping shear stress.

For first yield in warping torsion, equation 10.79 reduces to

$$\sigma_w / f_y \le 1$$

Thus

$$\sigma_{w,max} \le 275 \text{ N/mm}^2.$$

Using Section 10.9.4, a design torque of $T = 5$ kNm causes a design warping normal stress of $\sigma_{w,max} = 192.6$ N/mm^2. Thus the design warping torsion torque is

$$T_W = 5 \times 275/192.6 = 7.14 \text{ kNm}.$$

This is less than the value of 10.05 kNm determined for uniform torsion, and so the design torque resistance is 7.14 kNm.

10.9.6 Example 6 – plastic collapse analysis

Problem. Determine the plastic collapse torque of the cantilever of example 5.

Uniform torsion plastic collapse.
  Using equation 10.29,

$$T_{tp} = (2 \times 209.3 \times 15.6^2/2 + (533.1 - 2 \times 15.6) \times 10.1^2/2)$$
$$\times 159 \text{ Nmm} = 12.15 \text{ kNm}$$

Using Figure 10.20, $\alpha_t T = T_{tp} = 12.15$ kNm.

Warping torsion plastic collapse.
Using equation 10.60, $M_{fp} = 275 \times 209.3^2 \times 15.6/4$ Nmm $= 46.98$ kNm.
Using Figure 10.20, $\alpha_w T = M_{fp}d_f/L = 46.98 \times 517.5/5000 = 4.86$ kNm.

Non-uniform torsion plastic collapse.
Using equation 10.64, $\alpha_x T = 12.15 + 4.86 = 17.01$ kNm.

10.9.7 Example 7 – plastic design

Problem. Determine the plastic design torque capacity of the cantilever of example 5.

Section classification. Using equation 10.65,

$$t_f = 15.6 \text{ mm}, \quad f_y = 275 \text{ N/mm}^2$$
$$\varepsilon = \sqrt{(235/275)} = 0.924$$
$$c_f/(t_f\varepsilon) = (209.3/2 - 10.1/2 - 12.7)/(15.6 \times 0.924) = 6.0 < 9$$

and so the section is Class 1, and plastic design can be used.

Plastic design torque resistance.

Using the result of section 10.9.6, $T = 17.01$ kNm,
which is significantly higher than the first yield design torque resistance of 7.14 kNm determined in Section 10.9.5.

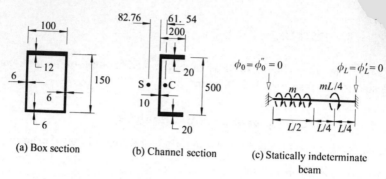

(a) Box section

(b) Channel section

(c) Statically indeterminate beam

Figure 10.38 Examples 8–13.

10.10 Unworked examples

10.10.1 Example 8 – uniform torsion of a box section

Determine the torsion section constant I_t for the box section shown in Figure 10.38a, and find the maximum elastic shear stress caused by a uniform torque of 10 kNm.

10.10.2 Example 9 – warping torsion section constant of a channel section

Determine the warping torsion section constant I_w for the channel section shown in Figure 10.38b.

10.10.3 Example 10 – approximation for the maximum twist rotation

The steel beam shown in Figure 10.38c is 5 m long, has the cross-section shown in Figure 10.38b and a torque per unit length of $m = 1000$ Nm/m. Determine approximate values of the maximum angle of twist rotation by assuming

(a) that the beam is in uniform torsion $(EI_w \equiv 0)$, or
(b) that the beam is in warping torsion $(GI_t \equiv 0)$, or
(c) the approximation of equation 10.56.

10.10.4 Example 11 – accurate twist rotation

If the steel beam shown in Figure 10.26 is 5 m long and has the cross-section shown in Figure 10.38b, determine an accurate value for the maximum angle of twist rotation.

resistance **Example 12 – elastic stresses**

the solutions of examples 9 and 11 to determine

(a) the maximum nominal uniform torsion shear stress,
(b) the maximum warping shear stress, and
(c) the maximum warping normal stress.

10.10.6   Example 13 – design

Determine the maximum design torque per unit length that should be permitted on the beam of examples 11 and 12 if the beam has a yield stress of $f_y = 275$ N/mm^2.

References

1. Kollbrunner, C.F. and Basler, K. (1969) *Torsion in Structures,* 2nd edition, Springer-Verlag, Berlin.
2. Timoshenko, S.P. and Goodier, J.N. (1970) *Theory of Elasticity,* 3rd edition, McGraw-Hill, New York.
3. El Darwish, I.A. and Johnston, B.G. (1965) Torsion of structural shapes, *Journal of the Structural Division, ASCE,* 91, No. ST1, pp. 203–27.
4. Davison, B. and Owens, G.W. (eds) (2005) *Steel Designers' Manual,* 6th edition, Blackwell Scientific Publications, Oxford.
5. Terrington, J.S. (1968) Combined bending and torsion of beams and girders, *Publication 31* (First Part), British Construction Steelwork Association, London.
6. Nethercot, D.A., Salter, P.R., and Malik, A.S. (1989) Design of members subject to combined bending and torsion, *SCI Publication 057,* The Steel Construction Institute, Ascot.
7. Hancock, G.J. and Harrison, H.B. (1972) A general method of analysis of stresses in thin-walled sections with open and closed parts, *Civil Engineering Transactions, Institution of Engineers, Australia,* CE14, No. 2, pp. 181–8.
8. Papangelis, J.P. and Hancock, G.J. (1975) *THIN-WALL – Cross Section Analysis and Finite Strip Buckling Analysis of Thin-walled Structures – Users Manual,* Centre for Advanced Structural Engineering, University of Sydney.
9. Von Karman, T. and Chien, W-Z. (1946) Torsion with variable twist, *Journal of the Aeronautical Sciences,* 13, No. 10, pp. 503–10.
10. Smith, F.A., Thomas, F.M., and Smith, J.0. (1970) Torsion analysis of heavy box beams in structures, *Journal of the Structural Division, ASCE,* 96, No. ST3, pp. 613–35.
11. Galambos, T.V. (1968) *Structural Members and Frames,* Prentice-Hall, Englewood Cliffs, New Jersey.
12. Kitipornchai, S. and Trahair, N.S. (1972) Elastic stability of tapered I-beams, *Journal of the Structural Division, ASCE,* 98, No. ST3, pp. 713–28.
13. Kitipornchai, S. and Trahair, N.S. (1975) Elastic behaviour of tapered monosymmetric I-beams, *Journal of the Structural Division, ASCE,* 101, No. ST8, pp. 1661–78.
14. Vlasov, V.Z. (1961) *Thin-walled Elastic Beams,* 2nd edition, Israel Program for Scientific Translations, Jerusalem.
15. Zbirohowski-Koscia, K. (1967) *Thin-walled Beams,* Crosby Lockwood and Son, Ltd, London.

16. Piaggio, H.T.H. (1962) *Differential Equations,* 3rd edition, G. Bell, London.
17. Trahair, N.S. and Pi, Y.-L. (1996) Simplified torsion design of compact I-beams, *Steel Construction, Australian Institute of Steel Construction,* 30, No. 1, pp. 2–19.
18. Trahair, N.S. (1999) Plastic torsion analysis of mono- and point-symmetric beams, *Journal of Structural Engineering, ASCE,* 125, No. 2, pp. 175–82.
19. Dinno, K.S. and Merchant, W. (1965) A procedure for calculating the plastic collapse of I-sections under bending and torsion, *The Structural Engineer,* 43, No. 7, pp. 219–221.
20. Farwell, C.R. and Galambos, T.V. (1969) Non-uniform torsion of steel beams in inelastic range, *Journal of the Structural Division, ASCE,* 95, No. ST12, pp. 2813–29.
21. Aalberg, A. (1995) An experimental study of beam-columns subjected to combined torsion, bending, and axial actions, *Dr.ing. thesis,* Department of Civil Engineering, Norwegian Institute of Technology, Trondheim.
22. Pi, Y.L. and Trahair, N.S. (1995) Inelastic torsion of steel I-beams, *Journal of Structural Engineering, ASCE,* 121, No. 4, pp. 609–20.
23. Trahair, N.S. (2005) Non-linear elastic non-uniform torsion, *Journal of Structural Engineering, ASCE,* 131, No. 7, pp. 1135–42.
24. British Standards Institution (2006) *Eurocode 3 – Design of Steel Structures – Part 1–5: Plated structural elements,* BSI, London.
25. Pi, Y.L. and Trahair, N.S. (1994) Inelastic bending and torsion of steel I-beams, *Journal of Structural Engineering, ASCE,* 120, No. 12, pp. 3397–417.
26. Subcommittee on Box Girders of the ASCE-AASHO Task Committee on Flexural Members (1971) Progress report on steel box bridges, *Journal of the Structural Division, ASCE,* 97, No. ST4, pp. 1175–86.
27. Vacharajittiphan, P. and Trahair, N.S. (1974) Warping and distortion at I-section joints, *Journal of the Structural Division, ASCE,* 100, No. ST3, pp. 547–64.
28. Goodier, J.N. and Barton, M.V. (1944) The effects of web deformation on the torsion of I-beams, *Journal of Applied Mechanics, ASME,* 2, pp. A-35–A-40.
29. Kubo, G.G., Johnston, B.G., and Eney, W.J. (1956) Non-uniform torsion of plate girders, *Transactions, ASCE,* 121, pp. 759–85.
30. Wright, R.N., Abdel-Samad, S.R., and Robinson, A.R. (1968) BEF analogy for analysis of box girders, *Journal of the Structural Division, ASCE,* 94, No. ST7, pp. 1719–43.
31. Hetenyi, M. (1946) *Beams on Elastic Foundations,* University of Michigan Press.